AF389207

Sustainable Assessment in Supply Chain and Infrastructure Management

Sustainable Assessment in Supply Chain and Infrastructure Management

Editors

Golam Kabir
Sanjoy Kumar Paul
Syed Mithun Ali

MDPI • Basel • Beijing • Wuhan • Barcelona • Belgrade • Manchester • Tokyo • Cluj • Tianjin

Editors
Golam Kabir
University of Regina
Canada

Sanjoy Kumar Paul
University of Technology Sydney
Australia

Syed Mithun Ali
Bangladesh University of
Engineering and Technology
Bangladesh

Editorial Office
MDPI
St. Alban-Anlage 66
4052 Basel, Switzerland

This is a reprint of articles from the Special Issue published online in the open access journal *Sustainability* (ISSN 2071-1050) (available at: https://www.mdpi.com/journal/sustainability/special_issues/Infrastructure_Management).

For citation purposes, cite each article independently as indicated on the article page online and as indicated below:

LastName, A.A.; LastName, B.B.; LastName, C.C. Article Title. *Journal Name* **Year**, *Volume Number*, Page Range.

ISBN 978-3-0365-4519-6 (Hbk)
ISBN 978-3-0365-4520-2 (PDF)

Contents

About the Editors . . . . **vii**

Golam Kabir, Sanjoy Kumar Paul and Syed Mithun Ali
Sustainable Assessment in Supply Chain and Infrastructure Management
Reprinted from: *Sustainability* **2022**, *14*, 6787, doi:10.3390/su14116787 . . . . **1**

Ananna Paul, Nagesh Shukla, Sanjoy Kumar Paul and Andrea Trianni
Sustainable Supply Chain Management and Multi-Criteria Decision-Making Methods: A
Systematic Review
Reprinted from: *Sustainability* **2021**, *13*, 7104, doi:10.3390/su13137104 . . . . **5**

Md. Rabbi, Syed Mithun Ali, Golam Kabir, Zuhayer Mahtab and Sanjoy Kumar Paul
Green Supply Chain Performance Prediction Using a Bayesian Belief Network
Reprinted from: *Sustainability* **2020**, *12*, 1101, doi:10.3390/su12031101 . . . . **33**

Priyabrata Chowdhury and Rezaul Shumon
Minimizing the Gap between Expectation and Ability: Strategies for SMEs to Implement Social
Sustainability Practices
Reprinted from: *Sustainability* **2020**, *12*, 6408, doi:10.3390/su12166408 . . . . **53**

Jiseong Noh, Jong Soo Kim and Seung-June Hwang
A Multi-Item Replenishment Problem with Carbon Cap-and-Trade under Uncertainty
Reprinted from: *Sustainability* **2020**, *12*, 4877, doi:10.3390/su12124877 . . . . **69**

Jeong Hugh Han, Yingli Wang and Mohamed Naim
Narrowing the Gaps: Assessment of Logistics Firms' Information Technology Flexibility for
Sustainable Growth
Reprinted from: *Sustainability* **2020**, *12*, 4372, doi:10.3390/su12114372 . . . . **85**

Malisa Dukić and Margareta Zidar
Sustainability of Investment Projects with Energy Efficiency and Non-Energy Efficiency Costs:
Case Examples of Public Buildings
Reprinted from: *Sustainability* **2021**, *13*, 5837, doi:10.3390/su13115837 . . . . **107**

Abdulaziz Alghamdi, Guangji Hu, Husnain Haider, Kasun Hewage and Rehan Sadiq
Benchmarking of Water, Energy, and Carbon Flows in Academic Buildings: A Fuzzy
Clustering Approach
Reprinted from: *Sustainability* **2020**, *12*, 4422, doi:10.3390/su12114422 . . . . **123**

Mrinal Kanti Sen, Subhrajit Dutta and Golam Kabir
Flood Resilience of Housing Infrastructure Modeling and Quantification Using a Bayesian Belief
Network
Reprinted from: *Sustainability* **2021**, *13*, 1026, doi:10.3390/su13031026 . . . . **149**

**Majed Alinizzi, Husnain Haider, Meshal Almoshaogeh, Fawaz Alharbi, Saleh M. Alogla and
Gamal A. Al-Saadi**
Sustainability Assessment of Construction Technologies for Large Pipelines on Urban
Highways: Scenario Analysis using Fuzzy QFD
Reprinted from: *Sustainability* **2020**, *12*, 2648, doi:10.3390/su12072648 . . . . **173**

Ngandu Balekelayi and Solomon Tesfamariam
Geoadditive Quantile Regression Model for Sewer Pipes Deterioration Using Boosting
Optimization Algorithm
Reprinted from: *Sustainability* **2020**, *12*, 8733, doi:10.3390/su12208733 . . . . **193**

Tzeu-Chen Han and Chih-Min Wang
Shipping Bunker Cost Risk Assessment and Management during the Coronavirus Oil Shock
Reprinted from: *Sustainability* **2021**, *13*, 4998, doi:10.3390/su13094998 . . . . **217**

Anastasia Roukouni, Heide Lukosch, Alexander Verbraeck and Rob Zuidwijk
Let the Game Begin: Enhancing Sustainable Collaboration among Actors in Innovation
Ecosystems in a Playful Way
Reprinted from: *Sustainability* **2020**, *12*, 8494, doi:10.3390/su12208494 . . . . **229**

Aydın Özdemir, Hakan Kitapçı, Mehmet Şahin Gök and Erşan Ciğerim
Efficiency Assessment of Operations Strategy Matrix in Healthcare Systems of US States Amid
COVID-19: Implications for Sustainable Development Goals
Reprinted from: *Sustainability* **2021**, *13*, 11934, doi:10.3390/su132111934 . . . . **247**

About the Editors

Golam Kabir

Golam Kabir is an Associate Professor in the Faculty of Engineering and Applied Science (Industrial Systems Engineering program) at the University of Regina, Canada. He is also an Adjunct Assistant Professor of the Department of Mechanical, Automotive and Materials Engineering, University of Windsor, Canada. He worked as an NSERC Postdoctoral Fellow at the University of Waterloo, Canada and Postdoctoral Research and Teaching Fellow at the University of British Columbia (UBC). He received his Ph.D. from the University of British Columbia (UBC), Canada. He completed his Masters and Bachelor degrees from the Department of Industrial and Production Engineering, Bangladesh University of Engineering and Technology (BUET), Bangladesh. He has published more than 90 referred journal and conference papers and was involved in multiple NSERC-funded projects. His major research interests include system risk, reliability, resilience assessment, interdependent network resilience analytics, sustainable system analytics, and data-driven decision making.

Sanjoy Kumar Paul

Sanjoy Kumar Paul is a Senior Lecturer in Operations and Supply Chain Management and Program Director of the Master of Strategic Supply Chain Management Programs at the University of Technology Sydney (UTS), Sydney, Australia. He has published more than 110 papers in top-tier journals and conferences, including European Journal of Operational Research, Transportation Research Part E, International Journal of Production Economics, Business Strategy and the Environment, Computers & Operations Research, Journal of Business Research, International Journal of Production Research, Annals of Operations Research, Journal of Management in Engineering, Journal of Cleaner Production, Computers & Industrial Engineering, Journal of Retailing and Consumer Services, and so on. He is also an editor and active reviewer of many reputed journals. His research interests include supply chain risk management, sustainability, and applied operations research. Dr. Paul has received several recognitions and awards in his career, including top 2% of scientists (based on the year 2020) in author databases of standardized citation indicators, ASOR Rising Star Award, Excellence in Early Career Research Award, the Stephen Fester prize for most outstanding thesis, and high-impact publications award for publishing articles in top-tier journals.

Syed Mithun Ali

Syed Mithun Ali is a Professor in the Department of Industrial and Production Engineering in Bangladesh University of Engineering and Technology (BUET). He holds a PhD majoring in supply chain management from the Nagoya Institute of Technology, Japan. He received both his B.Sc. and M.Sc. Engineering in Industrial and Production Engineering (IPE) from BUET. He has published more than 80 referred journal and conference papers. His research appears in Journal of Cleaner Production, Computers and Industrial Engineering, Sustainable Production and Consumption, Environmental Impact Assessment Review, Industrial Management and Data Systems, International Journal of Production Research, International Journal of Production Economics, among others. His current research interests include supply chain risk and resilience management, circular economy, data-driven/expert-driven decision analysis, optimization and machine learning approaches, capability management, project-driven supply chains, and tying that to supply chain sustainability.

sustainability

Editorial

Sustainable Assessment in Supply Chain and Infrastructure Management

Golam Kabir [1,*], **Sanjoy Kumar Paul** [2] **and Syed Mithun Ali** [3]

[1] Industrial Systems Engineering, University of Regina, Regina, SK S4S 0A2, Canada
[2] UTS Business School, University of Technology Sydney, Sydney, NSW 2007, Australia; sanjoy.paul@uts.edu.au
[3] Department of Industrial and Production Engineering, Bangladesh University of Engineering and Technology, Dhaka 1000, Bangladesh; mithun@ipe.buet.ac.bd
* Correspondence: golam.kabir@uregina.ca

Citation: Kabir, G.; Paul, S.K.; Ali, S.M. Sustainable Assessment in Supply Chain and Infrastructure Management. *Sustainability* **2022**, *14*, 6787. https://doi.org/10.3390/su14116787

Received: 23 May 2022
Accepted: 30 May 2022
Published: 1 June 2022

Publisher's Note: MDPI stays neutral with regard to jurisdictional claims in published maps and institutional affiliations.

Assessing sustainability in supply chain and infrastructure management is important for any organization in the competitive business environment or public domain. Organizations are currently trying to develop sustainable strategies through preparedness, response, and recovery because of increased competitive, regulatory, and community pressure [1]. Sustainability, in the context of supply chain, implies that companies identify, assess, and manage impacts and risks in all the echelons of the supply chain, considering upstream and downstream activities [2]. Considering the wider adoption and development of sustainability principles across the globe, there is a real need to develop a meaningful and more focused understanding of sustainability in supply chain management and infrastructure management practices. This Special Issue aimed to gather contributions on sustainable assessment in supply chain and infrastructure management. This Special Issue publishes 13 papers which provide a broad overview of the current knowledge on sustainable supply chain and infrastructure management.

To evaluate and understand the effectiveness of sustainable and green supply chain management, indicators must be carefully defined and monitored, including environmental, social, and economic aspects [3]. Sustainable supply chain management is addressed in five papers from different perspectives. Paul et al. (contribution one) analyzed existing research, identified research gaps, and proposed new future research opportunities in the area of sustainable supply chain management by applying multi-criteria decision-making (MCDM) methods. Rabbi et al. (contribution two) identified eleven green supply chain performance indicators and developed a Bayesian belief network (BBN) model to predict the overall environmental performance. However, it is always challenging for small- and medium-sized enterprises (SMEs) to adopt and practice social sustainability due to the lack of resources. To make SMEs socially sustainable, Chowdhury and Shumon (contribution three) described various situations and provided strategies outlining the implications for SMEs and their stakeholders. As global warming has become a critical issue, it is essential for companies to increase their efforts to control carbon emissions in green supply chain management (GSCM) activities. Noh et al. (contribution four) addressed the multi-item replenishment problem with carbon cap-and-trade for GSCM under limited resources, including limited storage capacity, budget, and carbon cap-and-trade regulation. For sustainable growth and to provide the best value from a logistics firm, Han et al. (contribution five) provided an analytical tool that measures the required and actual levels of information technology flexibility.

On the other hand, sustainable infrastructure management can be defined as the ability of infrastructure to meet the requirements of the present without sacrificing the ability of future generations to address their needs [4]. The complexity of the issues regarding sustainable infrastructure management drove managers and professionals in the field of asset management to seek different solutions and address different topics linked to sustainable infrastructure asset management [5]. Five papers address problems related to the sustainable infrastructure asset management. Đukić and Zidar (contribution six) focused on the sustainability of energy efficiency projects for public buildings considering both energy and non-energy efficiency investment costs. They assessed the sustainability of several projects in Serbia and Croatia and performed a cost-benefit analysis using the European Commission methodology. Public buildings such as higher education institutions are responsible for a substantial portion of energy consumption and anthropogenic greenhouse gas (GHG) emissions. Alghamdi et al. (contribution seven) proposed a fuzzy clustering approach to classify academic buildings in higher educational institutions. The authors benchmarked their environmental performance in terms of water, energy, and carbon flows. To ensure community safety and sustainability, it is needed to develop resilient housing infrastructure. For this, Sen et al. (contribution eight) developed a Bayesian belief network (BBN)-based model to assess the reliability, recovery, and resilience of housing infrastructure against flood hazards. They tested the model in a real community in Northeast India.

Proactive management is required for the effective maintenance and inspection of infrastructures. The performance of one infrastructure can affect other types of infrastructures. For example, urban highways frequently face disruptions due to the construction and maintenance of buried infrastructure such as potable water, wastewater, and stormwater. Alinizzi et al. (contribution nine) performed a sustainability assessment of construction technologies for large pipelines on urban highways. The developed framework evaluates various traffic detoured scenarios and trenchless technology scenarios based on all three dimensions of sustainability. Balekelayi and Tesfamariam (contribution ten) applied a Bayesian geo-additive quantile regression approach to estimate the deterioration of wastewater pipes. The proposed approach is suitable for prioritizing inspections and provides knowledge for future installations.

Logistics and transport systems are also critical for sustainable development. It is important to develop risk management strategies that enable logistics, transport, and shipping companies to handle fuel price fluctuations, reduce unnecessary fuel cost risks, and improve financial management. Three papers addressed these issues. Han et al. (contribution eleven) performed shipping bunker cost risk assessment and management during the coronavirus oil shock. Their study indicates that the best strategy is to install scrubbers on existing ships to purify their exhaust gas and choose natural gas-based marine fuel for new ships. Roukouni et al. (contribution twelve) developed truck platooning and multi-sided digital platforms games for barge transportation, both improving the sustainability of hinterland transportation. Besides these studies, Özdemir et al. (contribution thirteen) assessed the efficiency of the operations strategy matrix in the healthcare system amid COVID-19. They considered strategic decision areas such as supply network, capacity, process technology, and development and organization) to assess competitive priorities including cost, delivery, quality, and flexibility of different U.S. states.

To summarize, various issues have been addressed in this Special Issue from different aspects of these contributions. We believe that this Special Issue offered some solutions and also raised some questions for further research and development toward sustainable supply chain and infrastructure management.

List of Contributors

1. Paul, A., Shukla, N., Paul, S.K., Trianni, A. Sustainable Supply Chain Management and Multi-Criteria Decision-Making Methods: A Systematic Review.
2. Rabbi, M., Ali, S.M., Kabir, G., Mahtab, Z., Paul, S.K. Green Supply Chain Performance Prediction Using a Bayesian Belief Network.
3. Chowdhury, P., Shumon, R. Minimizing the Gap between Expectation and Ability: Strategies for SMEs to Implement Social Sustainability Practices.
4. Noh, J., Kim, J.S., Hwang, S.J. A Multi-Item Replenishment Problem with Carbon Cap-and-Trade under Uncertainty.
5. Han, J.H., Wang, Y., Naim, M. Narrowing the Gaps: Assessment of Logistics Firms' Information Technology Flexibility for Sustainable Growth.

6. Đukić, M., Zidar, M. Sustainability of Investment Projects with Energy Efficiency and Non-Energy Efficiency Costs: Case Examples of Public Buildings.
7. Alghamdi, A., Hu, G., Haider, H., Hewage, K., Sadiq, R. Benchmarking of Water, Energy, and Carbon Flows in Academic Buildings: A Fuzzy Clustering Approach.
8. Sen, M.K., Dutta, S., Kabir, G. Flood Resilience of Housing Infrastructure Modeling and Quantification Using a Bayesian Belief Network.
9. Alinizzi, M., Haider, H., Almoshaogeh, M., Alharbi, F., Alogla, S.M., Al-Saadi, G.A. Sustainability Assessment of Construction Technologies for Large Pipelines on Urban Highways: Scenario Analysis using Fuzzy QFD.
10. Balekelayi, N., Tesfamariam, S. Geoadditive Quantile Regression Model for Sewer Pipes Deterioration Using Boosting Optimization Algorithm.
11. Han, T.C., Wang, C.M., Shipping Bunker Cost Risk Assessment and Management during the Coronavirus Oil Shock.
12. Roukouni, A., Lukosch, H., Verbraeck, A., Zuidwijk, R. Let the Game Begin: Enhancing Sustainable Collaboration among Actors in Innovation Ecosystems in a Playful Way.
13. Özdemir, A., Kitapçı, H., Gök, M.Ş., Ciğerim, E. Efficiency Assessment of Operations Strategy Matrix in Healthcare Systems of US States Amid COVID-19: Implications for Sustainable Development Goals.

Funding: This research received no external funding.

References

1. Raian, S.; Ali, S.M.; Sarker, M.R.; Sankaranarayanan, B.; Kabir, G.; Paul, S.K.; Chakrabortty, R.K. Assessing sustainability risks in the supply chain of the textile industry under uncertainty. *Resour. Conserv. Recycl.* **2022**, *177*, 105975. [CrossRef]
2. Rajeev, A.; Pati, R.K.; Padhi, S.S.; Govindan, K. Evolution of sustainability in supply chain management: A literature review. *J. Clean. Prod.* **2017**, *162*, 299–314. [CrossRef]
3. Sharma, V.K.; Chandna, P.; Bhardwaj, A. Green supply chain management related performance indicators in agro industry: A review. *J. Clean. Prod.* **2017**, *141*, 1194–1208. [CrossRef]
4. Gardoni, P. (Ed.) *Routledge Handbook of Sustainable and Resilient Infrastructure*; Routledge: New York, NY, USA, 2019.
5. Nielsen, L.; Faber, M.H. Impacts of sustainability and resilience research on risk governance, management and education. *Sustain. Resilient Infrastruct.* **2021**, *6*, 339–384. [CrossRef]

 *sustainability*

Review

Sustainable Supply Chain Management and Multi-Criteria Decision-Making Methods: A Systematic Review

Ananna Paul [1,*], Nagesh Shukla [1], Sanjoy Kumar Paul [2,*] and Andrea Trianni [1]

[1] School of Information, Systems and Modelling, University of Technology Sydney, Sydney, NSW 2007, Australia; Nagesh.Shukla@uts.edu.au (N.S.); Andrea.Trianni@uts.edu.au (A.T.)

[2] UTS Business School, University of Technology Sydney, Sydney, NSW 2007, Australia

* Correspondence: ananna.du@gmail.com (A.P.); sanjoy.paul@uts.edu.au (S.K.P.)

Abstract: Multi-criteria decision-making (MCDM) methods are smart tools to deal with numerous criteria in decision-making. These methods have been widely applied in the area of sustainable supply chain management (SSCM) because of their computational capabilities. This paper conducts a systematic literature review on MCDM methods applied in different areas of SSCM. From the literature search, a total of 106 published journal articles have been selected and analyzed. Both individual and integrated MCDM methods applied in SSCM are reviewed and summarized. In addition, contributions, methodological focuses, and findings of the reviewed articles are discussed. It is observed that MCDM methods are widely used for analyzing barriers, challenges, drivers, enablers, criteria, performances, and practices of SSCM. In recent years, studies have focused on integrating more than one MCDM method to highlight methodological contributions in SSCM; however, in the literature, limited research papers integrate multiple MCDM methods in the area of SSCM. Most of the published articles integrate only two MCDM methods, and integration with other methods, such as optimization and simulation techniques, is missing in the literature. This review paper contributes to the literature by analyzing existing research, identifying research gaps, and proposing new future research opportunities in the area of sustainable supply chain management applying MCDM methods.

Keywords: literature review; multi-criteria decision-making; MCDM methods; sustainable supply chain management; SSCM

Citation: Paul, A.; Shukla, N.; Paul, S.K.; Trianni, A. Sustainable Supply Chain Management and Multi-Criteria Decision-Making Methods: A Systematic Review. *Sustainability* **2021**, *13*, 7104. https://doi.org/10.3390/su13137104

Academic Editor: Alessio Ishizaka

Received: 20 May 2021
Accepted: 21 June 2021
Published: 24 June 2021

Publisher's Note: MDPI stays neutral with regard to jurisdictional claims in published maps and institutional affiliations.

1. Introduction

In this competitive era, every business is part of a supply chain which involves efficient and effective movement of products or services from suppliers through to customers via manufacturers, distributors, and retailers. A typical supply chain involves multiple businesses, resources, people, technologies, and information for buying, manufacturing, distributing, storing, and selling products [1]. Several activities within a supply chain present direct social, environmental, and economic impacts [2]. These impacts are referred to as the triple bottom line (TBL) in sustainable supply chain literature. Social impact includes modern slavery, gender discrimination, unfair wages, child labor, and so on [3,4]. Environmental impact includes emission of carbon dioxide, polluting water and the environment, global warming, and so on [5,6]. Economic impact includes the return of investment, impact on profit, and productivity [2]. Considering their significant impact on society, the environment, and the economy, every supply chain is now taking steps to ensure sustainability.

Sustainable supply chain management (SSCM) integrates the economic, social, and environmental goals of the supply chain to improve long-term performance [7], evaluating and monitoring business performance against social, environmental, and economic dimensions [2]. Any good social and environmental performance with economic performance ensures better sustainability; however, ensuring all three performances are good creates the best sustainable supply chain [8]. Some recent studies have considered the triple bottom line (TBL) aspect of supply chain sustainability [9–15].

Examples of social sustainability include ensuring fair policies, ethical practices, equal opportunities, diversity, and so on [16–18]. Several papers in the literature focused on different social sustainability dimensions in supply chains, such as wages, child labor, equal opportunities, discrimination, ethics, corruption, health safety, diversity, equity, human rights, labor practice, training, and slavery [16,17,19–21]. A summary of social sustainability in SSCM literature is presented in Table 1. Empirical research, together with the application of different MCDM methods, were widely used to identify and analyze the social dimension of SSCM (see Table 1). From the contributions presented in Table 1, one can note that most of the research studies analyzing social sustainability focused on barriers, enablers, criteria in service, and manufacturing supply chains.

When a supply chain is environmentally sustainable, it is known as a green supply chain [22]. Examples of an environmentally sustainable supply chain include the treatment of waste, recycling, environmental education and training, green purchasing, green manufacturing, and green design [23,24]. In recent studies in this area, MCDM methods were widely applied (see Table 2). Looking at Table 2, most of the research studies focused on evaluating or analyzing factors, indicators, criteria, practices, performances, and suppliers in green supply chains. Different characteristics, including recycling, remanufacturing, greenhouse gas emissions, waste management, environmental education and training, green design, green/cleaner production, green purchasing, green logistics/distribution, and energy consumption are considered [22–29]. We have summarized the different characteristics of environmental sustainability and their source studies in Table 2.

Table 1. Different characteristics of social sustainability studied under SSCM literature.

Reference	Characteristic Name													Contribution	Methodology
	Wages	Child Labor	Equal Opportunity	Discrimination	Ethics	Corruption	Health-Safety	Diversity	Equity	Human Right	Labor Practice	Training	Slavery		
[19]	√	√	√	√	√		√	√	√	√	√	√		Identification and analysis of different dimensions of social sustainability in supply chains in India	Semi-structured interview
[16]	√	√					√	√	√	√	√			Analyzing forces for adopting social sustainability in emerging Indian and Portuguese economies	Empirical study
[30]	√	√				√				√	√		√	Analyzing modern slavery in supply chains perspective of United Kingdom from the clothing and textile sector	Secondary data analysis
[31]	√	√	√	√			√	√	√	√	√	√		Analyzing relationships between enablers to the social sustainability	ISM-MICMAC
[18]							√			√	√	√		Selecting supplier bases social sustainable criteria	Grey BWM–grey TODIM
[21]	√	√		√	√	√	√	√	√	√		√	√	Investigating integrated aspects of social sustainability	Empirical study
[4]	√	√		√		√	√					√		Analyzing enablers in social sustainability in footwear supply chains	BWM
[17]	√	√	√		√	√	√			√		√	√	Addressing social sustainability in supplier selection processes	Exploratory case study
[32]					√		√		√	√		√		Analyzing dimensions of social sustainability in healthcare supply chains	Stochastic exponential distribution model
[33]							√			√	√	√		Investigating social sustainability criteria	BWM
[34]			√		√		√		√			√		Identifying motivators, barriers, and enablers of social sustainability	Empirical study
[20]	√	√	√	√	√		√			√	√	√	√	Developing a taxonomy of supply chain social sustainability practices	Empirical study

Table 2. Different characteristics of environmental sustainability studied under SSCM literature.

Reference	Recycling	Remanufacturing	Circular Economy	Greenhouse Gas Emission	Waste Treatment/ Management	Use of Natural Resources	Environmental Education and Training	Green Design	Green/Cleaner Production	Green Purchasing	Green Logistics/Distribution	Energy Consumption	Contribution	Methodology
[25]	✓			✓			✓	✓		✓			Identifying critical dimensions and factors in green supply chains	DEMATEL and cast study
[26]	✓	✓		✓	✓			✓	✓	✓	✓		Evaluating indicators in green supply chains	Fuzzy VIKOR
[22]	✓			✓	✓	✓		✓	✓	✓		✓	Evaluating suppliers in green supply chain	Literature review
[29]	✓	✓		✓	✓			✓		✓		✓	Analyzing critical green supply chain practices	HPA approach
[23]	✓			✓	✓	✓	✓	✓	✓	✓	✓	✓	Analyzing criteria for green supply chains	Fuzzy DEMATEL
[28]	✓		✓	✓	✓			✓	✓	✓	✓	✓	Developing an assessment framework for green supply chain management	Conceptual study
[27]			✓					✓	✓	✓			Evaluating performance of green supply chain management	Fuzzy inference system
[24]	✓	✓		✓	✓		✓	✓		✓	✓	✓	Evaluating green suppliers	TOPSIS

Examples of economic sustainability include cost reduction, on-time delivery, reliability, and quality [11]. Sustainable supply chains simultaneously assess supply chain performance in terms of social, environmental, and economic aspects.

In the last few years, a good number of studies have been conducted on different dimensions of SSCM, including a number of review papers, such as a:

- review of green supply chain management [35–37];
- review of different theories in sustainable supply chains [38];
- review of the evolution of and future challenges in sustainable supply chain management [8,39–41];
- review of trends and future directions in social aspects of sustainable supply chains [42];
- review of SSCM in global supply chain context [43];
- review of drivers in SSCM [44]; and
- review of MCDM methods in green supply chains which focuses only on the environmental dimension of supply chain management [22,45].

In order to become more sustainable, supply chains should implement sustainable practices, with a certain impact on various TBL areas; however, decision makers need to consider multiple criteria to evaluate suppliers, practices, success factors, drivers, and challenges in SSCM in smart ways. For this purpose, MCDM methods have been widely applied in the area of SSCM [46]. In spite of having a reasonable number of contributing articles which applied different MCDM methods in SSCM, earlier literature is lacking a review on different MCDM methods applied to SSCM areas considering the social, environmental, and economic dimensions. In this paper, we aim to fill this gap by conducting a literature review on different MCDM methods applied in SSCM, contributing to the literature by analyzing existing studies systemically and proposing a future research framework in the area of MCDM methods in SSCM.

The paper is organized as follows. In Section 2, the scope of the literature review is described. The review of both individual and integrated MCDM methods is conducted in Section 3. Section 4 explains the bibliometric analysis for published articles. Section 5 summarizes the review and research gaps. Finally, conclusions and future research directions are presented in Section 6.

2. Scope of the Literature Review

The Scopus database was used to collect the relevant articles with the following phrases in the article's title, abstract, and keywords: "sustainable supply chain" and "multi-criteria decision making" or "multi-criteria decision analysis" or "MCDM". From the preliminary search of the literature, most of the studies in the area of SSCM modelling were found to be published since 2010, which is shown in Figure 1. Based on this observation, in this paper, the literature on MCDM methods applied in the area of SSCM is reviewed from 2010 to 2020.

After the preliminary search in Scopus, the search database was refined using the following criteria:

- Document type: article
- Source type: journals
- Year: 2010–2020
- Language: English

Other databases, such as the Web of Science and Google Scholar, were used to enhance the search. After a first screening of the articles (by title and abstract), the final subset of 106 relevant manuscripts for review was created. The inclusion criteria were articles focused on any dimension of supply chain sustainability and the search phrases appeared in the body text. The exclusion criterion was one or more keywords presented in the text or reference list without discussing supply chain sustainability using MCDM methods. After finalizing the list of articles, a deep review was conducted of the applications of different

MCDM methods in SSCM, and a bibliometric analysis was carried out within the set of finalized articles.

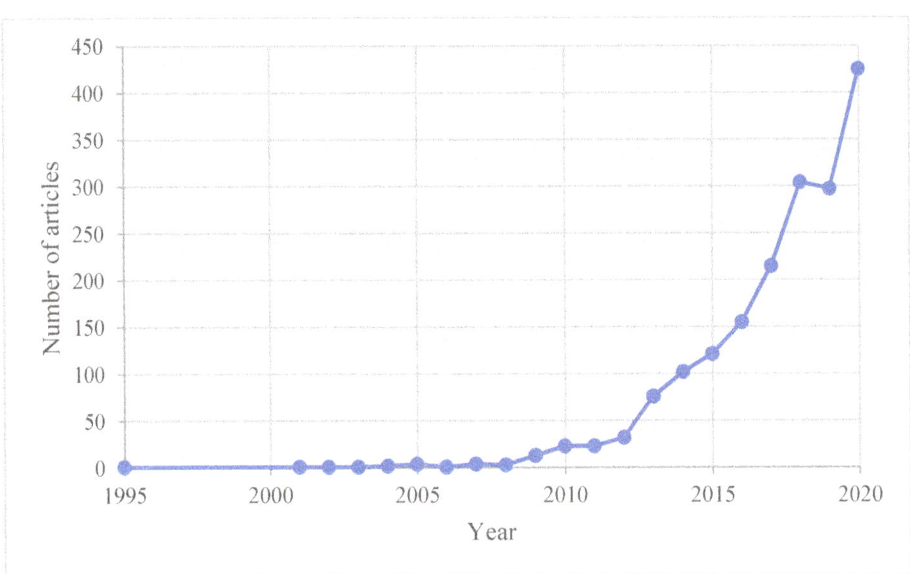

Figure 1. Number of articles published on SSCM modelling (Source: Scopus).

3. Review of Applications of MCDM Methods in SSCM

This section reviews the articles on MCDM methods applied in SSCM areas. The following sub-sections review the applications of both individual and integrated MCDM methods in detail.

3.1. Applications of Individual MCDM Methods

From the literature search, 59 articles applied individual MCDM methods in SSCM. The names of the methods and their abbreviated terms are as follows:

i.   Decision-making trial and evaluation laboratory (DEMATEL) and Fuzzy/Grey DEMATEL
ii. Analytical hierarchy process (AHP) and Fuzzy AHP
iii. The technique for order of preference by similarity to ideal solution (TOPSIS) and Fuzzy TOPSIS
iv. Best–worst method (BWM)
v.   VIseKriterijumska Optimizacija I Kompromisno Resenje (VIKOR) and Fuzzy VIKOR
vi. Rough set
vii. Elimination et choix traduisant la realité (ELECTRE) and Fuzzy ELECTRE
viii. Analytical network process (ANP)
ix. Rough strength-relation analysis method (RSRAM)
x.   Rough simple additive weighting (RSAW)
xi. Interpretive structural modelling (ISM)
xii. Preference ranking organization method for enriched evaluation (PROMETHEE)

3.1.1. DEMATEL and Fuzzy/Grey DEMATEL

DEMATEL and Fuzzy/Grey DEMATEL are the most applied methods in SSCM. Between 2010 and 2020, a total of 15 articles have been published on this method. Among these, six articles applied DEMATEL, four applied grey DEMATEL, and five articles applied fuzzy DEMATEL.

In particular, applications of DEMATEL include the identification and analysis of success factors for sustainability initiatives (grey [47]), sustainable food supply chain management [48], green supply chain practices (fuzzy [49]), SSCM for Industry 4.0 [50], and implementing green supply chain management [51]. In addition, DEMATEL was used in a number of studies to analyze and evaluate barriers or challenges for sustainable development [52], remanufacturing (grey [53]), and green supply chain [54]. A few studies also analyzed drivers for sustainable consumption and production adoption applying grey DEMATEL [55] and drivers to ICT for sustainability initiatives in supply chains using a fuzzy one [56]. DEMATEL was used in other applications including the analysis of criteria and alternatives in sustainable supply chains (grey [57]), evaluation of influential indicators for adopting sustainable supply chains [58], analysis of causal relationships between practices and performance in green supply chains (fuzzy [59]), assessing performance in green supply chains considering economic, logistics, operational, organizational, and marketing aspects (fuzzy [23]), and selection of suppliers based on multiple criteria (fuzzy [60]).

3.1.2. AHP and Fuzzy AHP

AHP is one of the most widely applied MCDM methods in SSCM. Eleven articles applied AHP or Fuzzy AHP. Among these, six articles applied AHP, and the remaining five articles applied the fuzzy AHP method.

Six studies applied AHP to evaluate barriers to adopting sustainable consumption and production initiatives [61], to analyze criteria for improving effectiveness in green supply chain management implementation [62], to analyze challenges for industry 4.0 initiatives toward SSCM [63], to evaluate pressures to implement GSCM [64], to evaluate manufacturing practices for sustainability [65], and to analyze drivers for sustainable manufacturing processes [66]. Fuzzy AHP was also used to identify and analyze risks in green supply chains [67], analyze success factors for sustainable food supply chain management [68], evaluate indicators of SSCM [69], assess the supply chain performance based on sustainability criteria [70], and evaluate European countries for renewable energy sectors [71].

3.1.3. TOPSIS and Fuzzy TOPSIS

Earlier studies applied TOPSIS (two articles) or fuzzy TOPSIS (six articles) in the context of SSCM. The applications encompass the suppliers' evaluation and selection in sustainable and green supply chains based on multiple criteria. These criteria include applications of TOPSIS in selecting sustainable suppliers [72,73] and applications of fuzzy TOPSIS in evaluating green supplier performance [74,75], evaluating sustainable and green suppliers [24,76,77], and assessing areas for improvement in implementing green supply chain initiatives [78].

Researchers applied TOPSIS and Fuzzy TOPSIS to select suppliers and performance in sustainable or green supply chains based on identified multi-criteria.

3.1.4. BWM

The eight articles which have applied BWM in SSCM include an assessment of sustainability in green supply chains in an emerging economy [79], assessment of social sustainability in supply chains [33], evaluation of external forces for sustainable supply chains in the context of the oil and gas industries [80], analysis of enablers for social sustainability in an emerging economy [4], evaluation and prioritization of criteria for sustainable innovation [13], analysis of product-package alternatives in food supply chains [81], ranking sustainable suppliers [82], and analyzing barriers for sustainable supply chain innovation [83].

3.1.5. VIKOR and Fuzzy VIKOR

Five articles applied VIKOR or fuzzy VIKOR in SSCM. These articles include evaluating green supply chain management practices using fuzzy VIKOR [26], selecting development programs for green suppliers using fuzzy VIKOR theory [84], evaluating green environmental factors in reverse logistics using fuzzy VIKOR [85,86], and assessing green supply chain initiatives using a probabilistic linguistic VIKOR method [87].

3.1.6. Rough Set

The Rough set method has been applied in SSCM to select suppliers with sustainability [88], analyzing relationships between organizational attributes, supplier development programs, and performance in green supply chains [89], evaluating a selection, performance measurement, and program development tool in green supply chains [90], and measuring SSCM performances [91].

3.1.7. ELECTRE and Fuzzy ELECTRE

ELECTRE and fuzzy ELECTRE have been applied in SSCM to classify suppliers in the manufacturing industry using the ELETCRE TRI-nC method [92] and to evaluate supplier performance in green supply chains using the fuzzy ELECTRE method [93].

3.1.8. ANP

Two studies applied the ANP method in SSCM. The applications include selecting suppliers for managing sustainability [94] and selecting suppliers integrating the triple-bottom-line aspect [95].

3.1.9. Rough Strength-Relation Analysis Method, RSAW, ISM, and PROMETHEE

One article applied the Rough strength-relation analysis method for analyzing risk factors in SSCM [96], the RSAW for sustainable supplier selection [97], the ISM for ranking of barriers in SSCM [98], and the PROMETHEE for analyzing alternatives of biomass [99].

3.1.10. Summary of Applications of Individual Methods

Researchers applied DEMATEL and Fuzzy/Grey DEMATEL, AHP, and BWM mostly for analyzing success factors, barriers and challenges, drivers, and enablers for different aspects of SSCM. Success factors are the important factors decision makers should consider to ensure success in different dimensions of SSCM. Barriers and challenges are the causes preventing the success of any dimension of SSCM. Drivers and enablers are the aspects driving toward the achievement of sustainable performance within any dimension of supply chain sustainability. The different MCDM methods applied to analyze and prioritize success factors, barriers and challenges, and drivers and enablers in SSCM are summarized in Tables 3–5, respectively.

Table 3. Application of MCDM methods to analyze success factors.

Analyzed Success Factors in SSCM	Reference	Method
Green design, recovering and recycling, green purchasing, environmental performance, supplier collaboration, and regulation	[49]	Fuzzy DEMATEL
Government regulations and standards, top management commitment, environmental certifications, adoption of new technology and processes, reverse logistics, and training of suppliers and employees	[51]	DEMATEL
Logistics integration, social development, and environmental development	[50]	DEMATEL

Table 3. *Cont.*

Analyzed Success Factors in SSCM	Reference	Method
Technology development and process innovation, training, reverse logistics and waste minimization, ecological considerations in organizations' policies and missions, green design and purchasing, societal considerations, ethical and safe practices, and community welfare and development	[47]	Grey DEMATEL
Climatic change, implementing green practice, governance and cooperation, technological innovation, and government regulation	[48]	DEMATEL
Proper use of irrigation, demographic and environmental conditions, risk analysis, government policies, and food packaging	[68]	Fuzzy AHP

Table 4. Applications of MCDM methods to analyze barriers and challenges.

Analyzed Barriers and Challenges in SSCM	Reference	Method
Lack of sufficient governmental policies, poor infrastructure, low level of integration, skill shortage, and poor quality of raw materials	[52]	DEMATEL
Lack of channels to collect used products, imperfect legal system, consumption attitude, customer willingness to return the products, uncertainty in demand of remanufactured product, uncertainty in quality, and quantity and timing of returned products	[53]	Grey DEMATEL
Lack of environmental regulation, lack of potential liability, high cost of disposal of hazardous materials, poor environmental performance, lack of information, lack of governmental support, high cost for renewable energy, lack of new technology, insufficient societal pressure, poor legislation, lack of adoption of green practices, health and safety issues, employment stability, less profit in remanufacturing, lack of adequate training, and lack of management support	[54].	DEMATEL
Lack of support from management, lack of innovative methods, lack of technology developments, communication gap, lack of rewards and encouragement programs, lack of governmental regulations, lack of promotion of ethical and safe practices, reluctance of consumers toward sustainable development practices, lack of promotion of sustainable products, and lack of knowledge among stakeholders	[61]	AHP
Low understanding of industry 4.0 implications, poor research and development of industry 4.0 adoption, legal issues, low management support and dedication, lack of global standards and data-sharing protocols, security issues, lack of governmental support and policies, and financial constraints	[63]	AHP
Technological, regulatory, social, cultural, organizational, market, and networking barriers	[83]	BWM

Table 5. Applications of MCDM methods for analyzing drivers and enablers.

Analyzed Drivers and Enablers in SSCM	Reference	Method
Top management role and support, government support systems and subsidies, information systems network design, socio-environmental impacts of the products, culture related factors, approach to ICT to adopt sustainability, understanding the nature of sustainability, security and support services, and human expertise	[56]	Fuzzy DEMATEL
Management support, dedication and involvement, educating suppliers and vendors, understanding the customer requirements about sustainability, governmental policies and regulations, information flow and sharing among supply chain members, competency and skill of workforce, integration of social, environmental, and economic advantages, and understanding the importance of sustainability	[55]	Grey DEMATEL
Market capabilities, compliance with regulations, green purchasing, green innovation, environmental conservation, education and training, and employee welfare	[66]	AHP
Commitment to continual improvement and pollution prevention, commitment to comply with legislation, framework for setting and reviewing environmental goals, legal and other requirements, environmental objectives and targets, environmental education and training, green teamwork, best practices, identification of culture, monitoring culture change, quantity of waste released at each stage, and communication between top management and employees	[62]	AHP
Waste management, reuse and recycle, renewable energy usage, resource utilization, land, air and water pollution, government regulations, and use of hazardous materials	[79]	BWM
Wages and benefits, customer requirements, workplace health and safety practices, food, housing, and sanitation, child labor or forced labor, commitment of top management, education and training of employees, non-discrimination, anti-corruption, and working hours	[4]	BWM
Sustainable product cost reduction, financial availability for innovation, enhanced sustainability value to customers, investment in R&D for sustainable products, designing sustainable products, green logistics capabilities development, green manufacturing, environment management commitment, conducting regular environmental audits, enhancing the social image of the organization, corporate social responsibility initiatives, cultural, social values and norms, occupational health, and safety and rights of the employees	[13]	BWM

Researchers applied TOPSIS, Fuzzy TOPSIS, VIKOR, Rough Set, and ANP to analyze and evaluate suppliers and practices in sustainable or green supply chains based on sustainable criteria. These studies are summarized in Table 6.

Table 6. Summary of applications in analyzing and evaluating suppliers and practices.

Sustainable Criteria Considered	Application Area	Reference	Method
Pollution controls, pollution prevention, environmental management system, resource consumption, employment practices, health and safety, local communities influence, stakeholders influence, cost, quality, and innovation	Supplier selection in sustainable supply chain	[73]	TOPSIS
Cost reduction activities, products' quality improvement, increase in supply flexibility, green design of products, green purchasing, green production, internal management support for green development, green logistics, provision for health and safety, protection of employee's rights, human rights, and fair-trading and against corruption	Supplier selection in sustainable supply chain	[72]	TOPSIS
Quality of products, service performance, cost, environmental efficiency, green image, pollution reduction, green competencies, health and safety, and employment practices	Supplier selection in sustainable supply chain	[77]	Fuzzy TOPSIS
Cost, financial capability, flexibility, innovation, service capability, environmental management system, green image, greenhouse gas emission, reuse/recycling, pollution control, energy and resource consumption, economic welfare and growth, social responsibility, job safety and labor health, the interest and rights of employees, and job opportunities	Supplier selection in sustainable supply chain	[76]	Fuzzy TOPSIS
Green design, green purchasing, green production, green warehousing, green transportation, and green recycling	Green practice evaluation	[26]	Fuzzy VIKOR
Cost, resource usage, energy usage, water consumption, emission and waste generation, green manufacturing, product design, transportation, warehouse and procurement, and reverse logistics	Evaluation of green supplier development program	[84]	VIKOR
Cost, quality, time, flexibility, innovation, culture, technology, relationships, pollution control and prevention, resource consumption, health and safety, employment practices, and local community influence	Supplier selection in sustainable supply chain	[88]	Rough Set
Quality, price, on-time delivery, lead time, flexibility, community initiatives, ethical behavior, health and safety, diversity, waste reduction, recycling, and reverse logistics	Supplier selection in sustainable supply chain	[95]	ANP

3.2. Applications of Integrated MCDM Methods

A total of 47 articles applied integrated MCDM methods in SSCM. Among these, AHP or Fuzzy AHP were most widely integrated with other methods such as DEMATEL, ELECTRE, ISM, TOPSIS, VIKOR, and SOWIA, followed by TOPSIS or Fuzzy TOPSIS with

FPP, Rough set, CRITIC, and VIKOR. Researchers have applied more integrated MCDM methods in recent years, making a significant methodological contribution; this section summarizes such studies.

AHP and Fuzzy AHP are mostly integrated with TOPSIS or fuzzy TOPSIS and VIKOR and fuzzy VIKOR. AHP-TOPSIS is widely applied in selecting sustainable or green suppliers, evaluating third-party logistics (3PL) service providers, and prioritizing solutions and responses in different aspects of SSCM [100–105]. AHP-VIKOR (with their fuzziness) integrated method was mostly applied for selecting a sustainable supplier and management practices in green supply chain management [106–108]. Other integrations of AHP or fuzzy AHP with DEMATEL or fuzzy DEMATEL, ELECTRE or fuzzy ELECTRE, ISM, and SOWIA were applied in analyzing success factors [109], barriers [110], enablers [111], and strategy decisions [112] in SSCM or green supply chain management.

ANP is mostly integrated with quality function deployment (QFD) to analyze supplier selection and environmental sustainability, and for designing sustainable supply chains [113–116]. Other integrations of ANP with VIKOR [117] and grey rational analysis (GRA) [118] were applied in green/sustainable supplier evaluation.

BWM or fuzzy BWM is mostly integrated with VIKOR or fuzzy VIKOR for evaluating transportation service providers and outsourcing partners based on sustainable criteria [119,120]. Other applications of integrated BWM or fuzzy BWM include evaluating dimensions of human resources in green supply chains using BWM-DEMATEL [121], selecting sustainable suppliers in manufacturing supply chains by integrating BWM and an alternative queuing method (AQM) [122] and selecting sustainable suppliers using integrated BWM and combined compromise solution [123].

TOPSIS or Fuzzy TOPSIS is mostly integrated with VIKOR or fuzzy VIKOR, fuzzy preference programming (FPP), Rough set, and criteria importance through intercriteria correlation (CRITIC). TOPSIS-VIKOR (and their fuzziness) integrated methods [124,125] were applied in selecting third-party reverse logistics service providers and classifying rural areas based on social sustainability criteria. TOPSIS-VIKOR-GRA (integrating three methods) was applied in analyzing locations for remanufacturing plants based on multiple criteria [126]. Other applications of integrated TOPSIS or fuzzy TOPSIS include evaluating supply chain practices by integrating TOPSIS and Rough set [127], analyzing risk factors in SSCM using TOPSIS-CRITIC [128], and selecting sustainable suppliers using TOPSIS-FPP [129].

Other integrated methods, such as ELECTRE with VIKOR, were applied in environmental performance evaluation [130]; DEMATEL with MABAC was applied in sustainable freight transport systems [131]; RSAW with MABAC applied in sustainable supplier selection [132]; factor relationship (FARE) with MABAC for selecting third-party logistics provider [133]; step-wise weight assessment ratio analysis (SWARA) and fuzzy complex proportional assessment of alternatives (COPRAS) were used for analyzing risks and solutions in sustainable manufacturing supply chains [134]; and fuzzy entropy and fuzzy multi-attribute utility were applied for sustainable performance measure in supply chain [135].

In summary, most of the integrated MCDM methods in SSCM were used for evaluating or analyzing suppliers, service providers, barriers, enablers, success factors, and evaluating performance. A summary of different integrated MCDM methods applied in SSCM is presented in Table 7.

Table 7. Summary of integrated MCDM methods applied in SSCM.

Method Name	Integrated with														References	Area of Application
	DEMATEL/ Fuzzy/Grey DEMATEL	ELECTRE/ Fuzzy ELECTRE	ISM	TOPSIS/ Fuzzy TOPSIS	VIKOR/ Fuzzy VIKOR	SOWIA	GRA	QFD	Rough Set	CRITIC	FPP	MABAAC	AQM	TODIM		
	√														[109]	Evaluating success factors of green supply chain
		√													[110]	Analyzing barriers to green supply chain management
			√												[111]	Analyzing enablers in SSCM
AHP/Fuzzy AHP				√											[100–105,123, 136,137]	Selecting sustainable/green supplier, prioritizing solutions for reverse logistics, prioritizing the responses to manage risks, third party logistics (3PL) selection, and analyzing key factors for supply chain sustainability
					√										[106–108]	Evaluating green supply chain management practices, and sustainable supplier selection
				√		√									[112]	Analyzing supply chain strategy decisions
ANP/Fuzzy ANP							√								[118]	Green supplier selection
								√							[113–116]	Analyzing environmental sustainability, designing a sustainable maritime supply chain, global logistics service provider, and sustainable supplier selection

Table 7. *Cont.*

Method Name	Integrated with														References	Area of Application
	DEMATEL/ Fuzzy/Grey DEMATEL	ELECTRE/ Fuzzy ELECTRE	ISM	TOPSIS/ Fuzzy TOPSIS	VIKOR/ Fuzzy VIKOR	SOWIA	GRA	QFD	Rough Set	CRITIC	FPP	MABAAC	AQM	TODIM		
					✓										[117]	Sustainable supplier evaluation
	✓				✓										[138]	Sustainable supplier selection
	✓		✓												[139]	Investigating agri-produce sustainable supply chains
	✓			✓											[140]	Sustainable supplier selection
BWM/Fuzzy BWM	✓														[121,141]	Evaluating human resource dimensions of green supply chain
					✓										[119,120]	Evaluating sustainable transportation service providers, sustainable outsourcing partner selection
													✓		[122]	Sustainable supplier selection in watch manufacturing
					✓									✓	[142]	Evaluating measurement for sustainable supply chain finance
TOPSIS/Fuzzy TOPSIS					✓										[124,125, 143]	Third-party reverse logistics provider selection, classification of rural areas based on social sustainability indicators
					✓		✓								[126]	Location for remanufacturing plant
									✓						[127]	Green supply chain practices evaluation

Table 7. *Cont.*

Method Name	Integrated with														References	Area of Application
	DEMATEL/ Fuzzy/Grey DEMATEL	ELECTRE/ Fuzzy ELECTRE	ISM	TOPSIS/ Fuzzy TOPSIS	VIKOR/ Fuzzy VIKOR	SOWIA	GRA	QFD	Rough Set	CRITIC	FPP	MABAAC	AQM	TODIM		
										√					[128,144]	Evaluation of sustainable supply chain risk management
											√				[129]	Sustainable supplier selection
							√								[145]	Sustainable supplier selection for building materials
ELECTRE					√										[130]	Supply chain environmental performance evaluation
DEMATEL												√			[131]	Sustainable freight transport system evaluation

4. Bibliometric Analysis on MCDM Methods Applied to SSCM

This section presents a bibliometric analysis of MCDM methods applied to SSCM. From the finalized literature search (see Section 2), we can note a lack of reviews, with only five studies (about 4.5%) reviewing particular topics such as: (i) green supplier evaluation and selection [22]; (ii) modelling approaches in SSCM [146]; (iii) MCDM approaches in green supply chains [45]; (iv) hybrid MCDM for general sustainability [147]; and (v) sustainable supplier selection [148]. In total, 106 contributing articles (about 95.5%) applied MCDM methods to better understand SSCM issues; this means, Figure 2 presents the keyword network obtained from the keywords used in each of the contributing articles. It is evident that supply chain management, decision-making, sustainable development, sustainability, and green supply chains, environmental management, and sustainable supply chains are the top keywords. Figures 3 and 4 present the citation networks of selected contributing papers based on source journals and authors, respectively. The Journal of Cleaner Production and International Journal of Production Economics are two leading cited journals. Govindan, K., and Mangla, S.K. are two leading cited authors.

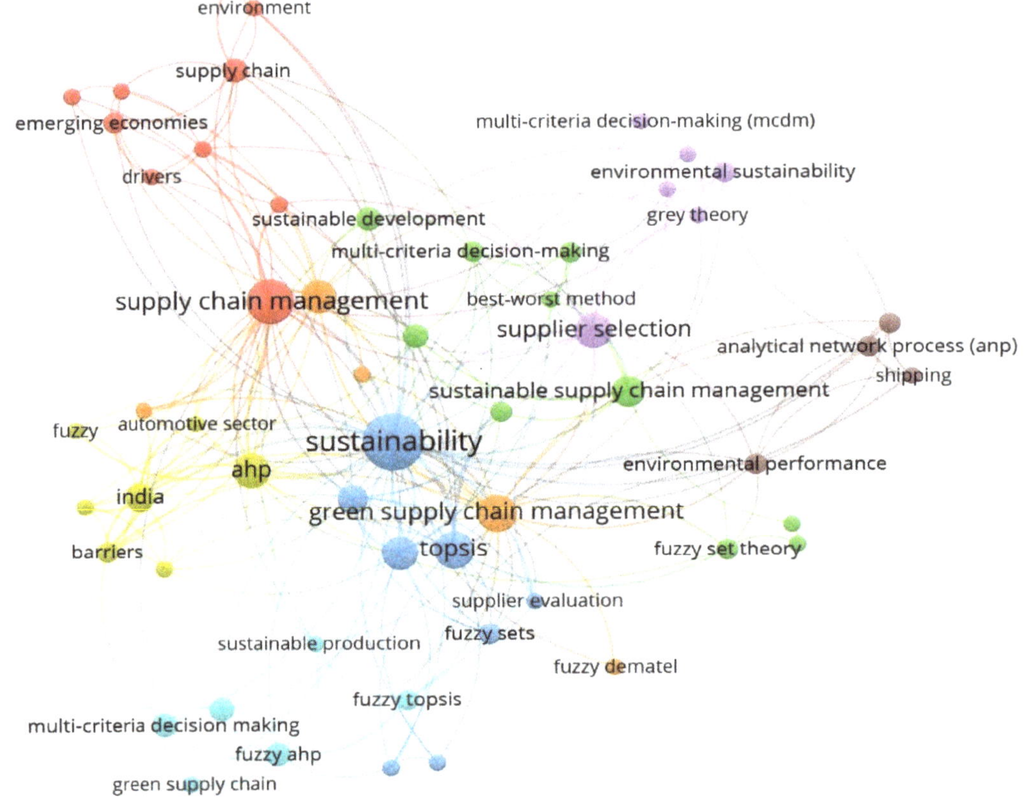

Figure 2. Co-occurrence of keywords used in the selected contributing papers (source: VOSviewer).

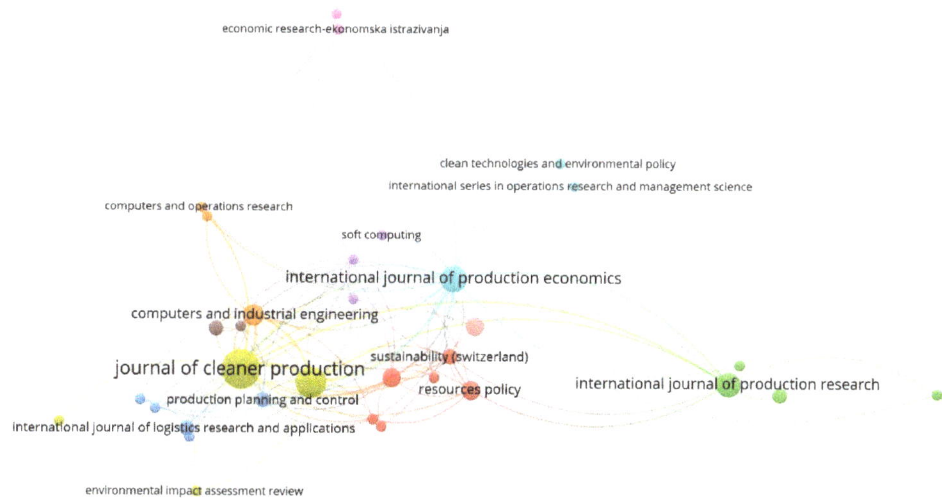

Figure 3. Co-occurrence of citation network from the source journals of the selected articles (source: VOSviewer).

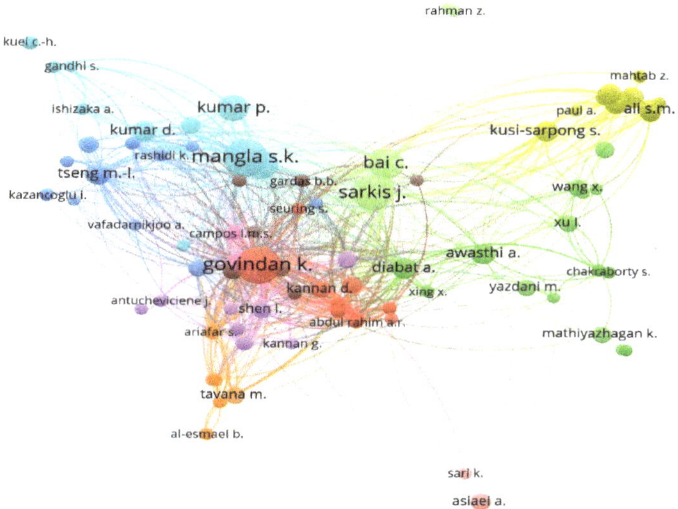

Figure 4. Co-occurrence of citation network from the authors of the selected articles (source: VOSviewer).

The number of contributing articles published from 2010 to 2020 is shown in Figure 5. From 2015, researchers started publishing an increasing number of articles applying MCDM methods in SSCM. The Journal of Cleaner Production (Publisher: Elsevier, Amsterdam, The Netherlands) has published the highest number of articles (18 articles), followed by Resource, Conservation and Recycling (Publisher: Elsevier), the International Journal of Production Economics (Publisher: Elsevier), Sustainability (Publisher: MDPI, Basel, Switzerland), and the International Journal of Production Research (Publisher: Taylor and Francis, Oxfordshire, UK). The number of articles published in each of these journals is presented in Figure 6.

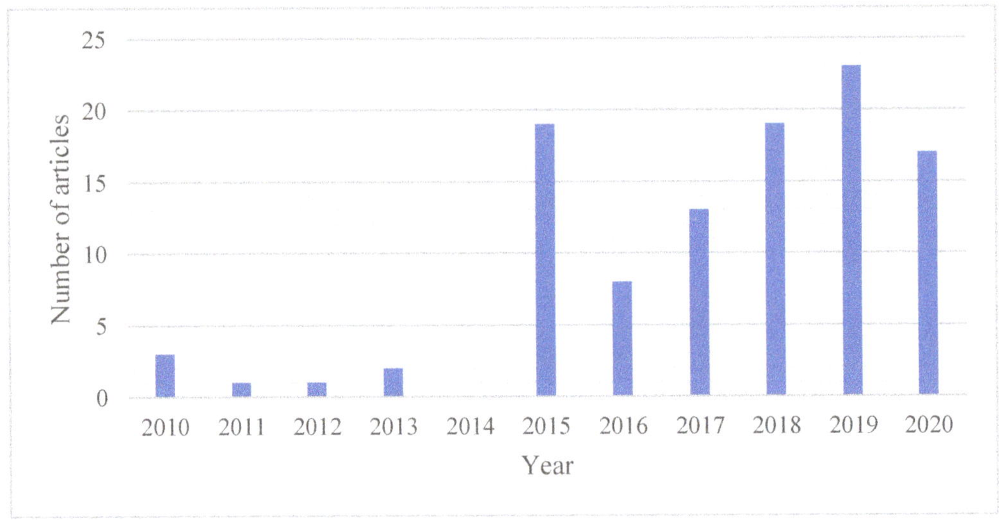

Figure 5. Number of articles published from 2010 to 2020.

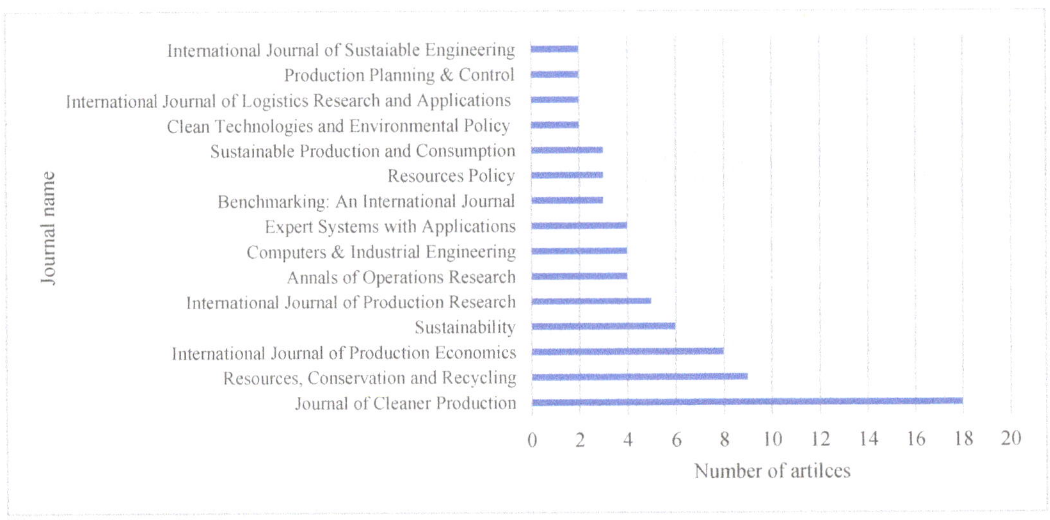

Figure 6. Number of articles published in different journals (N $\geq$ 2).

Figure 7 presents the authors who published the most articles in the area of MCDM for SSCM. Both Govindan, K. and Mangla, S.K. are at the top of the list, with 10 published articles, followed by Sarkis, J., Bai, C., and Luthra, S. with 9, 7, and 6 articles, respectively.

The affiliation by authors' institutions is also presented in Figure 8. Syddansk Universitet (University of Southern Denmark) is at the top of the list with 12 articles, followed by the Indian Institute of Technology Roorkee (India) and Dalian University of Technology (China). The affiliated countries which published most articles are presented in Figure 9. India is at the top of the list with 27, followed by China, the United States, and Denmark with 19, 15, and 12 articles, respectively.

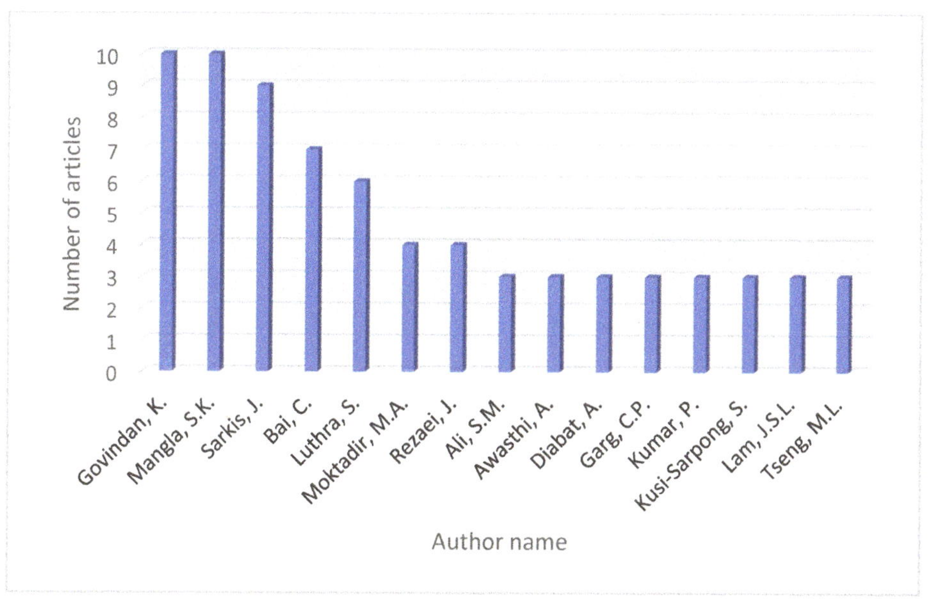

Figure 7. Articles published by different authors (N ≥ 3).

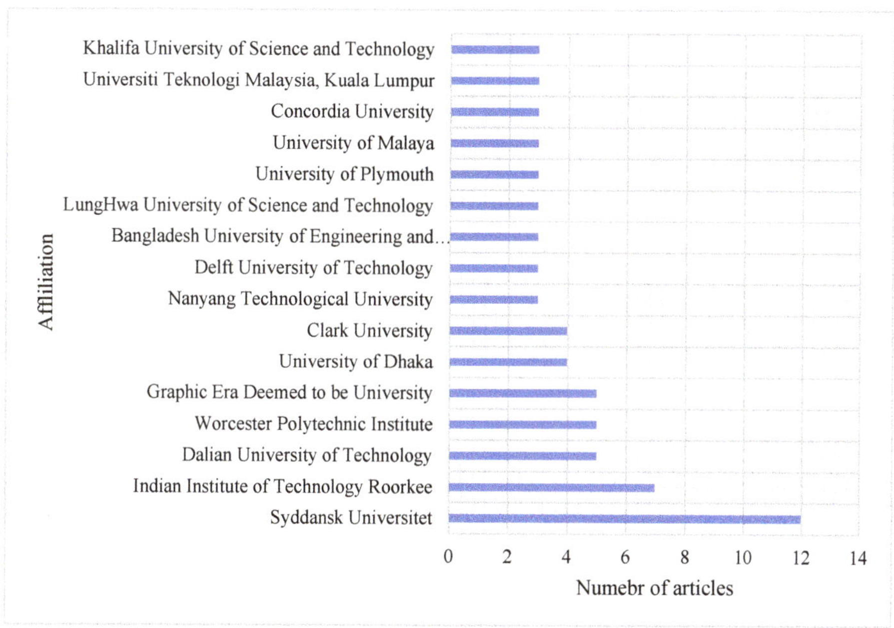

Figure 8. Most affiliated institutions (N ≥ 3).

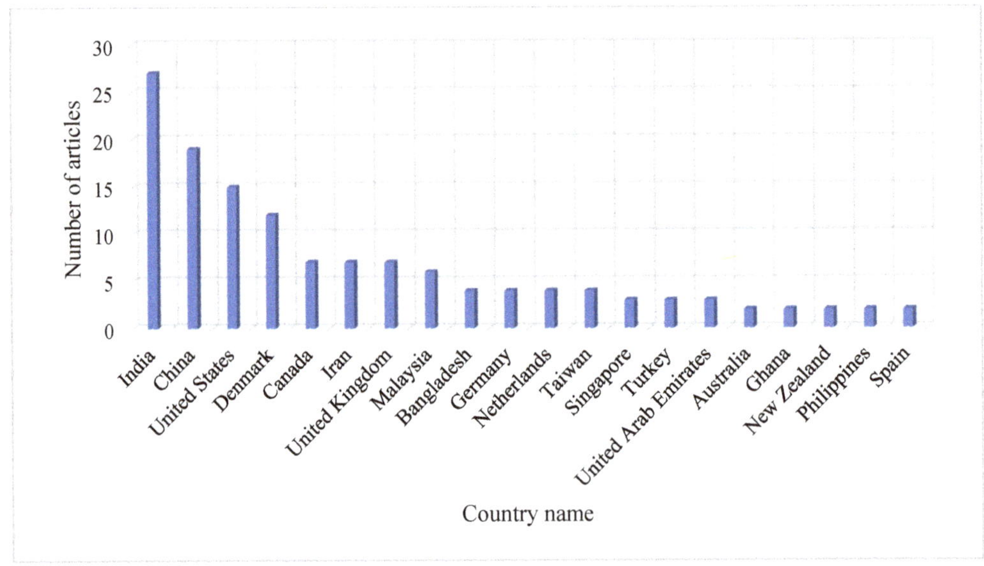

Figure 9. Source countries for publications (N ≥ 3).

Fifty-nine articles (about 56%) used individual MCDM methods for problem analysis, while the remaining 47 articles (about 44%) used integrated (two or more methods) methods. Figure 10 shows that integrated or hybrid MCDM methods were mostly applied in SSCM. Within individual MCDM methods: DEMATEL, AHP, and TOPSIS are the top three methods applied in SSCM literature.

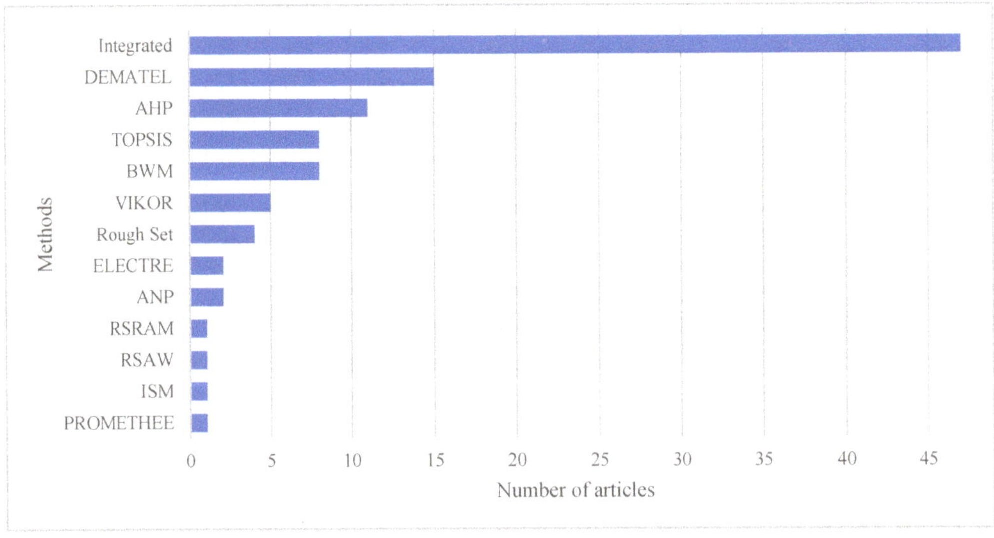

Figure 10. Distribution of contributing articles under different MCDM methods.

5. Summary of the Review and Research Gaps

The following key points present a summary of the extensive literature review and its findings.

Firstly, Section 3 described the application of different MCDM methods in SSCM, showing most studies applied an individual or two integrated MCDM methods, with a lack of studies integrating three or more MCDM methods for analyzing different dimensions of sustainability in supply chain management. Only a few studies integrated three methods (see Table 7); for example, AHP integrated with SOWIA and TOPSIS [112], TOPSIS integrated with GRA and VIKOR [126], and ANP integrated with DEMATEL and TOPSIS [140].

Secondly, in SSCM, most of the studies contributed to advancing the methodology by integrating different MCDM methods. MCDM methods were integrated within AHP, ANP, BWM, TOPSIS, ELECTRE, DEMATEL, ISM, VIKOR, SOWIA, GRA, QFD, Rough set, CRTIC, FPP, MABAAC, and AQM along with their fuzzy sets (see Table 7). It was observed that both single and integrated MCDM methods were applied, but there is a lack of studies which integrate MCDM methods with other operations research methods to improve the accuracy in decision making. For example, the integration of MCDM methods with mathematical modelling and optimization could help to obtain optimal decisions in SSCM.

Thirdly, most of the research considered social sustainability, environmental sustainability, and/or triple bottom line in the scope of their studies. Tables 6 and 7 presented the areas of application of individual and integrated MCDM methods. The application areas of MCDM methods were limited to different aspects of sustainability. Most of the studies integrated MCDM methods in models for green supply chains (see Table 7). Most of the studies on individual MCDM methods applied them in evaluating factors, barriers, challenges, drivers, enablers, and suppliers in different SSCM areas (see Tables 3–6).

Fourthly, most SSCM researchers limit their studies to analyzing factors, barriers, challenges, drivers, enablers, and suppliers (see Section 3). There are fewer studies that integrate sustainable strategies and performance with barriers/challenges and drivers/enablers using integrated MCDM methods.

Fifthly, from Section 3, most of the studies explained their results using individual or integrated MCDM methods. There are limited studies which compared the results between different MCDM methods. For example, there is only one study comparing fuzzy DEA and fuzzy TOPIS in sustainable supplier selection [76].

Finally, earlier literature largely overlooked small and medium enterprises (SMEs) and the emerging economy as study contexts for MCDM applied in SSCM. A small number of articles considered SMEs as their context, analyzing drivers for integrated lean-green manufacturing [149], innovation ability for supplier selection [150], determinants for cloud computing adoption [151], and entrepreneurship policies [152].

6. Conclusions and Future Research Directions

This paper presented a literature review on MCDM methods applied in different dimensions of SSCM. Although there have been many papers published in the area of SSCM, there are very few review papers published. A systematic literature analysis and review to identify different domains of supply chain sustainability and applications of different MCDM methods was lacking. Contributing to the academic discussion on this topic, this is the first effort to analyze both individual and integrated MCDM methods applied in different areas of supply chain sustainability.

Findings revealed the following important aspects of previous studies.

i.   Most of the research applied either individual or integrated methods of two MCDM techniques. All of the integrated MCDM methods applied were carried out in recent years, i.e., after 2015.

ii. Since 2015, environmental and social sustainability have been garnering the attention of researchers (see Tables 1 and 2). In recent years, several MCDM methods, such as ISM-TOPSIS [31], grey BWM-grey TODIM [18], and BWM [4,33] were applied in social sustainability while DEMATEL [25], fuzzy DEMATEL [23], fuzzy VIKOR [26], and TOPSIS [24] were applied in environmental sustainability areas.

iii. Most of the individual methods analyzed barriers, drivers, enablers, challenges, success factors, forces, and criteria in SSCM and green supply chains (see Tables 3–6).

iv. Integrated MCDM methods were applied in evaluating and analyzing sustainable suppliers and different alternatives in SSCM (see Table 7).

v. There are very few studies which integrated three or more MCDM methods and applied them in SSCM [112,126].

In conclusion, we would like to acknowledge some study limitations and propose some future research directions. At present, the study has been limited to three databases: Scopus (primary), Web of Science, and Google Scholar. Therefore, potential additional sources of information and knowledge such as conference proceedings or books not indexed in the three databases have not been included. Further, the research was limited to documents published in English exclusively, therefore potentially excluding local contributions redacted in other languages. Additionally, as there is limited research on the integration of three or more MCDM methods, it would be interesting to develop new integrated methodologies with three or more MCDM methods for decision-making in SSCM. Moreover, the fuzzy BWM method is a newly developed tool that can optimize a decision based on multiple criteria [153]. Currently, few studies apply BWM and fuzzy BWM in SSCM [4,13,79,119–122]. In the future, BWM and Fuzzy BWM could be applied widely in SSCM. In addition, the literature is lacking contributions able to integrate barriers and challenges with strategies and analyzes using an integrated MCDM method. In this regard, further research could explore the following research question: 'How can sustainable strategies be analyzed along with barriers and challenges to extend the current studies using MCDM methods?'

In the current literature, there is no study that correlated success factors, enablers, and drivers with sustainability performance analyzed using integrated MCDM methods. Hence, future research could address how barriers, challenges, drivers, enablers, and/or success factors can be linked with strategies and performance in one study, so as to develop a unique research framework using MCDM methods. In addition, it would be significantly new to the literature to integrate different aspects of SSCM with other supply chain areas such as risk management, lean supply chains, quality management, and supply chain network design. For example, the recent coronavirus (COVID-19) pandemic has impacted supply chains significantly [154]. It would be worth investigating the impacts of large-scale disruptions such as the COVID-19 pandemic in supply chain sustainability. Different MCDM methods can be applied to investigate different aspects of impacts on sustainable practices.

Our review pointed out that though service supply is an important area, research on applications of MCDM methods in sustainable service supply chains is very limited [130]. Different dimensions of sustainability in service supply chain management could be explored using MCDM methods.

In the future, another research direction could be integrating MCDM methods with operations research techniques, such as mathematical modelling and optimization, to make decisions more accurate. For example, integration of MCDM methods with mathematical modelling and optimization techniques could help to take optimal and accurate decision making in SSCM. Moreover, as there is only one study comparing the results of MCDM methods [76], it is necessary to develop benchmark problems and compare the results obtained from different MCDM methods used in SSCM. It would also be interesting to compare the sustainability results between supply chains from developed and emerging economies.

Finally, as SMEs and emerging economies play significant roles in local and global supply chains [155,156], it is important to analyze the sustainability of supply chains of SMEs and emerging economies. Usually, SMEs and emerging economies have fewer resources to deal with supply chain sustainability challenges. Hence, it is essential to analyze their supply chain sustainability and develop policies for improving sustainability practices. However, in the literature of MCDM methods and SSCM, SMEs and emerging economies were widely ignored. It would be interesting to develop decision-making tools using MCDM methods for analyzing different aspects of supply chain sustainability in the context of SMEs and emerging economies.

Author Contributions: Conceptualization, A.P.; methodology, A.P.; formal analysis, A.P.; investigation, A.P.; data curation, A.P.; writing—original draft preparation, A.P.; writing—review and editing, N.S., S.K.P., and A.T.; visualization, A.P. and S.K.P.; supervision, N.S. and A.T.; project administration, N.S. and A.T. All authors have read and agreed to the published version of the manuscript.

Funding: This research received no external funding.

Institutional Review Board Statement: Not applicable.

Informed Consent Statement: Informed consent was obtained from all subjects involved in the study.

Data Availability Statement: Not Applicable.

Acknowledgments: This research is supported by an Australian Government Research Training Program Scholarship.

Conflicts of Interest: The authors declare no conflict of interest.

References

1. Paul, S.K.; Sarker, R.; Essam, D. A quantitative model for disruption mitigation in a supply chain. *Eur. J. Oper. Res.* **2017**, *257*, 881–895. [CrossRef]
2. Mota, B.; Gomes, M.I.; Carvalho, A.; Barbosa-Povoa, A.P. Towards supply chain sustainability: Economic, environmental and social design and planning. *J. Clean. Prod.* **2015**, *105*, 14–27. [CrossRef]
3. Giannakis, M.; Papadopoulos, T. Supply chain sustainability: A risk management approach. *Int. J. Prod. Econ.* **2016**, *171*, 455–470. [CrossRef]
4. Munny, A.A.; Ali, S.M.; Kabir, G.; Moktadir, M.A.; Rahman, T.; Mahtab, Z. Enablers of social sustainability in the supply chain: An example of footwear industry from an emerging economy. *Sustain. Prod. Consum.* **2019**, *20*, 230–242. [CrossRef]
5. Green, K.W.; Zelbst, P.J.; Meacham, J.; Bhadauria, V.S. Green supply chain management practices: Impact on performance. *Supply Chain Manag. Int. J.* **2012**, *17*, 290–305. [CrossRef]
6. de Vries, G.J.; Ferrarini, B. What Accounts for the Growth of Carbon Dioxide Emissions in Advanced and Emerging Economies? The Role of Consumption, Technology and Global Supply Chain Participation. *Ecol. Econ.* **2017**, *132*, 213–223. [CrossRef]
7. Wittstruck, D.; Teuteberg, F. Understanding the Success Factors of Sustainable Supply Chain Management: Empirical Evidence from the Electrics and Electronics Industry. *Corp. Soc. Responsib. Environ. Manag.* **2012**, *19*, 141–158. [CrossRef]
8. Carter, C.; Easton, P.L. Sustainable supply chain management: Evolution and future directions. *Int. J. Phys. Distrib. Logist. Manag.* **2011**, *41*, 46–59. [CrossRef]
9. Shou, Y.; Shao, J.; Lai, K.-h.; Kang, M.; Park, Y. The impact of sustainability and operations orientations on sustainable supply management and the triple bottom line. *J. Clean. Prod.* **2019**, *240*, 118280. [CrossRef]
10. Padhi, S.S.; Pati, R.K.; Rajeev, A. Framework for selecting sustainable supply chain processes and industries using an integrated approach. *J. Clean. Prod.* **2018**, *184*, 969–984. [CrossRef]
11. Govindan, K.; Khodaverdi, R.; Jafarian, A. A fuzzy multi criteria approach for measuring sustainability performance of a supplier based on triple bottom line approach. *J. Clean. Prod.* **2013**, *47*, 345–354. [CrossRef]
12. Burki, U.; Ersoy, P.; Dahlstrom, R. Achieving triple bottom line performance in manufacturer-customer supply chains: Evidence from an emerging economy. *J. Clean. Prod.* **2018**, *197*, 1307–1316. [CrossRef]
13. Kusi-Sarpong, S.; Gupta, H.; Sarkis, J. A supply chain sustainability innovation framework and evaluation methodology. *IJPR* **2018**, *57*, 1990–2008. [CrossRef]
14. Sarkis, J.; Dhavale, D.G. Supplier selection for sustainable operations: A triple-bottom-line approach using a Bayesian framework. *Int. J. Prod. Econ.* **2015**, *166*, 177–191. [CrossRef]
15. Ahi, P.; Searcy, C. Assessing sustainability in the supply chain: A triple bottom line approach. *Appl. Math. Model.* **2015**, *39*, 2882–2896. [CrossRef]
16. Mani, V.; Gunasekaran, A. Four forces of supply chain social sustainability adoption in emerging economies. *Int. J. Prod. Econ.* **2018**, *199*, 150–161. [CrossRef]
17. Cole, R.; Aitken, J. Selecting suppliers for socially sustainable supply chain management:post-exchange supplier development activities as pre-selection requirements. *Prod. Plan. Control* **2019**, *30*, 1184–1202. [CrossRef]

18. Bai, C.; Kusi-Sarpong, S.; Badri Ahmadi, H.; Sarkis, J. Social sustainable supplier evaluation and selection: A group decision-support approach. *IJPR* **2019**, 1–22. [CrossRef]
19. Mani, V.; Gunasekaran, A.; Papadopoulos, T.; Hazen, B.; Dubey, R. Supply chain social sustainability for developing nations: Evidence from India. *Resour. Conserv. Recycl.* **2016**, *111*, 42–52. [CrossRef]
20. Mani, V.; Gunasekaran, A.; Delgado, C. Supply chain social sustainability: Standard adoption practices in Portuguese manufacturing firms. *Int. J. Prod. Econ.* **2018**, *198*, 149–164. [CrossRef]
21. Mani, V.; Gunasekaran, A.; Delgado, C. Enhancing supply chain performance through supplier social sustainability: An emerging economy perspective. *Int. J. Prod. Econ.* **2018**, *195*, 259–272. [CrossRef]
22. Govindan, K.; Rajendran, S.; Sarkis, J.; Murugesan, P. Multi criteria decision making approaches for green supplier evaluation and selection: A literature review. *J. Clean. Prod.* **2015**, *98*, 66–83. [CrossRef]
23. Kazancoglu, Y.; Kazancoglu, I.; Sagnak, M. Fuzzy DEMATEL-based green supply chain management performance. *Ind. Manag. Data Syst.* **2018**, *118*, 412–431. [CrossRef]
24. dos Santos, B.M.; Godoy, L.P.; Campos, L.M.S. Performance evaluation of green suppliers using entropy-TOPSIS-F. *J. Clean. Prod.* **2019**, *207*, 498–509. [CrossRef]
25. Wu, H.-H.; Chang, S.-Y. A case study of using DEMATEL method to identify critical factors in green supply chain management. *Appl. Math. Comput.* **2015**, *256*, 394–403. [CrossRef]
26. Rostamzadeh, R.; Govindan, K.; Esmaeili, A.; Sabaghi, M. Application of fuzzy VIKOR for evaluation of green supply chain management practices. *Ecol. Indic.* **2015**, *49*, 188–203. [CrossRef]
27. Pourjavad, E.; Shahin, A. The Application of Mamdani Fuzzy Inference System in Evaluating Green Supply Chain Management Performance. *Int. J. Fuzzy Syst.* **2017**, *20*, 901–912. [CrossRef]
28. Kazancoglu, Y.; Kazancoglu, I.; Sagnak, M. A new holistic conceptual framework for green supply chain management performance assessment based on circular economy. *J. Clean. Prod.* **2018**, *195*, 1282–1299. [CrossRef]
29. Islam, M.S.; Tseng, M.-L.; Karia, N.; Lee, C.-H. Assessing green supply chain practices in Bangladesh using fuzzy importance and performance approach. *Resour. Conserv. Recycl.* **2018**, *131*, 134–145. [CrossRef]
30. Stevenson, M.; Cole, R. Modern slavery in supply chains: A secondary data analysis of detection, remediation and disclosure. *Supply Chain Manag. Int. J.* **2018**, *23*, 81–99. [CrossRef]
31. Mani, V.; Agrawal, R.; Sharma, V. Social sustainability in the supply chain: Analysis of enablers. *Manag. Res. Rev.* **2015**, *38*, 1016–1042. [CrossRef]
32. Khosravi, F.; Izbirak, G. A stakeholder perspective of social sustainability measurement in healthcare supply chain management. *Sustain. Cities Soc.* **2019**, *50*, 101681. [CrossRef]
33. Badri Ahmadi, H.; Kusi-Sarpong, S.; Rezaei, J. Assessing the social sustainability of supply chains using Best Worst Method. *Resour. Conserv. Recycl.* **2017**, *126*, 99–106. [CrossRef]
34. Hussain, M.; Ajmal, M.M.; Gunasekaran, A.; Khan, M. Exploration of social sustainability in healthcare supply chain. *J. Clean. Prod.* **2018**, *203*, 977–989. [CrossRef]
35. Tseng, M.-L.; Islam, M.S.; Karia, N.; Fauzi, F.A.; Afrin, S. A literature review on green supply chain management: Trends and future challenges. *Resour. Conserv. Recycl.* **2019**, *141*, 145–162. [CrossRef]
36. Maditati, D.R.; Munim, Z.H.; Schramm, H.-J.; Kummer, S. Recycling, A review of green supply chain management: From bibliometric analysis to a conceptual framework and future research directions. *Resour. Conserv. Recycl.* **2018**, *139*, 150–162. [CrossRef]
37. Fahimnia, B.; Sarkis, J.; Davarzani, H. Green supply chain management: A review and bibliometric analysis. *Green Supply Chain Manag. A Rev. Bibliometr. Anal.* **2015**, *162*, 101–114. [CrossRef]
38. Touboulic, A.; Walker, H. Theories in sustainable supply chain management: A structured literature review. *Int. J. Phys. Distrib. Logist. Manag.* **2015**, *45*, 16–42. [CrossRef]
39. Rajeev, A.; Pati, R.K.; Padhi, S.S.; Govindan, K. Evolution of sustainability in supply chain management: A literature review. *J. Clean. Prod.* **2017**, *162*, 299–314. [CrossRef]
40. Ansari, Z.N.; Kant, R. A state-of-art literature review reflecting 15 years of focus on sustainable supply chain management. *J. Clean. Prod.* **2017**, *142*, 2524–2543. [CrossRef]
41. Ghadimi, P.; Wang, C.; Lim, M.K. Sustainable supply chain modeling and analysis: Past debate, present problems and future challenges. *Resour. Conserv. Recycl.* **2019**, *140*, 72–84. [CrossRef]
42. Bubicz, M.E.; Barbosa-Póvoa, A.P.F.D.; Carvalho, A. Incorporating social aspects in sustainable supply chains: Trends and future directions. *J. Clean. Prod.* **2019**, *237*, 117500. [CrossRef]
43. Koberg, E.; Longoni, A. A systematic review of sustainable supply chain management in global supply chains. *J. Clean. Prod.* **2019**, *207*, 1084–1098. [CrossRef]
44. Saeed, M.A.; Kersten, W. Drivers of sustainable supply chain management: Identification and classification. *Sustainability* **2019**, *11*, 1137. [CrossRef]
45. Banasik, A.; Bloemhof-Ruwaard, J.M.; Kanellopoulos, A.; Claassen, G.D.H.; van der Vorst, J.G.A.J. Multi-criteria decision making approaches for green supply chains: A review. *Flex. Serv. Manuf. J.* **2016**, *30*, 366–396. [CrossRef]
46. Zandieh, M.; Aslani, B. A hybrid MCDM approach for order distribution in a multiple-supplier supply chain: A case study. *J. Ind. Inf. Integr.* **2019**, *16*, 100104. [CrossRef]

47. Luthra, S.; Mangla, S.K.; Shankar, R.; Prakash Garg, C.; Jakhar, S. Modelling critical success factors for sustainability initiatives in supply chains in Indian context using Grey-DEMATEL. *Prod. Plan. Control* **2018**, *29*, 705–728. [CrossRef]
48. Sharma, Y.K.; Mangla, S.K.; Patil, P.P.; Uniyal, S. Sustainable Food Supply Chain Management Implementation Using DEMATEL Approach. In *Advances in Health and Environment Safety*; Springer: Singapore, 2018; pp. 115–125.
49. Wu, K.-J.; Liao, C.-J.; Tseng, M.-L.; Chiu, A.S.F. Exploring decisive factors in green supply chain practices under uncertainty. *Int. J. Prod. Econ.* **2015**, *159*, 147–157. [CrossRef]
50. Bhagawati, M.T.; Manavalan, E.; Jayakrishna, K.; Venkumar, P. Identifying Key Success Factors of Sustainability in Supply Chain Management for Industry 4.0 Using DEMATEL Method. In Proceedings of the International Conference on Intelligent Manufacturing and Automation, Penang, Malaysia, 18–20 December 2018; pp. 583–591.
51. Gandhi, S.; Mangla, S.K.; Kumar, P.; Kumar, D. Evaluating factors in implementation of successful green supply chain management using DEMATEL: A case study. *Int. Strateg. Manag. Rev.* **2015**, *3*, 96–109. [CrossRef]
52. Gardas, B.B.; Raut, R.D.; Narkhede, B. Modelling the challenges to sustainability in the textile and apparel (T&A) sector: A Delphi-DEMATEL approach. *Sustain. Prod. Consum.* **2018**, *15*, 96–108.
53. Bhatia, M.S.; Srivastava, R.K. Analysis of external barriers to remanufacturing using grey-DEMATEL approach: An Indian perspective. *Resour. Conserv. Recycl.* **2018**, *136*, 79–87. [CrossRef]
54. Kaur, J.; Sidhu, R.; Awasthi, A.; Chauhan, S.; Goyal, S. A DEMATEL based approach for investigating barriers in green supply chain management in Canadian manufacturing firms. *IJPR* **2017**, *56*, 312–332. [CrossRef]
55. Luthra, S.; Govindan, K.; Mangla, S.K. Structural model for sustainable consumption and production adoption—A grey-DEMATEL based approach. *Resour. Conserv. Recycl.* **2017**, *125*, 198–207. [CrossRef]
56. Luthra, S.; Mangla, S.K.; Chan, F.T.S.; Venkatesh, V.G. Evaluating the Drivers to Information and Communication Technology for Effective Sustainability Initiatives in Supply Chains. *Int. J. Inf. Technol. Decis. Mak.* **2018**, *17*, 311–338. [CrossRef]
57. Su, C.-M.; Horng, D.-J.; Tseng, M.-L.; Chiu, A.S.F.; Wu, K.-J.; Chen, H.-P. Improving sustainable supply chain management using a novel hierarchical grey-DEMATEL approach. *J. Clean. Prod.* **2016**, *134*, 469–481. [CrossRef]
58. Li, Y.; Mathiyazhagan, K. Application of DEMATEL approach to identify the influential indicators towards sustainable supply chain adoption in the auto components manufacturing sector. *J. Clean. Prod.* **2018**, *172*, 2931–2941. [CrossRef]
59. Govindan, K.; Khodaverdi, R.; Vafadarnikjoo, A. Intuitionistic fuzzy based DEMATEL method for developing green practices and performances in a green supply chain. *Expert Syst. Appl.* **2015**, *42*, 7207–7220. [CrossRef]
60. Lin, K.-P.; Tseng, M.-L.; Pai, P.-F. Sustainable supply chain management using approximate fuzzy DEMATEL method. *Resour. Conserv. Recycl.* **2018**, *128*, 134–142. [CrossRef]
61. Luthra, S.; Mangla, S.K.; Xu, L.; Diabat, A. Using AHP to evaluate barriers in adopting sustainable consumption and production initiatives in a supply chain. *Int. J. Prod. Econ.* **2016**, *181*, 342–349. [CrossRef]
62. Shen, L.; Muduli, K.; Barve, A. Developing a sustainable development framework in the context of mining industries: AHP approach. *Resour. Policy* **2015**, *46*, 15–26. [CrossRef]
63. Luthra, S.; Mangla, S.K. Evaluating challenges to Industry 4.0 initiatives for supply chain sustainability in emerging economies. *Process Saf. Environ. Prot.* **2018**, *117*, 168–179. [CrossRef]
64. Mathiyazhagan, K.; Diabat, A.; Al-Refaie, A.; Xu, L. Application of analytical hierarchy process to evaluate pressures to implement green supply chain management. *J. Clean. Prod.* **2015**, *107*, 229–236. [CrossRef]
65. Gupta, S.; Dangayach, G.S.; Singh, A.K.; Rao, P.N. Analytic Hierarchy Process (AHP) Model for Evaluating Sustainable Manufacturing Practices in Indian Electrical Panel Industries. *Procedia Soc. Behav. Sci.* **2015**, *189*, 208–216. [CrossRef]
66. Shankar, K.; Kumar, P.; Kannan, D. Analyzing the Drivers of Advanced Sustainable Manufacturing System Using AHP Approach. *Sustainability* **2016**, *8*, 824. [CrossRef]
67. Mangla, S.K.; Kumar, P.; Barua, M.K. Risk analysis in green supply chain using fuzzy AHP approach: A case study. *Resour. Conserv. Recycl.* **2015**, *104*, 375–390. [CrossRef]
68. Sharma, Y.K.; Yadav, A.K.; Mangla, S.K.; Patil, P.P. Ranking the Success Factors to Improve Safety and Security in Sustainable Food Supply Chain Management Using Fuzzy AHP. *Mater. Today Proc.* **2018**, *5*, 12187–12196. [CrossRef]
69. Kumar, D.; Garg, C.P. Evaluating sustainable supply chain indicators using fuzzy AHP. *Benchmarking Int. J.* **2017**, *24*, 1742–1766. [CrossRef]
70. Mejías, A.M.; Bellas, R.; Pardo, J.E.; Paz, E. Traceability management systems and capacity building as new approaches for improving sustainability in the fashion multi-tier supply chain. *Int. J. Prod. Econ.* **2019**, *217*, 143–158. [CrossRef]
71. Mastrocinque, E.; Ramírez, F.J.; Honrubia-Escribano, A.; Pham, D.T. An AHP-based multi-criteria model for sustainable supply chain development in the renewable energy sector. *Expert Syst. Appl.* **2020**, *150*, 113321. [CrossRef]
72. Li, J.; Fang, H.; Song, W. Sustainable supplier selection based on SSCM practices: A rough cloud TOPSIS approach. *J. Clean. Prod.* **2019**, *222*, 606–621. [CrossRef]
73. Bai, C.; Sarkis, J. Integrating Sustainability into Supplier Selection: A Grey-Based Topsis Analysis. *Technol. Econ. Dev. Econ.* **2018**, *24*, 2202–2224. [CrossRef]
74. Shen, L.; Olfat, L.; Govindan, K.; Khodaverdi, R.; Diabat, A. A fuzzy multi criteria approach for evaluating green supplier's performance in green supply chain with linguistic preferences. *Resour. Conserv. Recycl.* **2013**, *74*, 170–179. [CrossRef]
75. Rouyendegh, B.D.; Yildizbasi, A.; Üstünyer, P. Intuitionistic Fuzzy TOPSIS method for green supplier selection problem. *Soft Comput.* **2020**, *24*, 2215–2228. [CrossRef]

76. Rashidi, K.; Cullinane, K. A comparison of fuzzy DEA and fuzzy TOPSIS in sustainable supplier selection: Implications for sourcing strategy. *Expert Syst. Appl.* **2019**, *121*, 266–281. [CrossRef]
77. Memari, A.; Dargi, A.; Akbari Jokar, M.R.; Ahmad, R.; Abdul Rahim, A.R. Sustainable supplier selection: A multi-criteria intuitionistic fuzzy TOPSIS method. *J. Manuf. Syst.* **2019**, *50*, 9–24. [CrossRef]
78. Wang, X.; Chan, H.K. A hierarchical fuzzy TOPSIS approach to assess improvement areas when implementing green supply chain initiatives. *IJPR* **2013**, *51*, 3117–3130. [CrossRef]
79. Suhi, S.A.; Enayet, R.; Haque, T.; Ali, S.M.; Moktadir, M.A.; Paul, S.K. Environmental sustainability assessment in supply chain: An emerging economy context. *Env. Impact Assess. Rev.* **2019**, *79*, 106306. [CrossRef]
80. Wan Ahmad, W.N.K.; Rezaei, J.; Sadaghiani, S.; Tavasszy, L.A. Evaluation of the external forces affecting the sustainability of oil and gas supply chain using Best Worst Method. *J. Clean. Prod.* **2017**, *153*, 242–252. [CrossRef]
81. Rezaei, J.; Papakonstantinou, A.; Tavasszy, L.; Pesch, U.; Kana, A. Sustainable product-package design in a food supply chain: A multi-criteria life cycle approach. *Packag. Technol. Sci.* **2019**, *32*, 85–101. [CrossRef]
82. Jafarzadeh Ghoushchi, S.; Khazaeili, M.; Amini, A.; Osgooei, E. Multi-criteria sustainable supplier selection using piecewise linear value function and fuzzy best-worst method. *J. Intell. Fuzzy Syst.* **2019**, *37*, 2309–2325. [CrossRef]
83. Gupta, H.; Kusi-Sarpong, S.; Rezaei, J. Barriers and overcoming strategies to supply chain sustainability innovation. *Resour. Conserv. Recycl.* **2020**, *161*, 104819. [CrossRef]
84. Awasthi, A.; Kannan, G. Green supplier development program selection using NGT and VIKOR under fuzzy environment. *Comput. Ind. Eng.* **2016**, *91*, 100–108. [CrossRef]
85. Haji Vahabzadeh, A.; Asiaei, A.; Zailani, S. Green decision-making model in reverse logistics using FUZZY-VIKOR method. *Resour. Conserv. Recycl.* **2015**, *103*, 125–138. [CrossRef]
86. Haji Vahabzadeh, A.; Asiaei, A.; Zailani, S. Reprint of "Green decision-making model in reverse logistics using FUZZY-VIKOR method". *Resour. Conserv. Recycl.* **2015**, *104*, 334–347. [CrossRef]
87. Zhang, X.; Xing, X. Probabilistic Linguistic VIKOR Method to Evaluate Green Supply Chain Initiatives. *Sustainability* **2017**, *9*, 1231. [CrossRef]
88. Bai, C.; Sarkis, J. Integrating sustainability into supplier selection with grey system and rough set methodologies. *Int. J. Prod. Econ.* **2010**, *124*, 252–264. [CrossRef]
89. Bai, C.; Sarkis, J. Green supplier development: Analytical evaluation using rough set theory. *J. Clean. Prod.* **2010**, *18*, 1200–1210. [CrossRef]
90. Bai, C.; Sarkis, J.; Wei, X. Addressing key sustainable supply chain management issues using rough set methodology. *Manag. Res. Rev.* **2010**, *33*, 1113–1127. [CrossRef]
91. Bai, C.; Sarkis, J. Performance Measurement and Evaluation for Sustainable Supply Chains using Rough Set and Data Envelopment Analysis. In *Sustainable Supply Chains*; Springer: New York, NY, USA, 2012; pp. 223–241.
92. Costa, A.S.; Govindan, K.; Figueira, J.R. Supplier classification in emerging economies using the ELECTRE TRI-nC method: A case study considering sustainability aspects. *J. Clean. Prod.* **2018**, *201*, 925–947. [CrossRef]
93. Kumar, P.; Singh, R.K.; Vaish, A. Suppliers' green performance evaluation using fuzzy extended ELECTRE approach. *Clean Technol. Environ. Policy* **2016**, *19*, 809–821. [CrossRef]
94. Lin, C.; Madu, C.N.; Kuei, C.-h.; Tsai, H.-L.; Wang, K.-n. Developing an assessment framework for managing sustainability programs: A Analytic Network Process approach. *Expert Syst. Appl.* **2015**, *42*, 2488–2501. [CrossRef]
95. Faisal, M.N.; Al-Esmael, B.; Sharif, K.J. Supplier selection for a sustainable supply chain. *Benchmarking Int. J.* **2017**, *24*, 1956–1976. [CrossRef]
96. Song, W.; Ming, X.; Liu, H.-C. Identifying critical risk factors of sustainable supply chain management: A rough strength-relation analysis method. *J. Clean. Prod.* **2017**, *143*, 100–115. [CrossRef]
97. Stević, Ž.; Durmić, E.; Gajić, M.; Pamučar, D.; Puška, A. A Novel Multi-Criteria Decision-Making Model: Interval Rough SAW Method for Sustainable Supplier Selection. *Information* **2019**, *10*, 292. [CrossRef]
98. Raut, R.; Gardas, B.B.; Narkhede, B. Ranking the barriers of sustainable textile and apparel supply chains. *Benchmarking: Int. J.* **2019**, *26*, 371–394. [CrossRef]
99. Pehlken, A.; Wulf, K.; Grecksch, K.; Klenke, T.; Tsydenova, N. More Sustainable Bioenergy by Making Use of Regional Alternative Biomass? *Sustainability* **2020**, *12*, 7849. [CrossRef]
100. Sirisawat, P.; Kiatcharoenpol, T. Fuzzy AHP-TOPSIS approaches to prioritizing solutions for reverse logistics barriers. *Comput. Ind. Eng.* **2018**, *117*, 303–318. [CrossRef]
101. Singh, R.K.; Gunasekaran, A.; Kumar, P. Third party logistics (3PL) selection for cold chain management: A fuzzy AHP and fuzzy TOPSIS approach. *AnOR* **2017**, *267*, 531–553. [CrossRef]
102. Mangla, S.K.; Kumar, P.; Barua, M.K. Prioritizing the responses to manage risks in green supply chain: An Indian plastic manufacturer perspective. *Sustain. Prod. Consum.* **2015**, *1*, 67–86. [CrossRef]
103. Freeman, J.; Gary Graham, D.; Chen, T. Green supplier selection using an AHP-Entropy-TOPSIS framework. *Supply Chain Manag. Int. J.* **2015**, *20*, 327–340. [CrossRef]
104. Azimifard, A.; Moosavirad, S.H.; Ariafar, S. Selecting sustainable supplier countries for Iran's steel industry at three levels by using AHP and TOPSIS methods. *Resour. Policy* **2018**, *57*, 30–44. [CrossRef]
105. Mohammed, A.; Harris, I.; Govindan, K. A hybrid MCDM-FMOO approach for sustainable supplier selection and order allocation. *Int. J. Prod. Econ.* **2019**, *217*, 171–184. [CrossRef]

106. Sari, K. A novel multi-criteria decision framework for evaluating green supply chain management practices. *Comput. Ind. Eng.* **2017**, *105*, 338–347. [CrossRef]
107. Luthra, S.; Govindan, K.; Kannan, D.; Mangla, S.K.; Garg, C.P. An integrated framework for sustainable supplier selection and evaluation in supply chains. *J. Clean. Prod.* **2017**, *140*, 1686–1698. [CrossRef]
108. Awasthi, A.; Govindan, K.; Gold, S. Multi-tier sustainable global supplier selection using a fuzzy AHP-VIKOR based approach. *Int. J. Prod. Econ.* **2018**, *195*, 106–117. [CrossRef]
109. Gandhi, S.; Mangla, S.K.; Kumar, P.; Kumar, D. A combined approach using AHP and DEMATEL for evaluating success factors in implementation of green supply chain management in Indian manufacturing industries. *Int. J. Logist. Res. Appl.* **2016**, *19*, 537–561. [CrossRef]
110. Uddin, S.; Ali, S.M.; Kabir, G.; Suhi, S.A.; Enayet, R.; Haque, T. An AHP-ELECTRE framework to evaluate barriers to green supply chain management in the leather industry. *Int. J. Sustain. Dev. World Ecol.* **2019**, 1–20. [CrossRef]
111. Kumar, D.; Rahman, Z.; Huang, Z.; Wang, K. Analyzing enablers of sustainable supply chain: ISM and Fuzzy AHP approach. *J. Model. Manag.* **2017**, *12*, 498–524. [CrossRef]
112. Sreekumar, V.; Rajmohan, M. Supply chain strategy decisions for sustainable development using an integrated multi-criteria decision-making approach. *Sustain. Dev.* **2019**, *27*, 50–60. [CrossRef]
113. Tavana, M.; Yazdani, M.; Di Caprio, D. An application of an integrated ANP–QFD framework for sustainable supplier selection. *Int. J. Logist. Res. Appl.* **2016**, *20*, 254–275. [CrossRef]
114. Lam, J.S.L.; Lai, K.-h. Developing environmental sustainability by ANP-QFD approach: The case of shipping operations. *J. Clean. Prod.* **2015**, *105*, 275–284. [CrossRef]
115. Lam, J.S.L.; Dai, J. Environmental sustainability of logistics service provider: An ANP-QFD approach. *Int. J. Logist. Manag.* **2015**, *26*, 313–333. [CrossRef]
116. Lam, J.S.L. Designing a sustainable maritime supply chain: A hybrid QFD–ANP approach. *Transp. Res. Part E Logist. Transp. Rev.* **2015**, *78*, 70–81. [CrossRef]
117. Liu, K.; Liu, Y.; Qin, J. An integrated ANP-VIKOR methodology for sustainable supplier selection with interval type-2 fuzzy sets. *Granul. Comput.* **2018**, *3*, 193–208. [CrossRef]
118. Hashemi, S.H.; Karimi, A.; Tavana, M. An integrated green supplier selection approach with analytic network process and improved Grey relational analysis. *Int. J. Prod. Econ.* **2015**, *159*, 178–191. [CrossRef]
119. Paul, A.; Moktadir, M.A.; Paul, S.K. An innovative decision-making framework for evaluating transportation service providers based on sustainable criteria. *IJPR* **2020**, *58*, 7334–7352. [CrossRef]
120. Garg, C.P.; Sharma, A. Sustainable outsourcing partner selection and evaluation using an integrated BWM–VIKOR framework. *Env. Dev. Sustain.* **2018**, 1–29. [CrossRef]
121. Kumar, A.; Mangla, S.K.; Luthra, S.; Ishizaka, A. Evaluating the human resource related soft dimensions in green supply chain management implementation. *Prod. Plan. Control* **2019**, *30*, 699–715. [CrossRef]
122. Liu, H.-C.; Quan, M.-Y.; Li, Z.; Wang, Z.-L. A new integrated MCDM model for sustainable supplier selection under interval-valued intuitionistic uncertain linguistic environment. *Inf. Sci.* **2019**, *486*, 254–270. [CrossRef]
123. Jain, N.; Singh, A.R.; Upadhyay, R.K. Sustainable supplier selection under attractive criteria through FIS and integrated fuzzy MCDM techniques. *Int. J. Sustain. Eng.* **2020**, *13*, 441–462. [CrossRef]
124. Papathanasiou, J.P.; Nikolaos, P.; Bournaris, T.; Manos, B. A Decision Support System for Multiple Criteria Alternative Ranking Using TOPSIS and VIKOR: A Case Study on Social Sustainability in Agriculture. In *Decision Support Systems VI—Addressing Sustainability and Societal Challenges, Lecture Notes in Business Information Processing*; Springer International Publishing: Basel, Switzerland, 2016; pp. 1–13.
125. Bai, C.; Sarkis, J. Integrating and extending data and decision tools for sustainable third-party reverse logistics provider selection. *Comput. Oper. Res.* **2019**, *110*, 188–207. [CrossRef]
126. Bhatia, M.S.; Dora, M.; Jakhar, S.K. Appropriate location for remanufacturing plant towards sustainable supply chain. *AnOR* **2019**, 1–22. [CrossRef]
127. Kusi-Sarpong, S.; Bai, C.; Sarkis, J.; Wang, X. Green supply chain practices evaluation in the mining industry using a joint rough sets and fuzzy TOPSIS methodology. *Resour. Policy* **2015**, *46*, 86–100. [CrossRef]
128. Rostamzadeh, R.; Ghorabaee, M.K.; Govindan, K.; Esmaeili, A.; Nobar, H.B.K. Evaluation of sustainable supply chain risk management using an integrated fuzzy TOPSIS- CRITIC approach. *J. Clean. Prod.* **2018**, *175*, 651–669.
129. Fallahpour, A.; Udoncy Olugu, E.; Nurmaya Musa, S.; Yew Wong, K.; Noori, S. A decision support model for sustainable supplier selection in sustainable supply chain management. *Comput. Ind. Eng.* **2017**, *105*, 391–410. [CrossRef]
130. Chithambaranathan, P.; Subramanian, N.; Gunasekaran, A.; Palaniappan, P.K. Service supply chain environmental performance evaluation using grey based hybrid MCDM approach. *Int. J. Prod. Econ.* **2015**, *166*, 163–176. [CrossRef]
131. Yazdani, M.; Pamucar, D.; Chatterjee, P.; Chakraborty, S. Development of a decision support framework for sustainable freight transport system evaluation using rough numbers. *IJPR* **2019**, 1–27. [CrossRef]
132. Matić, B.; Jovanović, S.; Das, D.K.; Zavadskas, E.K.; Stević, Ž.; Sremac, S.; Marinković, M. A New Hybrid MCDM Model: Sustainable Supplier Selection in a Construction Company. *Symmetry* **2019**, *11*, 353.
133. Roy, J.; Pamučar, D.; Kar, S. Evaluation and selection of third party logistics provider under sustainability perspectives: An interval valued fuzzy-rough approach. *AnOR* **2020**, *293*, 669–714. [CrossRef]

134. Ansari, Z.N.; Kant, R.; Shankar, R. Evaluation and ranking of solutions to mitigate sustainable remanufacturing supply chain risks: A hybrid fuzzy SWARA-fuzzy COPRAS framework approach. *Int. J. Sustain. Eng.* **2020**, *13*, 473–494. [CrossRef]
135. Erol, I.; Sencer, S.; Sari, R. A new fuzzy multi-criteria framework for measuring sustainability performance of a supply chain. *Ecol. Econ.* **2011**, *70*, 1088–1100. [CrossRef]
136. Muhammad, N.; Fang, Z.; Shah, S.A.A.; Akbar, M.A.; Alsanad, A.; Gumaei, A.; Solangi, Y.A. A Hybrid Multi-Criteria Approach for Evaluation and Selection of Sustainable Suppliers in the Avionics Industry of Pakistan. *Sustainability* **2020**, *12*, 4744. [CrossRef]
137. Sharma, R.K.; Singh, P.K.; Sarkar, P.; Singh, H. A hybrid multi-criteria decision approach to analyze key factors affecting sustainability in supply chain networks of manufacturing organizations. *Clean Technol. Environ. Policy* **2020**, *22*, 1871–1889. [CrossRef]
138. Phochanikorn, P.; Tan, C. A New Extension to a Multi-Criteria Decision-Making Model for Sustainable Supplier Selection under an Intuitionistic Fuzzy Environment. *Sustainability* **2019**, *11*, 5413. [CrossRef]
139. Chauhan, A.; Kaur, H.; Yadav, S.; Jakhar, S.K. A hybrid model for investigating and selecting a sustainable supply chain for agri-produce in India. *AnOR* **2019**, *290*, 621–642. [CrossRef]
140. Tirkolaee, E.B.; Mardani, A.; Dashtian, Z.; Soltani, M.; Weber, G.-W. A novel hybrid method using fuzzy decision making and multi-objective programming for sustainable-reliable supplier selection in two-echelon supply chain design. *J. Clean. Prod.* **2020**, *250*, 119517. [CrossRef]
141. Yazdani, M.; Torkayesh, A.E.; Chatterjee, P. An integrated decision-making model for supplier evaluation in public healthcare system: The case study of a Spanish hospital. *J. Enterp. Inf. Manag.* **2020**, *33*, 965–989. [CrossRef]
142. Abdel-Basset, M.; Mohamed, R.; Sallam, K.; Elhoseny, M. A novel decision-making model for sustainable supply chain finance under uncertainty environment. *J. Clean. Prod.* **2020**, *269*, 122324. [CrossRef]
143. Rajesh, R. Sustainable supply chains in the Indian context: An integrative decision-making model. *Technol. Soc.* **2020**, *61*, 101230. [CrossRef]
144. Abdel-Basset, M.; Mohamed, R. A novel plithogenic TOPSIS- CRITIC model for sustainable supply chain risk management. *J. Clean. Prod.* **2020**, *247*, 119586. [CrossRef]
145. Chen, C. A New Multi-Criteria Assessment Model Combining GRA Techniques with Intuitionistic Fuzzy Entropy-Based TOPSIS Method for Sustainable Building Materials Supplier Selection. *Sustainability* **2019**, *11*, 2265. [CrossRef]
146. Seuring, S. A review of modeling approaches for sustainable supply chain management. *Decis. Support Syst.* **2013**, *54*, 1513–1520. [CrossRef]
147. Zavadskas, E.K.; Govindan, K.; Antucheviciene, J.; Turskis, Z. Hybrid multiple criteria decision-making methods: A review of applications for sustainability issues. *Econ. Res. Ekon. Istraživanja* **2016**, *29*, 857–887. [CrossRef]
148. Schramm, V.B.; Cabral, L.P.B.; Schramm, F. Approaches for supporting sustainable supplier selection—A literature review. *J. Clean. Prod.* **2020**, *273*, 123089. [CrossRef]
149. Gandhi, N.S.; Thanki, S.J.; Thakkar, J.J. Ranking of drivers for integrated lean-green manufacturing for Indian manufacturing SMEs. *J. Clean. Prod.* **2018**, *171*, 675–689. [CrossRef]
150. Gupta, H.; Barua, M.K. A novel hybrid multi-criteria method for supplier selection among SMEs on the basis of innovation ability. *Int. J. Logist. Res. Appl.* **2017**, *21*, 201–223. [CrossRef]
151. Raut, R.D.; Gardas, B.B.; Narkhede, B.E.; Narwane, V.S. To investigate the determinants of cloud computing adoption in the manufacturing micro, small and medium enterprises. *Benchmarking Int. J.* **2019**, *26*, 990–1019. [CrossRef]
152. Tsai, W.-H.; Lee, P.-L.; Shen, Y.-S.; Hwang, E.T.Y. A combined evaluation model for encouraging entrepreneurship policies. *AnOR* **2011**, *221*, 449–468. [CrossRef]
153. Rezaei, J. Best-worst multi-criteria decision-making method: Some properties and a linear model. *Omega* **2016**, *64*, 126–130. [CrossRef]
154. Chowdhury, P.; Paul, S.K.; Kaisar, S.; Moktadir, M.A. COVID-19 pandemic related supply chain studies: A systematic review. *Transp. Res. E Logist Transp. Rev.* **2021**, *148*, 102271. [CrossRef]
155. Sabuj, S.U.; Ali, S.M.; Hasan, K.W.; Paul, S.K. Contextual relationships among key factors related to environmental sustainability: Evidence from an emerging economy. *Sustain. Prod. Consum.* **2021**, *27*, 86–99. [CrossRef]
156. Ali, S.M.; Paul, S.K.; Chowdhury, P.; Agarwal, R.; Fathollahi-Fard, A.M.; Jabbour, C.J.C.; Luthra, S. Modelling of supply chain disruption analytics using an integrated approach: An emerging economy example. *Expert Syst. Appl.* **2021**, *173*, 114690.

 *sustainability*

Article

Green Supply Chain Performance Prediction Using a Bayesian Belief Network

Md. Rabbi [1], Syed Mithun Ali [1], Golam Kabir [2,*], Zuhayer Mahtab [3] and Sanjoy Kumar Paul [4]

[1] Department of Industrial and Production Engineering, Bangladesh University of Engineering and Technology, Dhaka-1000, Bangladesh; mdrabbi@pg.ipe.buet.ac.bd (M.R.); mithun@ipe.buet.ac.bd (S.M.A.)

[2] Industrial Systems Engineering, University of Regina, Regina, SK S4S 0A2, Canada

[3] Department of Industrial and Production Engineering, Military Institute of Science and Technology, Dhaka-1216, Bangladesh; zuhayer.mahtab@ipe.mist.ac.bd

[4] UTS Business School, University of Technology Sydney, Sydney, NSW 2007, Australia; sanjoy.paul@uts.edu.au

* Correspondence: golam.kabir@uregina.ca; Tel.: +1-306-858-5271

Received: 20 December 2019; Accepted: 21 January 2020; Published: 4 February 2020

Abstract: Green supply chain management (GSCM) has emerged as an important issue to lessen the impact of supply chain activities on the natural environment, as well as reduce waste and achieve sustainable growth of a company. To understand the effectiveness of GSCM, performance measurement of GSCM is a must. Monitoring and predicting green supply chain performance can result in improved decision-making capability for managers and decision-makers to achieve sustainable competitive advantage. This paper identifies and analyzes various green supply chain performance measures and indicators. A probabilistic model is proposed based on a Bayesian belief network (BBN) for predicting green supply chain performance. Eleven green supply chain performance indicators and two green supply chain performance measures are identified through an extensive literature review. Using a real-world case study of a manufacturing industry, the methodology of this model is illustrated. Sensitivity analysis is also performed to examine the relative sensitivity of green supply chain performance to each of the performance indicators. The outcome of this research is expected to help managers and practitioners of GSCM improve their decision-making capability, which ultimately results in improved overall organizational performance.

Keywords: green supply chain; performance measurement; Bayesian belief network; sustainability

---

1. Introduction

Recently, supply chain performance prediction has received increased attention from academics and practitioners [1]. To predict the supply chain performance, the development of supply chain performance metrics is crucial. Developing a performance measurement system to empower the coordination mechanism for mutual decision-making has become a vital issue in supply chain management [2]. This mutual decision-making process can be used to combine the goals of independent participants and integrate their individual activities so as to optimize the performance of the whole supply chain [3]. Hoole [4] discussed that the performance measurement of a supply chain enables a company to employ more mature supply chain practices and, consequently, enables them to reduce cost faster than their less mature competitors. More accurately, there can be a variation of 5% to 6% of annual revenue in supply chain costs among competitor companies of the same industry. Therefore, it is important for a company to develop a performance measurement model to improve its operation [5,6].

Companies are presently trying to include environmental performance in the evaluation of the overall supply chain performance because of increased competitive, regulatory, and community pressures [3]. Companies need to minimize the environmental impact of their goods and services and, thus, they have to formulate as well as implement strategies. Consequently, it helps companies to compensate competitive, community and regulatory pressures and also helps them to achieve environmental sustainability [7–9]. The basic principles upon which a company's business is based can be reviewed and readjusted so that the company can project a green image. In addition, Bhattacharya et al. [10] discussed that it is important for a company to address environmental issues to develop a unique competitive advantage for increasing the value of its core business programs. In 1994, the Confederation of British Industries observed various elements that build a competitive advantage through environmental performance: market expectations, risk management, regulatory compliance and business efficiency are some of these elements [11–13]. To handle all these elements properly, researchers and practitioners use green supply chain management (GSCM) as an effective tool [14,15]. Thus, GSCM enables a company not only to increase its ecological efficiency but also to increase profit and market share. It also results in sustainable growth [7].

Green supply chain performance prediction is receiving more and more attention because of the recent progress in the area of GSCM. Though several metrics have been suggested for supply chain performance measurement [16–19], these metrics do not include all aspects of the green supply chain. So, more comprehensive environmental performance metrics need to be introduced. This study aims to identify these performance indicators and use them to predict green supply chain performance in different scenarios.

There are significant works completed on traditional supply chain performance measurement but very few of them focused on the environmental performance of the supply chain. Gunasekaran et al. [20] described various supply chain performance metrics. Though they did not include environmental performance metrics, they emphasized a more comprehensive study of these general measures. In addition, works completed on environmental performance measurement mostly considered qualitative performance measures and indicators. Hervani et al. [21] considered ISO14031 to develop the basic principle for green supply chain performance measurement, but no definite quantitative model was suggested in their study. So, all of the existing literature lacks developing and implementing probabilistic and quantitative techniques to predict green supply chain performance. Therefore, the research questions this study aims to answer are:

(a)   What are the performance metrics of a green supply chain?
(b)   How do the performance metrics affect the supply chain performance?
(c)   Which metrics have the most impact on GSCM performance?
(d)   How can a manager achieve a specified level of GSCM performance?

This study attempts to answer the above research questions by identifying the GSCM performance measures by reviewing the existing literature and taking expert opinion, developing a quantitative and probabilistic model using a Bayesian belief network (BBN). The effectiveness of the model is demonstrated using a real-world case study, as well as performing sensitivity and diagnostic analyses to determine the impact of various metrics on overall performance. The BBN-based green supply chain performance prediction model can consider the cause–effect relationships between different performance indicators and provide informed decisions effectively in cases of incomplete, imprecise and ambiguous information. The proposed BBN model is flexible enough to perform both diagnostic analysis or bottom-up inference, and predictive analysis or top-down inference.

The rest of the paper is organized as follows. Section 2 presents a review of the existing literature on traditional supply chain management performance measurement and GSCM performance measurement. Section 3 gives a brief overview of BBN. Section 4 presents the proposed research framework. Section 5 includes a BBN-based performance prediction model. Data collection and analysis, results, discussion of the findings and sensitivity analysis are also included in this section. Finally, Section 6 includes conclusions, managerial implications and recommendations.

2. Literature Review

In this section, a brief overview of GSCM and review of the related literature is provided. The literature review is divided into two parts. In the first section, the traditional supply chain performance measurement is discussed. In the second section, literature focusing on GSCM is discussed.

2.1. Green Supply Chain Management (GSCM)

GSCM is defined by incorporating the environmental aspect into the supply chain management that considers the effect and association of supply chain management to the surrounding environment [14,19,22,23]. The traditional supply chain focuses on the economic aspect of the supply chain, ignoring the environmental impact of the activities, whereas the green supply chain focuses on and tries to minimize the adverse impact of the supply chain activities on the environment. A green supply chain, however, must not just be environment friendly, it also must be economically viable [14]. According to Wilkreson [24], GSCM is not a cost center. Instead, it is an important business driver.

The rising scarcity of raw materials, environmental pollution and ever-increasing world population have placed utmost importance on GSCM [14,25]. It encompasses two major areas, green design and green operations [25]. Mishra et al. [19] expanded the repertoire of GSCM by adding green distribution and marketing to the definition. Green design is the incorporation of the environmental effect of a product throughout its life cycle in the design and development process [26]. Several tools exist that can help designers understand the impact of their product. Life-cycle assessment, design for environment principles and product stewardship are some of the tools [27]. Green operations include green manufacturing or remanufacturing, reverse logistics and waste management [28–30]. Reverse logistics is a crucial part of GSCM. Reverse logistics is a method of increasing environmental performance of a company by using the concept of the 3Re's (recycle, reuse and reduce the use of material) [10,19,29]. Swami and Shah [31] included the coordination of functional areas to share the responsibility for environmental performance as an important element of GSCM.

2.2. Traditional Supply Chain Management Performance Measurement

Though much work on performance measurement and management of internal organizational activities has been done, only a handful have focused on supply chain performance measurement [18]. Multiple echelon inventory-based supply chain models have generally considered various performance measures like cost, quality, delivery time, inventory levels, and environmental costs [32,33].

Comprehensive supply chain performance measurement has been the focus of some of the existing literature. Supplier performance evaluation and study of appropriate performance measures have received special attention from some researchers [34,35]. Most of these studies have focused on measuring supplier performance, and also focused on their roles in the supply chain. Beamon [36] observed the impact of the various elements on supply chain performance and recognized the inherent association between these elements and supply chain performance. Inventory system stock-out risk, the probability distribution of demand and transportation time are some of the major elements identified by the authors.

A. Gunasekaran et al. [20] described various supply chain performance metrics. Using these performance metrics, they also described sources. Though they did not include environmental performance metrics, they emphasized a comprehensive study of these general measures. As existing literature on traditional supply chain measurement fails to include environmental performance, several researchers have tried to incorporate environmental performance in their different studies. These studies will be reviewed in the next section.

2.3. GSCM Performance Measurement

The perception of ecological sustainability has been considered as a basis for studying management practices by several researchers [37–39]. They have acknowledged its applicability in both operational and strategic contexts. Greening of supply chains within various contexts has been studied and these contexts include product design [26], process design [25], manufacturing practices [40–42], purchasing [43,44], and a comprehensive combination of these factors [25,45].

Several researchers [10,25,46,47] emphasized that GSCM has emerged as an approach to build a competitive advantage and to fulfill the environmental requirements that are set by various regulatory bodies. Ahi and Searcy [48] proposed various metrics to measure environmental performance. Reducing the negative environmental impact (different types of pollution) and reducing the waste of resources (energy, materials, goods) were considered as overall objectives of a green supply chain by Hervani et al. [21]. The authors also noted that this reduction process should start from the extraction of raw material and should continue up to the consumption and shipment of products. Green supplier selection has attracted quite a bit of attention from researchers [49–51]. Several metrics for assessing supplier performance were identified by Kuo, Wang, and Tien [52]. These metrics include "green competencies", "current environment efficiency", "supplier's green image" and "net life-cycle cost". Actually, this supplier performance assessment procedure is a part of the green purchasing process. Yazdani, Chatterjee, Zavadskas, and Zolfani [53] used a Quality Function Development (QFD)-based multi-criteria decision-making approach for green supplier selection. Tang, Wei, and Gao [54] used the Muirhead Mean operator and dual Muirhead Mean (DMM) operator to process the interval-valued Pythagorean fuzzy numbers (IVPFNs), which they then used to solve a supplier selection problem. Jenssen and de Boer [55] incorporated life-cycle assessment in the green supplier selection problem. Xu, Shi, Cui, and Quan [56] used interval 2-tuple linguistic hybrid aggregation operators to select green suppliers.

Researchers have introduced tools such as the analytical hierarchy process (AHP), activity-based costing and design for environmental analysis; life-cycle analysis and balanced scorecard for Green Supply Chain Performance Measurement. Although a few tools can be directly implemented for evaluating the performance, the remaining others need to be modified. For example, a management tool known as ecological supply chain analysis (ECOSCAN) was developed by Faruk et al. [45] to observe the effect of environmental management across the supply chain. The life cycle analysis model is the basis of the ECOSCAN tool, which emphasizes the connection between life-cycle analysis and GSCM methods.

Handfield, Walton, Sroufe, and Melnyk [57] combined AHP with an extensive information system. This information system enables environmentally conscious purchasing. Pineda-Henson, Culaba, and Mendoza [58] used AHP to analyze the impact of environment by following the life-cycle assessment approach, which mainly considers the manufacturing operations. Handfield et al. considered only green purchasing and Pineda-Henson et al. considered a particular case study of pulp and paper manufacturing; however, none of them considered the overall green supply chain performance.

By reviewing the existing literature, eleven corresponding green supply chain performance indicators were identified. These eleven indicators were then classified into a hierarchy model which was inspired by the works of Maleki and Machado [33]. The indicators were first clustered into small groups, e.g., indicators related to water and energy consumption were put into the group 'consumption'. These groups were then classified into performance measures. Two performance measures were considered: business wastage, and emission and consumption. Business wastage, in the context of this paper, is defined as waste materials produced as a result of various business processes. On the other hand, emissions and consumption includes all those indicators related to various solid, liquid emission and different resource consumptions. It is to be noted that two indicators, i.e., 'Output amount of hazardous and toxic material' and 'Percentage of energy obtained from renewable sources' were not clustered into any group as there no performance indicators similar to them. These performance measures and indicators were used for developing a BBN model. Using these performance measures and indicators, the green supply chain performance in different scenarios can be predicted. These performance measures and indicators are presented in Table 1 along with corresponding references and notations that will be used throughout the paper.

Table 1. Green supply chain performance measures and indicators.

Measures	Groups	Indicators	Reference
Business Wastage related	Materials related.	Total flow quantity of scrap.	[59,60]
		Percentage of materials remanufactured.	[21,61]
		Percentage of materials recycled/re-used.	[15,62]
	Output amount of hazardous and toxic material.		[32,63]
	Waste related.	Amount of solid wastes.	[46,64]
		Amount of liquid wastes.	[46,64]
Emissions and consumption related	Emission.	Amount of greenhouse gas emissions.	[65]
		Air emission quality.	[66,67]
	Consumption.	Amount of water consumption.	[68,69]
		Amount of energy consumption.	[68,69]
	Percentage of energy obtained from renewable source.		[70,71]

3. Bayesian Belief Network

As mentioned before, Bayesian Belief Network is the method of choice for predicting the performance of green supply chain in this study. A Bayesian Belief Network, or Bayes net in short form, is a probabilistic graphical model that presents knowledge about an uncertain domain. Bayes net is an effective method to represent causality and conditional probabilities among various factors [72,73]. Moreover, this method is suitable when the factors are probabilistic in nature. According to several authors, such as Langseth and Portinale; Mahadevan, Zhang, and Smith; Maleki and Machado [33,74,75], BBN shows high performance in handling uncertainty. As the factors and the relationships among them are represented using nodes and edges, any model represented using this method is easier to understand for practitioners than any other techniques [72]. As the performance indicators used in this study are probabilistic and the state of performance measures are conditionally dependent upon the states of performance indicators, Bayesian Network has been chosen to predict environmental performance.

BBN consists of two parts $B = (G, \theta)$. The first part, "G", is a directed acyclic graph (DAG) which includes nodes and arcs. DAG presents the network visually where variables of the data set $X_1, \ldots \ldots ,$ X_n represent nodes and arcs indicate dependencies among nodes [76]. The second part of BBN is the conditional dependency distribution of θ where $\theta_{xi|\pi xi} = P_B(x_i|\pi_{xi})$ is the set of direct parent variables of x_i in G [77]. Using the joint probability distribution, the network B can be represented by:

$$P_B(X_1, \ldots., X_n) = \prod_{i=1}^{n} P_B(X_i \,|\pi_{Xi}) = \prod_{i=1}^{n} X_i \,|\pi_{Xi} \tag{1}$$

In a BBN, random variables are represented by nodes, and probabilistic dependencies among the corresponding random variables are represented by edges between the nodes. A BBN actually is a probabilistic model that can compute the posterior probability distribution of any unobserved stochastic variables, given the observation of complementary subset variables [78]. In a BBN, "backward" probability propagation is also possible and it is helpful to find the most probable scenario indicating the evidence set [79].

Inference in BBN is used to update the probability for a hypothesis as more evidence or information becomes available. Figure 1 illustrates a simple Bayesian network where "A" and "B" are parent nodes, "C" is a child node. There are two types of inference support: predictive and diagnostic support for a node X_i. Predictive support for node X_i is a top-down approach that considers evidence nodes connected to X_i through its parent nodes. On the other hand, diagnostic support for node X_i is a bottom-up approach that considers evidence nodes connected to X_i through its child node [80].

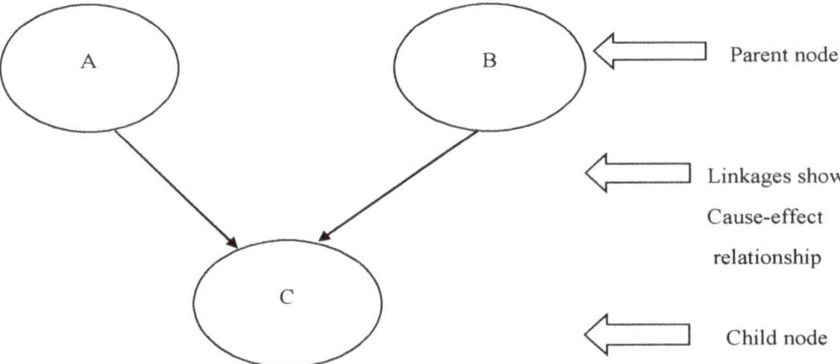

Figure 1. A simple Bayesian network.

4. Framework Development

The aim of this research work is to find out the most significant quantitative green supply chain performance measures and predict green supply chain performance in different scenarios. Figure 2 illustrates the methodology of this research. The proposed research consists of five steps as described below.

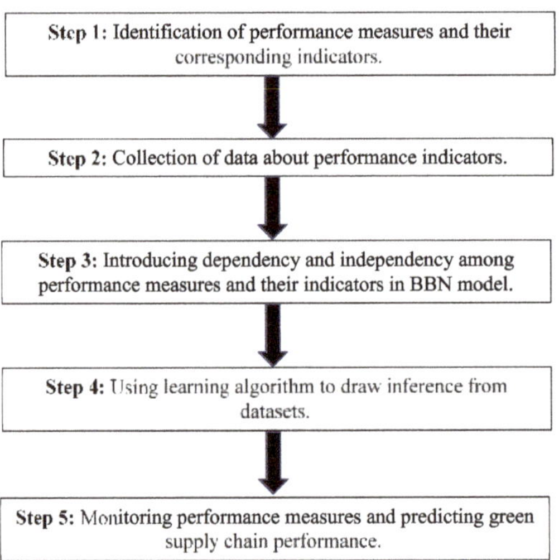

Figure 2. Proposed research framework.

Step 1: Identification of performance measures and their corresponding indicators

In the first step, an extensive list of green supply chain performance measures (GSC-PM) is generated based on the factors that have a significant influence on green supply chain performance. The most significant GSC-PM are identified through a literature review. For the literature review, papers published from 1995 to present day were chosen for the review timeline. Literature about green supply chain was searched for factors that the authors believe can significantly affect the environmental performance of a company. Papers about green purchasing, green consumption, green marketing, green manufacturing, green 3R (reduce, reuse, recycle) and supply chain performance measurement were also studied for green performance indicators. A total of 11 performance indicators were identified, which represent two major performance measures. These indicators and measures were identified by reviewing the existing literature.

Step 2: Collection of data about performance indicators

In this step, data about performance indicators were collected. These data were used for calculating the prior probability of each performance indicator in the BBN model and for learning in the BBN model.

Step 3: Introducing dependency and independency among performance measures and their indicators in the BBN model

In this step, dependency and independency among performance measures and their indicators are introduced. They determine the mutual influence between performance measures and their indicators.

Step 4: Using a learning algorithm to draw an inference from data sets

In this step, a learning algorithm is used to draw an inference from data which are collected during step 3. The inference is used to update the probability for a hypothesis (here, the hypothesis is whether the performance indicator will be in a satisfactory state or in an unsatisfactory state) as more evidence or information becomes available.

Step 5: Monitoring performance measures and predicting green supply chain performance by applying evidence to specific nodes in the BBN model

In this step, performance measures are monitored and green supply chain performance is predicted by applying evidence to specific nodes which represent performance indicators in the BBN model. Evidence will be different for different scenarios and, as a result, green supply chain performance will be different for different scenarios. The methodology of this research is illustrated in Figure 2.

5. BBN-Based Performance Prediction Model

5.1. Identification of Performance Measures and Their Corresponding Indicators

Using a thorough literature review, eleven performance indicators have been identified which will be used to build the model.

5.2. Data Collection

The developed methodology has been applied to a real-world case study and used to predict the green supply chain performance of this case company. Necessary data about performance indicators have been collected from this company. These data and information will help to determine the prior probability of the performance indicators being in the satisfactory state, and also the prior probability of performance indicators being in the unsatisfactory state. The data of "Amount of Solid Wastes" performance indicator are highlighted in Table 2 as an example.

Table 2. Data about 'Amount of Solid Wastes' performance indicator.

Time Period (Month)	Amount of Solid Wastes (Kg)	
1st	64,100	
2nd	60,000	
3rd	65,700	
4th	62,000	
5th	59,800	
6th	63,500	
7th	71,000	Recommended Level ≤ 68,000 kg
8th	60,500	
9th	73,000	
10th	70,500	
11th	69,000	
12th	72,000	
13th	68,500	
14th	66,000	
15th	61,000	
16th	66,700	
17th	69,700	
18th	63,300	
19th	67,600	
20th	64,500	

Using a histogram, collected data about performance indicators and their corresponding recommended levels are presented in Figure 3.

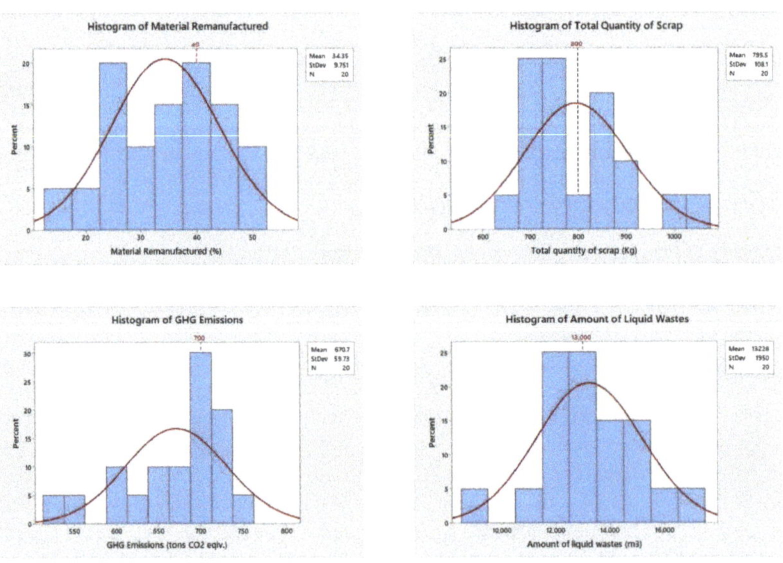

Figure 3. *Cont.*

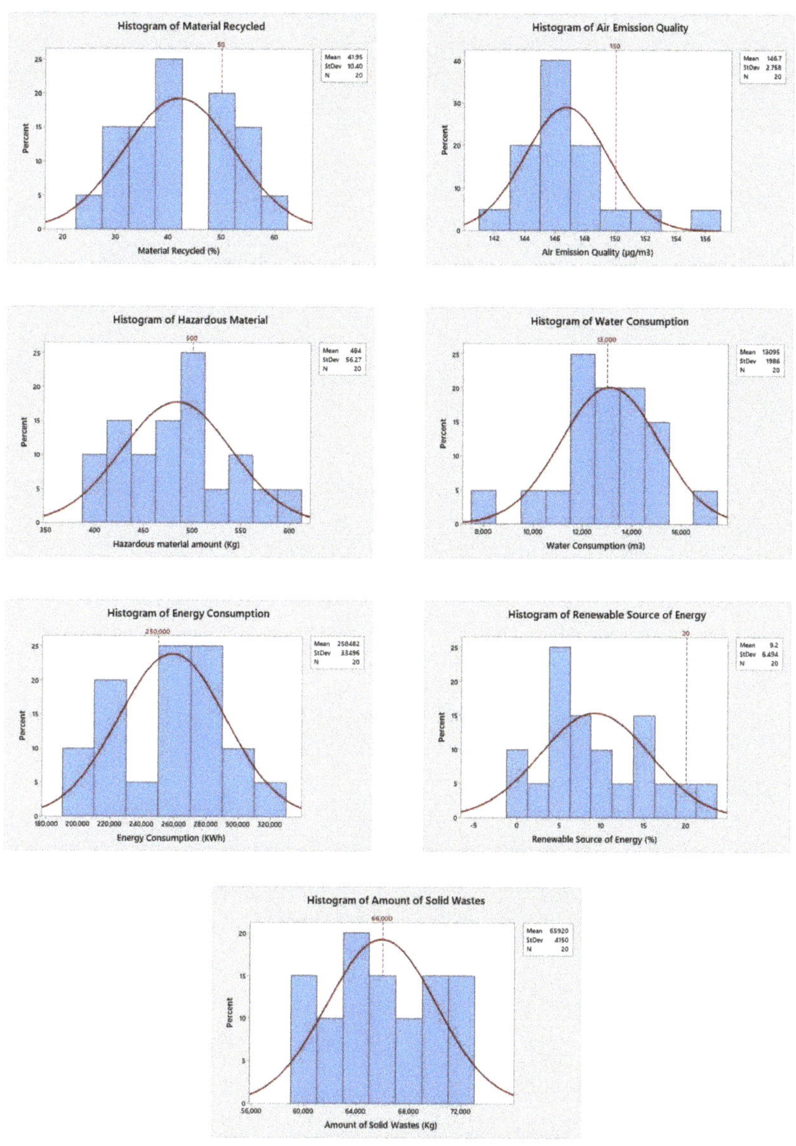

Figure 3. Histogram of performance indicators.

According to Table 2, the company has met the recommended level 13 out of 20 times. For this, the probability that performance indicator "Amount of Solid Wastes" will be in the Satisfactory state = ((13/20) × 100) % = 65% and Unsatisfactory state = ((7/20) × 100) % = 35%. Similarly, prior probabilities for other performance indicators have been calculated and are listed in Table 3.

Table 3. Prior probabilities for green supply chain performance indicators.

Performance Indicators	State Probability (%)	
	Satisfactory (S)	Unsatisfactory (U)
1. Total flow quantity of scrap.	60	40
2. Percentage of materials recycled/re-used.	35	65
3. Percentage of materials remanufactured.	40	60
4. Output amount of hazardous and toxic material.	60	40
5. Amount of solid wastes.	65	35
6. Amount of liquid wastes.	55	45
7. Amount of water consumption.	50	50
8. Amount of greenhouse gas emissions.	60	40
9. Air emission quality.	90	10
10. Amount of energy consumption.	35	65
11. Percentage of energy obtained from renewable sources.	10	90

5.3. BBN Model Development

The BBN model was developed using commercially available software Netica [81] and prior probabilities of performance indicators listed in Table 3 are presented in Figure 4. To learn the conditional probabilities in the network, Netica has counting algorithms, expectation-maximization (EM) and gradient descent [79]. As there are no hidden variables or incomplete data, Netica used counting algorithms for the model development.

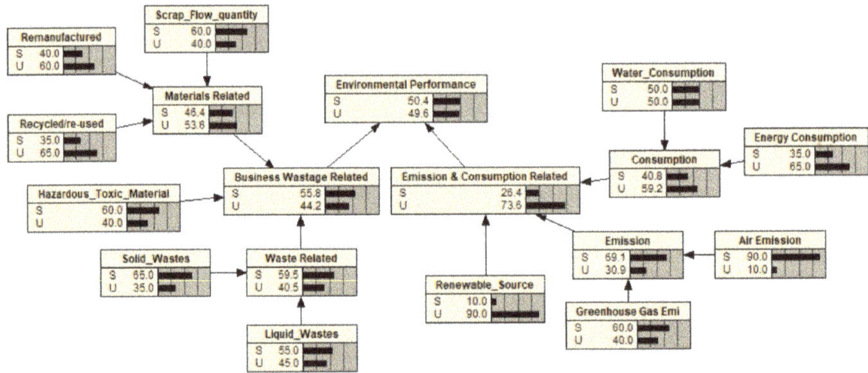

Figure 4. BBN Model for predicting Green Supply Chain performance.

From Figure 4, it can be found that there is a 52.4% probability that performance measure "Business Wastage" will be in a satisfactory state and 53.6% probability that performance measure "Emissions" will be in a satisfactory state and, finally, there is a 57.6% probability that Environmental Performance or Green Supply Chain Performance will be in a satisfactory state. These results are based on current prior probabilities of performance indicators which were collected from the data.

5.4. Model Validation

Both qualitative (extreme-condition test and scenario analysis) and quantitative (sensitivity analysis) validation approaches were performed for validation of the proposed model [82].

5.4.1. Extreme-Condition Test

To validate the model, two extreme conditions are considered in this paper. In extreme case 1, all the performance indicators are in a satisfactory condition. In extreme case 2, all the performance indicators are in an unsatisfactory condition. The results for extreme cases 1 and 2 are shown in Figures 5 and 6, respectively. From Figure 5, it can be found that, when all the performance indexes have a 100 percent probability of being satisfactory, the environmental performance has a 92.8 percent probability of being in a satisfactory condition. Similarly, according to Figure 6, when all the performance indicators have a 100 percent probability of being in an unsatisfactory condition, the environmental performance has only 4.15 percent probability of being in a satisfactory condition. So, the extreme-condition tests show that the proposed environmental performance model works according to expected model behavior.

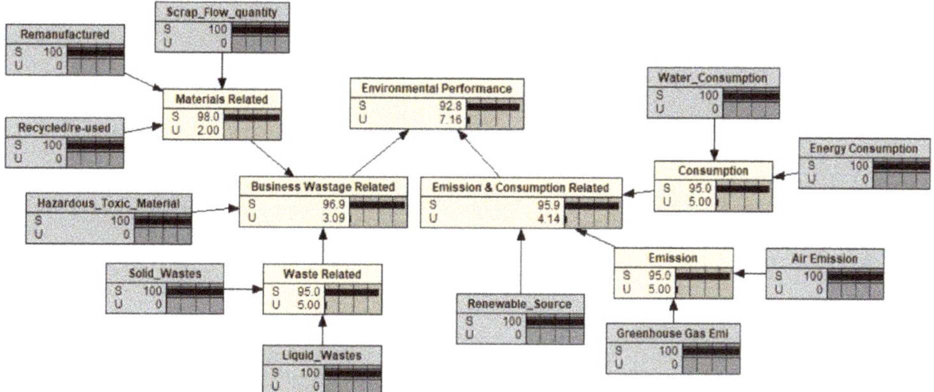

Figure 5. Extreme 1 case when all the performance indicators are in satisfactory states.

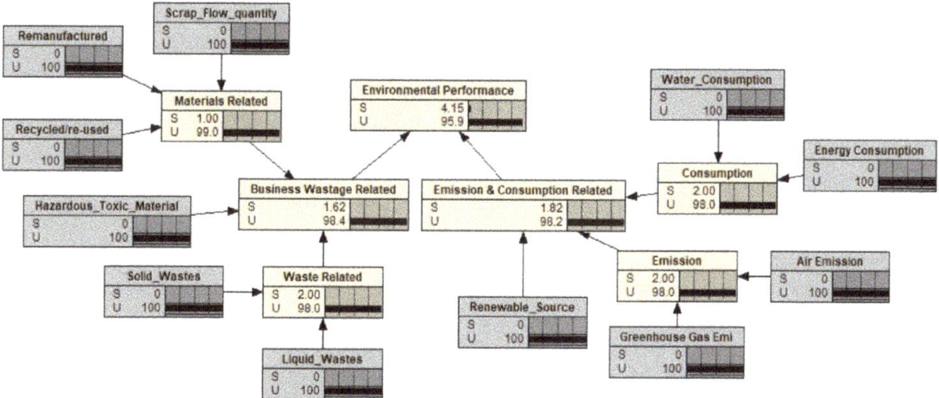

Figure 6. Extreme 2 case when all the performance indicators are in unsatisfactory states.

5.4.2. Scenario Analysis

In this analysis, different hypothetical scenarios are considered except the two extreme cases. Due to space limitations, only two performance factors, namely, *Scrap_Flow_quantity* and *Hazardous_Toxic_Material* have been considered here. Eight different scenarios are considered where the probability of a satisfactory state of these two performance factors is varied from 90 percent to 10 percent in increments of 10 percent. The results of the analysis are shown in Figure 7 and summarized in Table 4.

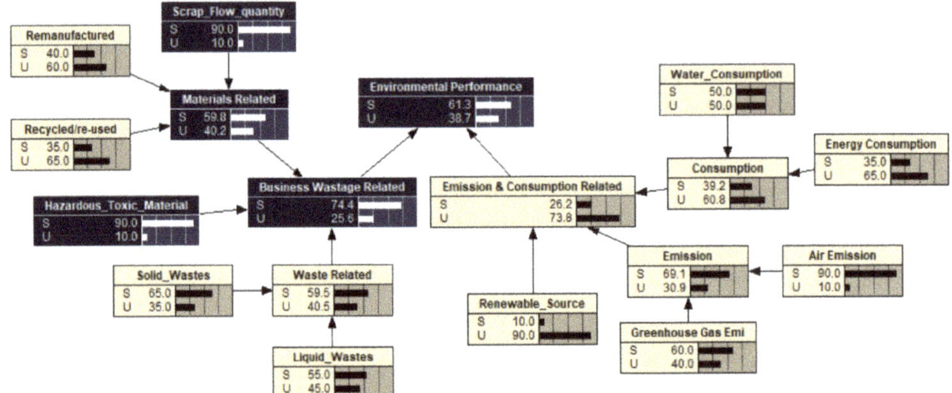

Figure 7. Scenario analysis of simultaneous change of *Scrap_Flow_quantity* and *Hazardous_Toxic_Material*.

Table 4. Scenario analysis results of the proposed BBN-based model.

Nodes	Sates	Conditional Probabilities of Different Scenarios							
		Sc-1	Sc-2	Sc-3	Sc-4	Sc-5	Sc-6	Sc-7	Sc-8
Scrap_Flow_quantity	Satisfactory	90	80	70	50	40	30	20	10
	Unsatisfactory	10	20	30	50	60	70	80	90
Hazardous_Toxic_Material	Satisfactory	90	80	70	50	40	30	20	10
	Unsatisfactory	10	20	30	50	60	70	80	90
Materials Related	Satisfactory	59.8	55.4	50.9	41.9	37.4	32.9	28.5	24
	Unsatisfactory	40.2	44.6	49.1	58.1	62.6	67.1	71.5	76
Business Wastage Related	Satisfactory	74.4	68.2	62	49.6	43.5	37.3	31.1	25
	Unsatisfactory	25.6	31.8	38	50.4	56.5	62.7	68.9	75
Environmental Performance	Satisfactory	61.3	57.6	54	46.6	43	39.3	35.7	32
	Unsatisfactory	38.7	42.4	46	53.4	57	60.7	64.3	68

In scenario 1, where *Scrap_Flow_quantity* and *Hazardous_Toxic_Material* are at 90 percent, Environmental Performance is at 61.3 percent. In scenario 2, where the probability of satisfactory states of performance indicators decreases to 80 percent, the nvironmental performance decreases to 57.6 percent. Scenario 3 depicts 70 percent satisfactory probability of performance indicators where *Environmental Performance* decreases to 54 percent. Similarly, from scenarios 4–8, where the probability of a satisfactory state of *Scrap_Flow_quantity* and *Hazardous_Toxic_Material* decreases, the probability of *Environmental Performance* being in a satisfactory state also decreases. In scenario 8, where the performance indicators are at 10 percent of the probability of being in a satisfactory state, the *Environmental Performance* has only a 32 percent probability of being in a satisfactory state. Thus, the scenarios show the expected model behavior.

All these eight scenarios represent the anticipated model behavior. In a similar way, the different combinations of the performance indicators are considered to generate different scenarios and their *Environmental Performance* probability distribution is tested to perform model validation.

5.4.3. Sensitivity Analysis

To identify the contribution of each individual input in the model output, a sensitivity analysis was performed. This analysis provides information about how slight variations in input parameters like water consumption and solid wastes can affect the model output, which in this case is Environmental Performance. Because the input variables, in this case, are discretized continuous parent nodes, using a variance reduction method is recommended [82]. However, the results of an entropy reduction method are also provided.

The variance reduction method calculates the variance reduction of the expected real value of a query node E (e.g., *Environmental_Performance*) due to a finding in a varying variable node I (e.g., *Recycled/re-used*, *Remanufactured*, *Renewable_Source*). The variance of the real value of E given evidence I, $V(e|i)$ is computed using the following equation [82,83]:

$$(e|i) = \sum_e p(e|i) \left[Y_e - E(e|i) \right]^2, \tag{2}$$

where e is the state of the query node E, i is the state of varying node I, $p(e|i)$ is the conditional probability of e given i, Y_e is the numeric value corresponding to state e and $E(e|i)$ is the expected value of E after the new finding i for node I.

Entropy reduction calculates the expected reduction in mutual information of E from a finding for variable I (Kabir et al., 2019). The formula is given below:

$$ER = H(E) - H(E|I) = \sum_e \sum_i P(e,i) \frac{\log_2[P(e,i)]}{P(e)P(i)}, \tag{3}$$

where $H(E)$ and $H(E|I)$ are the entropy before the new findings and after the new findings. The results of the sensitivity analysis are provided in Figure 8.

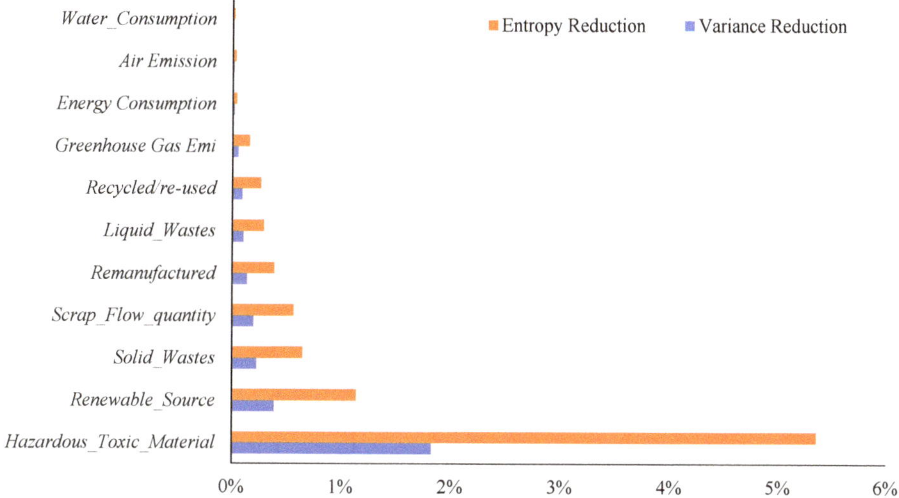

Figure 8. Sensitivity analysis of environmental performance node.

For query node *Environmental_Performance*, *Hazardous_Toxic_Material* has the highest contribution (1.830% variance reduction and 5.354% entropy reduction, respectively) followed by *Renewable_Source* (0.386% and 1.141%), *Solid_Wastes* (0.223% and 0.645%), *Scrap_Flow_quantity* (0.195% and 0.564%), *Remanufactured* (0.133% and 0.385%), *Liquid_Wastes* (0.099% and 0.287%), *Recycled/re-used* (0.090% and 0.260%), *Greenhouse Gas Emission* (0.053% and 0.153%) have medium effects on *Environmetal_Performance*. *Energy_Consumption* and *Air_Emission*, *Water_Consumption* have very low contributions. The variance reduction and entropy reduction for both are below 0.05%. The total contribution of parent nodes in variance reduction is 3.039% and for entropy reduction is 8.856%.

The result of sensitivity analysis allows the decision-maker to identify the input parameters that affect the output most and prioritize them in the decision-making. In the case of *Environmental*

Performance, the managers should first prioritize reducing toxic waste and focus on using renewable sources to improve environmental performance.

5.5. Diagnostic Analysis

Marginal probabilities of root or parent nodes can be determined by performing a diagnostic analysis [83]. The posterior probabilities conditioned to the aggregated risk can be identified using this analysis. The posterior probabilities of the parent nodes conditioned to aggregated risk are shown in Table 5. Table 5 shows the change in the probability of performance indicators in two extreme cases of Environmental Performance. All the probabilities for a satisfactory state decrease when Environmental Performance goes from a satisfactory state to an unsatisfactory state. On the other hand, the probabilities of an unsatisfactory state go up. This is in line with the expected model behavior.

Table 5. Posterior probabilities of parent nodes conditioned to *Environmental Performance*.

Parent Nodes	Posterior Probabilities of States	Environmental Performance	
		Satisfactory	Unsatisfactory
		1	1
Air Emission	Satisfactory	0.906	0.894
	Unsatisfactory	0.094	0.106
Energy Consumption	Satisfactory	0.361	0.339
	Unsatisfactory	0.639	0.661
Greenhouse Gas Emi	Satisfactory	0.622	0.577
	Unsatisfactory	0.378	0.423
Hazardous_Toxic_Material	Satisfactory	0.732	0.466
	Unsatisfactory	0.268	0.534
Liquid_Wastes	Satisfactory	0.581	0.518
	Unsatisfactory	0.419	0.482
Recycled/re-used	Satisfactory	0.378	0.321
	Unsatisfactory	0.622	0.679
Remanufactured	Satisfactory	0.436	0.364
	Unsatisfactory	0.564	0.636
Renewable_Source	Satisfactory	0.137	0.062
	Unsatisfactory	0.863	0.938
Scrap_Flow_Quantity	Satisfactory	0.643	0.556
	Unsatisfactory	0.357	0.444
Solid_Wastes	Satisfactory	0.695	0.605
	Unsatisfactory	0.305	0.395
Water_Consumption	Satisfactory	0.508	0.492
	Unsatisfactory	0.492	0.508

6. Conclusions, Implications and Limitations to This Study

In this study, a BBN-based probabilistic model is proposed for predicting green supply chain performance. Eleven green supply chain performance indicators and two green supply chain performance measures were identified through a review of the existing literature, and then BBN was used to develop the model that would predict the overall environmental performance. Inputting the satisfactory/unsatisfactory states of the performance indicators in the model will provide the decision maker with the overall green supply chain environmental performance state. This will allow the manager to see how the performance metrics can affect the overall green supply chain performance. To validate the model, an extreme condition test was used. Furthermore, performing a sensitivity analysis reveals the most important performance indicators and provides the decision makers with a ranking of the indicators in the order of importance. For the case company in this study, output amount of hazardous toxic material was found to be the most important indicator having the highest variance reduction and entropy, followed by percentage of energy obtained from renewable sources. Amount of solid wastes, total flow quantity of scraps, and percentage of materials remanufactured are also some important indicators with relatively high sensitivity. On the other hand, amount of water consumption, quality of air emission and amount of energy consumption were found to be the least important indicators with low sensitivity. The manager of this company can give relatively less focus to these factors without worsening the environmental performance too much. This model can also help a manager to achieve a prespecified level of performance. The model allows a manager to input the required level of environmental performance, and the model will determine at what level each environmental indicator needs to be.

This study makes several theoretical contributions in the field of green supply chains. First, this study identified the key performance indicators that affect the overall environmental performance across the supply chain of an organization. Second, this study proposes a BBN-based framework for the green or environmental performance prediction of a supply chain. The final contribution of this paper is that it investigates how the individual key performance indicators affect the overall environmental performance of a supply chain.

Using this model, managers and executives will be able to understand which performance indicators most affect green supply chain performance of their company. They will also be able to understand which performance indicators to focus on first, and where the resource should be allocated first to achieve the optimum level of environmental performance. As managers usually have limited resources at their disposal, prioritizing and optimizing resource allocation can greatly help in achieving environmental objectives of a supply chain. Using the current performance level of each performance indicator, they will be able to monitor current environmental performance, which will help them to understand the company's current relative position in the industry. Using diagnostic analysis, managers can determine the target performance indicator levels required to achieve satisfactory overall performance.

This study has some limitations. Due to limitations in collecting data, only one company's data has been used to test the model. Another limitation is that not all possible key environmental performance indicators were added to the proposed BBN model. This model only considered two states, namely satisfactory and unsatisfactory states for the BBN nodes.

There are several directions future researches on this topic can take. One direction can be adding more performance indicators. A more comprehensive model validation test can be performed by collecting data from multiple sources and performing multiple case studies. Another direction for future research can be incorporating multiple states instead of just satisfactory and unsatisfactory states, which can be an exciting direction for future research. IoT devices can be used to collect environmental data in real time which can be used as the input for an online algorithm. This algorithm can update the BBN model in real time for monitoring purposes. Other artificial intelligence-based tools such as knowledge systems, fuzzy AHP, or Bayesian Regression Trees can be developed for predicting green supply chain performance and the results can be compared.

Author Contributions: Conceptualization, M.R., S.M.A., G.K. and S.K.P.; methodology, M.R., S.M.A., G.K. and S.K.P.; software, M.R., S.M.A., G.K.; validation, M.R., S.M.A., G.K.; formal analysis, M.R., S.M.A., G.K. and S.K.P.; investigation, M.R., S.M.A., Z.M.; resources, M.R., S.M.A., Z.M.; data curation, M.R., S.M.A. and G.K.; writing—original draft preparation, M.R., S.M.A., G.K., Z.M. and S.K.P.; writing—review and editing, M.R., S.M.A., G.K., Z.M. and S.K.P.; visualization, M.R., S.M.A., G.K., Z.M.; supervision, S.M.A., G.K. and S.K.P.; project administration, S.M.A., G.K. and S.K.P. All authors have read and agreed to the published version of the manuscript.

Funding: This research received no external funding.

Conflicts of Interest: The authors declare no conflict of interest.

References

1. Gong, R.; Xue, J.; Zhao, L.; Zolotova, O.; Ji, X.; Xu, Y. A Bibliometric Analysis of Green Supply Chain Management Based on the Web of Science (WOS) Platform. *Sustainability* **2019**, *11*, 3459. [CrossRef]
2. Kim, B.; Oh, H. The impact of decision-making sharing between supplier and manufacturer on their collaboration performance. *Supply Chain Manag. Int. J.* **2005**, *10*, 223–236. [CrossRef]
3. Jian, J.; Guo, Y.; Jiang, L.; An, Y.; Su, J. A multi-objective optimization model for green supply chain considering environmental benefits. *Sustainability* **2019**, *11*, 5911. [CrossRef]
4. Hoole, R. Five ways to simplify your supply chain. *Supply Chain Manag. Int. J.* **2005**, *10*, 3–6. [CrossRef]
5. Panchal, G.; Jain, V. A review on select models for supply chain formation. *Int. J. Bus. Perform. Supply Chain Model.* **2011**, *3*, 113. [CrossRef]
6. Panicker, V.V.; Sridharan, R. Modelling supply chain decision problem with fixed charge—A review. *Int. J. Bus. Perform. Supply Chain Model.* **2011**, *3*, 195. [CrossRef]
7. Xing, G.; Xia, B.; Guo, J. Sustainable cooperation in the green supply chain under financial constraints. *Sustainability* **2019**, *11*, 5977. [CrossRef]
8. Sarkis, J. Manufacturing's role in corporate environmental sustainability—Concerns for the new millennium. *Int. J. Oper. Prod. Manag.* **2001**, *21*, 666–686. [CrossRef]
9. Lewis, H.; Gertsakis, J.; Grant, T.; Morelli, N.; Sweatman, A. *Design and Environment: A Global Guide to Designing Greener Goods*; Greenleaf Pub: Austin, TX, USA, 2001; ISBN 1874719438.
10. Bhattacharya, A.; Mohapatra, P.; Kumar, V.; Dey, P.K.; Brady, M.; Tiwari, M.K.; Nudurupati, S.S. Green supply chain performance measurement using fuzzy ANP-based balanced scorecard: A collaborative decision-making approach. *Prod. Plan. Control* **2014**, *25*, 698–714. [CrossRef]
11. Dvorsky, J.; Popp, J.; Virglerova, Z.; Kovács, S.; Oláh, J. Assessing the importance of market risk and its sources in the SME of the Visegrad Group and Serbia. *Adv. Decis. Sci.* **2018**, *22*, 1–22.
12. Oláh, J.; Virglerova, Z.; Popp, J.; Kliestikova, J.; Kovács, S. The Assessment of Non-Financial Risk Sources of SMES in the V4 Countries and Serbia. *Sustainability* **2019**, *11*, 4806. [CrossRef]
13. Oláh, J.; Kovács, S.; Virglerova, Z.; Lakner, Z.; Kovacova, M.; Popp, J. Analysis and Comparison of Economic and Financial Risk Sources in SMEs of the Visegrad Group and Serbia. *Sustainability* **2019**, *11*, 1853. [CrossRef]
14. Srivastava, S.K. Green supply-chain management: A state-of-the-art literature review. *Int. J. Manag. Rev.* **2007**, *9*, 53–80. [CrossRef]
15. Parsaeifar, S.; Bozorgi-Amiri, A.; Naimi-Sadigh, A.; Sangari, M.S. A game theoretical for coordination of pricing, recycling, and green product decisions in the supply chain. *J. Clean. Prod.* **2019**, *226*, 37–49. [CrossRef]
16. Lima-Junior, F.R.; Carpinetti, L.C.R. Predicting supply chain performance based on SCOR®metrics and multilayer perceptron neural networks. *Int. J. Prod. Econ.* **2019**, *212*, 19–38. [CrossRef]
17. Lima-Junior, F.R.; Carpinetti, L.C.R. Quantitative models for supply chain performance evaluation: A literature review. *Comput. Ind. Eng.* **2017**, *113*, 333–346. [CrossRef]
18. Maestrini, V.; Luzzini, D.; Maccarrone, P.; Caniato, F. Supply chain performance measurement systems: A systematic review and research agenda. *Int. J. Prod. Econ.* **2017**, *183*, 299–315. [CrossRef]
19. Mishra, D.; Gunasekaran, A.; Papadopoulos, T.; Hazen, B. Green supply chain performance measures: A review and bibliometric analysis. *Sustain. Prod. Consum.* **2017**, *10*, 85–99. [CrossRef]
20. Gunasekaran, A.; Patel, C.; Tirtiroglu, E. Performance measures and metrics in a supply chain environment. *Int. J. Oper. Prod. Manag.* **2001**, *21*, 71–87. [CrossRef]
21. Hervani, A.A.; Helms, M.M.; Sarkis, J. Performance measurement for green supply chain management. *Benchmarking* **2005**, *12*, 330–353. [CrossRef]
22. Lau, K.H. Benchmarking green logistics performance with a composite index. *Benchmarking* **2011**, *18*, 873–896. [CrossRef]
23. Testa, F.; Iraldo, F. Shadows and lights of GSCM (green supply chain management): Determinants and effects of these practices based on a multi-national study. *J. Clean. Prod.* **2010**, *18*, 953–962. [CrossRef]
24. Wilkerson, T. Can one green deliver another? *Harv. Bus. Rev.* **2005**, 3–4.

25. Cousins, P.D.; Lawson, B.; Petersen, K.J.; Fugate, B. Investigating green supply chain management practices and performance: The moderating roles of supply chain ecocentricity and traceability. *Int. J. Oper. Prod. Manag.* **2019**, *39*, 767–786. [CrossRef]

26. Zhu, W.; He, Y. Green product design in supply chains under competition. *Eur. J. Oper. Res.* **2017**, *258*, 165–180. [CrossRef]

27. Shohan, S.; Ali, S.M.; Kabir, G.; Ahmed, S.K.; Suhi, S.A.; Haque, T. Green supply chain management in the chemical industry: structural framework of drivers. *Int. J. Sustain. Dev. World Ecol.* **2019**, *26*, 752–768. [CrossRef]

28. Yu, W.; Ramanathan, R. An empirical examination of stakeholder pressures, green operations practices and environmental performance. *Int. J. Prod. Res.* **2015**, *53*, 6390–6407. [CrossRef]

29. Uddin, S.; Ali, S.M.; Kabir, G.; Suhi, S.A.; Enayet, R.; Haque, T. An AHP-ELECTRE framework to evaluate barriers to green supply chain management in the leather industry. *Int. J. Sustain. Dev. World Ecol.* **2019**, *26*, 732–751. [CrossRef]

30. Banerjee, M.; Mishra, M. Retail supply chain management practices in India: A business intelligence perspective. *J. Retail. Consum. Serv.* **2017**, *34*, 248–259. [CrossRef]

31. Swami, S.; Shah, J. Channel coordination in green supply chain management. *J. Oper. Res. Soc.* **2013**, *64*, 336–351. [CrossRef]

32. Azevedo, S.G.; Carvalho, H.; Cruz-Machado, V. A proposal of LARG Supply Chain Management Practices and a Performance Measurement System. *Int. J. e-Educ. e-Bus. e-Manag. e-Learn.* **2011**, *1*, 7–14. [CrossRef]

33. Maleki, M.; Machado, V.C. Supply chain performance monitoring using Bayesian network. *Int. J. Bus. Perform. Supply Chain Model.* **2013**, *5*, 177. [CrossRef]

34. Bai, C.; Sarkis, J. Green supplier development: A review and analysis. In *Handbook on the Sustainable Supply Chain*; Edward Elgar Publishing: Chatham, UK, 2019.

35. Vahdani, B.; Mousavi, S.M.; Tavakkoli-Moghaddam, R.; Hashemi, H. A new enhanced support vector model based on general variable neighborhood search algorithm for supplier performance evaluation: A case study. *Int. J. Comput. Intell. Syst.* **2017**, *10*, 293–311. [CrossRef]

36. Beamon, B.M. Measuring supply chain performance. *Int. J. Oper. Prod. Manag.* **1999**, *19*, 275–292. [CrossRef]

37. King, A.A.; Lenox, M.J. Lean and Green? an Empirical Examination of the Relationship Between Lean Production and Environmental Performance. *Prod. Oper. Manag.* **2009**, *10*, 244–256. [CrossRef]

38. Klassen, R.D.; McLaughlin, C.P. The Impact of Environmental Management on Firm Performance. *Manag. Sci.* **1996**, *42*, 1199–1214. [CrossRef]

39. Sarkis, J.; Rasheed, A. Greening the manufacturing function. *Bus. Horiz.* **1995**, *38*, 17–27. [CrossRef]

40. Rehman, M.A.; Seth, D.; Shrivastava, R.L. Impact of green manufacturing practices on organisational performance in Indian context: An empirical study. *J. Clean. Prod.* **2016**, *137*, 427–448. [CrossRef]

41. Vazquez-Brust, D.A.; Campos, L.M.S. Mapping lean manufacturing practices and green manufacturing practices in supply chains. In *Handbook on the Sustainable Supply Chain*; Edward Elgar Publishing: Chatham, UK, 2019; p. 291.

42. Choudhary, K.; Sangwan, K.S. Benchmarking Indian ceramic enterprises based on green supply chain management pressures, practices and performance. *Benchmarking Int. J.* **2018**, *25*, 3628–3653. [CrossRef]

43. Chekima, B.; Wafa, S.A.; Igau, O.A.; Chekima, S.; Sondoh, S.L., Jr. Examining green consumerism motivational drivers: Does premium price and demographics matter to green purchasing? *J. Clean. Prod.* **2016**, *112*, 3436–3450. [CrossRef]

44. Jaiswal, D.; Kant, R. Green purchasing behaviour: A conceptual framework and empirical investigation of Indian consumers. *J. Retail. Consum. Serv.* **2018**, *41*, 60–69. [CrossRef]

45. Faruk, A.C.; Lamming, R.C.; Cousins, P.D.; Bowen, F.E. Analyzing, mapping, and managing environmental impacts along supply chains. *J. Ind. Ecol.* **2002**, *5*, 13–36. [CrossRef]

46. Green, K.W., Jr.; Zelbst, P.J.; Meacham, J.; Bhadauria, V.S. Green supply chain management practices: Impact on performance. *Supply Chain Manag. Int. J.* **2012**, *17*, 290–305. [CrossRef]

47. Tseng, M.-L.; Islam, M.S.; Karia, N.; Fauzi, F.A.; Afrin, S. A literature review on green supply chain management: Trends and future challenges. *Resour. Conserv. Recycl.* **2019**, *141*, 145–162. [CrossRef]

48. Ahi, P.; Searcy, C. An analysis of metrics used to measure performance in green and sustainable supply chains. *J. Clean. Prod.* **2015**, *86*, 360–377. [CrossRef]

49.	Keshavarz Ghorabaee, M.; Amiri, M.; Zavadskas, E.K.; Antucheviciene, J. Supplier evaluation and selection in fuzzy environments: A review of MADM approaches. *Econ. Res. Istraživanja* **2017**, *30*, 1073–1118. [CrossRef]

50.	Maditati, D.R.; Munim, Z.H.; Schramm, H.-J.; Kummer, S. A review of green supply chain management: From bibliometric analysis to a conceptual framework and future research directions. *Resour. Conserv. Recycl.* **2018**, *139*, 150–162. [CrossRef]

51.	Sen, D.K.; Datta, S.; Mahapatra, S.S. Sustainable supplier selection in intuitionistic fuzzy environment: A decision-making perspective. *Benchmarking Int. J.* **2018**, *25*, 545–574. [CrossRef]

52.	Kuo, R.J.; Wang, Y.C.; Tien, F.C. Integration of artificial neural network and MADA methods for green supplier selection. *J. Clean. Prod.* **2010**, *18*, 1161–1170. [CrossRef]

53.	Yazdani, M.; Chatterjee, P.; Zavadskas, E.K.; Zolfani, S.H. Integrated QFD-MCDM framework for green supplier selection. *J. Clean. Prod.* **2017**, *142*, 3728–3740. [CrossRef]

54.	Tang, X.; Wei, G.; Gao, H. Models for multiple attribute decision making with interval-valued pythagorean fuzzy muirhead mean operators and their application to green suppliers selection. *Informatica* **2019**, *30*, 153–186. [CrossRef]

55.	Jenssen, M.M.; de Boer, L. Implementing life cycle assessment in green supplier selection: A systematic review and conceptual model. *J. Clean. Prod.* **2019**, *229*, 1198–1210. [CrossRef]

56.	Xu, X.-G.; Shi, H.; Cui, F.-B.; Quan, M.-Y. Green Supplier Evaluation and Selection Using Interval 2-Tuple Linguistic Hybrid Aggregation Operators. *Informatica* **2018**, *29*, 801–824. [CrossRef]

57.	Handfield, R.; Walton, S.V.; Sroufe, R.; Melnyk, S.A. Applying environmental criteria to supplier assessment: A study in the application of the Analytical Hierarchy Process. *Eur. J. Oper. Res.* **2002**, *141*, 70–87. [CrossRef]

58.	Pineda-Henson, R.; Culaba, A.B.; Mendoza, G.A. Evaluating environmental performance of pulp and paper manufacturing using the analytic hierarchy process and life-cycle assessment. *J. Ind. Ecol.* **2002**, *6*, 15–28. [CrossRef]

59.	Lin, R.-J.; Chen, R.-H.; Nguyen, T.-H. Green supply chain management performance in automobile manufacturing industry under uncertainty. *Procedia Soc. Behav. Sci.* **2011**, *25*, 233–245. [CrossRef]

60.	Rostamzadeh, R.; Govindan, K.; Esmaeili, A.; Sabaghi, M. Application of fuzzy VIKOR for evaluation of green supply chain management practices. *Ecol. Indic.* **2015**, *49*, 188–203. [CrossRef]

61.	Gunasekaran, A.; Subramanian, N.; Rahman, S. Green supply chain collaboration and incentives: Current trends and future directions. *Transp. Res. Part E: Logist. Transp. Rev.* **2015**, *74*, 1–10. [CrossRef]

62.	Li, Y. Research on the performance measurement of green supply chain management in China. *J. Sustain. Dev.* **2011**, *4*, 101. [CrossRef]

63.	Vachon, S. Green supply chain practices and the selection of environmental technologies. *Int. J. Prod. Res.* **2007**, *45*, 4357–4379. [CrossRef]

64.	Laosirihongthong, T.; Adebanjo, D.; Choon Tan, K. Green supply chain management practices and performance. *Ind. Manag. Data Syst.* **2013**, *113*, 1088–1109. [CrossRef]

65.	Shaw, S.; Grant, D.B.; Mangan, J. Developing environmental supply chain performance measures. *Benchmarking Int. J.* **2010**, *17*, 320–339. [CrossRef]

66.	Seman, N.A.A.; Zakuan, N.; Jusoh, A.; Arif, M.S.M.; Saman, M.Z.M. Green supply chain management: A review and research direction. *Int. J. Manag. Value Supply Chains* **2012**, *3*, 1–18. [CrossRef]

67.	Zhu, Q.; Sarkis, J.; Geng, Y. Green supply chain management in China: Pressures, practices and performance. *Int. J. Oper. Prod. Manag.* **2005**, *25*, 449–468. [CrossRef]

68.	Noci, G. Designing 'green'vendor rating systems for the assessment of a supplier's environmental performance. *Eur. J. Purch. Supply Manag.* **1997**, *3*, 103–114. [CrossRef]

69.	Shen, L.; Olfat, L.; Govindan, K.; Khodaverdi, R.; Diabat, A. A fuzzy multi criteria approach for evaluating green supplier's performance in green supply chain with linguistic preferences. *Resour. Conserv. Recycl.* **2013**, *74*, 170–179. [CrossRef]

70.	Cucchiella, F.; D'Adamo, I. Issue on supply chain of renewable energy. *Energy Convers. Manag.* **2013**, *76*, 774–780. [CrossRef]

71.	Kurien, G.P.; Qureshi, M.N. Performance measurement systems for green supply chains using modified balanced score card and analytical hierarchical process. *Sci. Res. Essays* **2012**, *7*, 3149–3161.

72.	Cai, Z.; Sun, S.; Si, S.; Yannou, B. Identifying product failure rate based on a conditional Bayesian network classifier. *Expert Syst. Appl.* **2011**, *38*, 5036–5043. [CrossRef]

73. Abolbashari, M.H.; Chang, E.; Hussain, O.K.; Saberi, M. Smart buyer: A Bayesian network modelling approach for measuring and improving procurement performance in organisations. *Knowl. Based Syst.* **2018**, *142*, 127–148. [CrossRef]

74. Langseth, H.; Portinale, L. Bayesian networks in reliability. *Reliab. Eng. Syst. Saf.* **2007**, *92*, 92–108. [CrossRef]

75. Mahadevan, S.; Zhang, R.; Smith, N. Bayesian networks for system reliability reassessment. *Struct. Saf.* **2001**, *23*, 231–251. [CrossRef]

76. Baesens, B.; Verstraeten, G.; Van den Poel, D.; Egmont-Petersen, M.; Van Kenhove, P.; Vanthienen, J. Bayesian network classifiers for identifying the slope of the customer lifecycle of long-life customers. *Eur. J. Oper. Res.* **2004**, *156*, 508–523. [CrossRef]

77. Abad-Grau, M.M.; Arias-Aranda, D. Operations strategy and flexibility: Modeling with Bayesian classifiers. *Ind. Manag. Data Syst.* **2006**, *106*, 460–484. [CrossRef]

78. Gambelli, D.; Bruschi, V. A Bayesian network to predict the probability of organic farms' exit from the sector: A case study from Marche, Italy. *Comput. Electron. Agric.* **2010**, *71*, 22–31. [CrossRef]

79. Neapolitan, R.E. Learning Bayesian Networks. *Mol. Biol.* **2003**, *6*, 674. [CrossRef]

80. Gopnik, A.; Tenenbaum, J.B. Bayesian networks, Bayesian learning and cognitive development. *Dev. Sci.* **2007**, *10*, 281–287. [CrossRef]

81. Norsys Software Corp. Netica Version 4.16. Canada: Norsys Software Corp. 2014. Available online: www.norsys.com/Accessed (accessed on 15 September 2017).

82. Kabir, G.; Balek, N.B.C.; Tesfamariam, S. Consequence-based framework for buried infrastructure systems: A Bayesian belief network model. *Reliab. Eng. Syst. Saf.* **2018**, *180*, 290–301. [CrossRef]

83. Kabir, G.; Cruz, A.M.; Suda, H.; Tesfamariam, S.; Giraldo, F.M. Earthquake-related Natech risk assessment using a Bayesian belief network model. *Struct. Infrastruct. Eng.* **2019**, 1–15. [CrossRef]

Article

Minimizing the Gap between Expectation and Ability: Strategies for SMEs to Implement Social Sustainability Practices

Priyabrata Chowdhury * and Rezaul Shumon

School of Accounting, Information Systems and Supply Chain, RMIT University, Melbourne, Victoria 3000, Australia; rezaul.shumon@rmit.edu.au
* Correspondence: priyabrata.chowdhury@rmit.edu.au

Received: 30 June 2020; Accepted: 7 August 2020; Published: 9 August 2020

Abstract: Traditionally, it is believed that small- and medium-sized enterprises (SMEs) do not have enough ability to adopt and persistently practice social sustainability. This is because SMEs are not capital-intensive companies and neither are their returns nor skills. At the same time, the wellbeing of the employees in SMEs cannot be ensured and sustainable development goals cannot be achieved without making SMEs socially sustainable, as they account for the majority of world businesses. Moreover, the expectation of the stakeholders and subsequent pressure on SMEs to practicing social sustainability remains. Such pressure from the stakeholders creates a "mismatch problem" between stakeholders' expectations and SMEs' abilities to adopt socially sustainable practices. This study aims to explore what factors are responsible for this "mismatch problem", and how SMEs can handle this mismatch to be socially sustainable firms. Based on a rigorous literature review, this study reveals that both internal issues, such as a lack of resources and awareness, and external issues, such as the non-existence of a tailored social sustainability standard for SMEs and lack of institutional support, are responsible for this gap. This study develops several propositions that highlight the requirements in various situations and provides strategies outlining the implications for SMEs and their stakeholders to make SMEs socially sustainable. Overall, this study discloses that cooperative support from stakeholders, especially during a disruption such as the COVID-19 pandemic, a finance mechanism, the development of awareness and human capital in SMEs, and a unified standard for SMEs are likely to improve social sustainability practices in SMEs.

Keywords: social sustainability; small- and medium-sized enterprises (SMEs); stakeholder support; social sustainability awareness

---

1. Introduction

There has been a widespread agreement that achieving social sustainability is critical for businesses, organizations, and society [1,2]. Big companies and corporations have traveled far towards attaining their goal of implementing social sustainability practices (SSPs). However, small- and medium-sized enterprises (SMEs) are falling significantly behind in terms of achieving social sustainability goals [3,4]. While the advancement of multi-national and big corporations in this regard is appreciable, nonetheless, the full benefit of social sustainability cannot be realized by society unless SMEs implement SSPs, as these smaller firms account for the majority of the businesses in all countries around the world. For example, more than 99 percent of European enterprises are SMEs, generating more than 85 percent of new employment in Europe [5]. Therefore, it is obvious that achieving social sustainability in SMEs will be vital for society as a whole. It has also been established that socially sustainable companies generally perform better than other companies that do not adequately maintain these practices, as these non-participating firms suffer from a negative reputation and a loss of brand value [6,7]. Therefore, practicing social sustainability is important for SMEs to maintain their competitiveness [8,9].

Although research on social sustainability in SMEs is not well explored and still at a nascent stage, a notion can be extracted from the literature that indicates SMEs' inability to adopt SSPs. Many factors can be cited as responsible for such an inability. For example, a lack of finance and capital investment, knowledge, skills, and awareness and the misperceptions of SME owners and managers regarding their role are predominantly mentioned in the literature [10,11]. No matter what the factors are, stakeholder expectations and subsequent pressures for SSPs in SMEs are inevitable. This creates a gap between the expectation of stakeholders and the ability of SMEs to implement SSPs in the supply chain or business environment. Therefore, it is important to comprehensively understand this gap, which has been termed in this study as a "matching problem", and to formulate appropriate strategies to reduce the gap. Due to the lack of focus on SMEs, the current body of literature is unable to provide such understandings. On the other hand, underlying factors in adopting SSPs are different in multinational or large companies and SMEs [4], suggesting that knowledge on social sustainability in the context of large firms is not readily applicable to SMEs. Therefore, studies are particularly needed in the context of SMEs to understand the "matching problem" and to formulate strategies to handle the problem.

In this study, therefore, we aim to provide an understanding of this issue. Based on a rigorous literature review, we first explore the factors responsible for creating the gap between stakeholder expectation and SMEs' abilities to adopt SSPs. Then, we find strategies to reduce the gap so that SMEs can be socially sustainable. In designing the strategies, we also outline the role of SMEs and various stakeholders, such as buyers, government, and NGOs involved in the implementation of social sustainability requirements in SMEs. By doing so, the study contributes to the literature on social sustainability, as well as on SMEs. More importantly, we believe the study findings can serve as a guide for SMEs and their stakeholders in formulating strategies to enhance SSPs in these smaller firms.

The rest of the paper is structured as follows. Section 2 outlines the concept and necessity of social sustainability in SMEs. Section 3 introduces the "mismatch problem" and Section 4 explores it further and develops propositions to understand why it exists in SMEs and how various factors contribute to handling this problem. In line with the propositions, Section 5 provides the strategies and outlines the role of SMEs and their stakeholders to make SMEs socially sustainable. The contributions of the study are highlighted in Section 6, and a concluding remark, along with the future research directions, are provided in Section 7.

2. Social Sustainability in SMEs

Supply chain social sustainability is concerned with the human side of sustainability [3,12,13]. It refers to the practices and ways firms address issues related to the health, safety, career progression initiatives, freedom, and welfare of the people associated with the supply chain [1]. In an attempt to specify the SSPs, Wolf [14] mentioned nine (9) main indicators: (1) a healthy and safe work environment, (2) an acceptable minimum wage, (3) the specification of maximum work hours, (4) freedom to join an employee union, (5) a policy for child labor, (6) suitable living conditions, (7) non-discrimination, (8) a clear policy for corporate disciplinary practices, and (9) a policy for forced labor. In addition to the organizational side, social sustainability is also concerned about advocating for the local community and culture. For example, according to Zhang and Zhang [15], social sustainability addresses issues relating to respecting, protecting, and advocating for native cultures and communities via providing benefits to neighboring residents and actively participating in various community functions. More specifically, firms can engage in several local programs on health, education, and sports via developing these services or collaborating with the existing local providers to help the low-income earners of the community [16]. From the social sustainability perspective, participating in these functions is generally considered as an intervention towards the enrichment of the society and community development [8,17].

Valuing and practicing SSPs are important for all firms, including SMEs. The importance for SMEs is twofold: (1) SMEs can improve various performances by practicing SSPs efficiently and (2) the active participation of SMEs in SSPs is required to reach overall social sustainability goals from the societal perspective. By being socially sustainable, SMEs can enjoy tangible performance improvements [18]. For example, Mani, Jabbour, and Mani [4] found that SSPs contribute to improvements in the supply chain performance of SMEs. Moreover, the study revealed that the higher the investment made by SMEs in SSPs, the better their supplier performance, customer performance, and operational performance. These SSPs also significantly improve the financial performance of SMEs and enhance customer satisfaction and employee satisfaction [16]. Moreover, SMEs' opportunities to improve innovation are increased when they can integrate SSPs with their core business policies and strategies [19], as well as with other operations of the organization [9]. In addition, the integration of workforce-oriented and society-oriented sustainability practices has a positive impact on the overall competitiveness of SMEs [8]. Lee, Che-Ha, and Alwi [9] also argued that SSPs enhance the competitive advantage of SMEs. Furthermore, the implementation of adequate SSPs enhances the reputation of SMEs in societies and communities [20].

The adoption of SSPs by SMEs is also important from the societal point of view. Complete social sustainability cannot be achieved if it is only practiced by large organizations as, in some economies, such as EU countries and Malaysia, more than 99 percent of businesses are SMEs [21,22]. In addition to adopting SSPs in workforce management, SMEs can play a substantial role in achieving complete sustainability via focusing on local development by employing from surrounding areas and solving local problems, as they generally manage niche social resources [19]. Moreover, SME operations have substantial impacts on societies due to their large numbers. Therefore, their active participation in SSPs is essential for overall social sustainability development [10,23].

While SSPs in SMEs are necessary from both SME and societal points of view, SMEs find it challenging to implement and generally consider these practices as elective [9,10]. They perceive that the potential benefits of SSPs are far less than the cost associated with their implementation [24]. Although they are not under the intense scrutiny of stakeholders, stakeholder's expectations remain a pressure for SMEs, which creates a mismatch between their abilities and expectations. Through an extensive literature review, this research has identified a "mismatch problem", which is yet to be investigated in the literature. The objective of this study is to materialize the mismatch problem and seek potential solutions to be carried out.

3. What Is the Mismatch Problem?

There is a growing concern among stakeholders about implementing sustainability practices in the SME sector. Although the perception of SMEs is that they have a little individual impact on the environment [25], collectively, their impacts are very high [26]. So, it is expected that this sector will step forward to make their operations sustainable. Stakeholders around the world are taking measures, such as new legislation, to force SMEs to behave ethically, collaborating with them by educating them on social sustainability issues, providing direct or indirect incentives or rewards, and so on. However, the concern within SMEs is how a single economic entity, especially a small-scale enterprise, can be engaged in the uptake of sustainability practices. It is known that SMEs have a lack of expertise and understanding with regard to strategies to address social and environmental issues [27]. This can further be elaborated into many other issues faced by SMEs, such as the cost of implementation and the limited capacity to absorb the cost within the volume of their operation. It is well known that cost is a major barrier for SMEs to show more proactive sustainable behavior, with mangers perceiving little financial benefit from environmental investments. This is even more exacerbated by little or no access to finances from banks and other investment corporations, as they tend to favor large organizations due to credit rating history and wealth. A lack of awareness regarding sustainability practices among SMEs is evident, especially in the least developed or developing countries, where cultural factors play a big role [10]. In many countries, children are employed in SMEs rather than going to school because the legislation does not work fully due to cultural issues. The health and safety of workers and workplaces are also among the things with the least priority. Despite all these issues limiting

SMEs' abilities to take up social sustainability issues, stakeholders' expectations are still increasing. Therefore, there is a clear mismatch between the expectations of stakeholders and the reality of SMEs' abilities to meet those expectations. We envisage that this mismatch between expectation and reality could hinder the progress of social sustainability implementation in SMEs and thus we investigate further. Accordingly, we aim to explore the following research questions:

i.   Why there is a gap between stakeholders' expectations and SSPs adopted by SMEs?
ii. What should be the strategies to minimize the gap?

4. Exploration of the Mismatch Problem in SMEs

4.1. Factors Responsible for the Gap in SMEs' SSPs

The literature suggests a number of reasons why there is a social sustainability gap in SMEs. Among them, a lack of various types of resources, such as finances, skills, staff, and time, are mentioned most frequently [9,10,28]. A lack of financial resources to invest in implementing sustainable practices is considered one of the main barriers to adopting SSPs in SMEs [27,29,30]. The initial investment is required for several SSPs, such as building infrastructure for a safe work environment. While SMEs have a lack of infrastructure, they also cannot only make compliance investments to build such infrastructure due to a lack of financial resources [19]. As such, SMEs are unable to implement several SSPs in their firms. Moreover, an efficient implementation of SSPs requires appropriate skills in the labor force and managers that may not be adequate in SMEs [19,31]. The lack of skills in SMEs is evident in the findings in Johnson and Schaltegger [10], who report that, even when SMEs are aware of the impacts and benefits of SSPs, they are still unable to implement these practices properly due to a lack of skills, knowledge, and expertise. In line with this, a recent study [32] also suggests that practitioners and leaders in SMEs need to be skilled to properly implement SSPs. Furthermore, SMEs generally have too few human resources (staff) to assign someone specifically to the job of taking care of social sustainability or sustainability issues. Due to wearing different hats within a business, SME practitioners also face time constraints in implementing social sustainability indicators [33]. For example, according to the Chamber of Commerce, Industry, Craft and Agriculture of Milan (2003), while more than 80 percent of SMEs with five to nine employees are not willing to implement SSPs, this percentage drops substantially when the results are analyzed for SMEs with 10 or more employees (cited in [24]).

In addition to a lack of resources, a lack of long-term orientation, and the misperceptions of SME owners and managers are mentioned as barriers to adopting SSPs. The investment made in improving SSPs does not provide sufficient return in the short to medium term [24]. However, SMEs care more about the investment that gives them immediate returns, as they do not value the investment that only provides benefits in the long term [34,35]. As a result, smaller firms perceive very few benefits of sustainability practices compared to larger firms [36]. Moreover, SMEs generally lack a consistent and proactive sustainability culture to promote SSPs. Given that such culture and orientation are important for all firms, including SMEs, in adopting both basic and advanced SSPs [37–39], a lack of these elements certainly creates obstacles in the implementation of these practices. In addition, while SMEs are characterized by a small number of employees, they also have a high degree of necrocracy, as they are mostly family businesses where conflict among family members is a common issue [40]. Due to these conflicts, it is difficult for firms to reach a consensus for implementing a social practice unless there exists an established culture and formal governance mechanism for SSPs [41,42]. The other barrier is that SMEs often perceive that their operations have no or minimal impacts on society [30]. However, research shows that the cumulative impacts of SME operations are very high, as most of the businesses across the world are SMEs [26]. These misperceptions are found to be counterproductive to adopting SSPs [10,11,23].

While SMEs' lack of resources and poor social sustainability orientations are mostly mentioned as the barriers to adopting SSPs in the literature, we dig deeper into the SME literature to understand if any other factors are also responsible for the gap between stakeholders' expectations and SMEs' SSPs. It is important to understand this because some SMEs successfully implement SSPs [43]. If some can, a common question is what makes the gap bigger for other companies. Moreover, both resources and orientation are internal issues of SMEs. Highlighting only these issues means that the blame is on SMEs for the social sustainability gap in their firms. The literature survey reveals that SMEs face problems in finding guidelines for implementing a social sustainability standard that is specifically designed for them [24]. This is a crucial factor for increasing SSPs in SMEs because many guidelines are set in the context of large multi-national companies and may not be applicable for SMEs [44].

The other issue that increases the gap is the lack of support from other stakeholders in implementing SSPs. Due to their smaller size, the business counterparts of SMEs, such as buyers and suppliers, often consider them less critical; hence, they provide limited support in implementing SSPs. Moreover, SMEs do not have enough bargaining power [23], which is required to solicit support from other supply chain partners in complying with social sustainability standards [45]. Due to limited bargaining power, SMEs are also not able to provide requirements to their supply chain partners to ensure supply chain-wide social sustainability. For example, Jorgensen and Knudsen [29] found that SMEs receive more social sustainability requirements from their buyers than they provide to their immediate suppliers. Due to the lack of support from their supply chain partners or other relevant stakeholders, SMEs tend to ignore SSPs.

As discussed, many issues are responsible for the social sustainability gap in SMEs. These include several internal issues, such as the resource, skill, and knowledge constraints of SMEs and the poor orientation culture of SMEs, as well as external issues, such as the complexity of the standards and a lack of stakeholder support. Therefore, we come up with the following proposition:

Proposition 1 (P1). *Both internal factors, such as a lack of resources and poor social sustainability orientation and culture, and external factors, such as a lack of standard and institutional support from various stakeholder groups, are responsible for the gap between stakeholders' expectations and SSPs adopted by SMEs.*

4.2. Role of Stakeholder Pressure and Support

Stakeholder pressure has been considered as a dominant factor for firms to implement sustainable practices in their organizations. However, as big organizations are under the intense scrutiny of stakeholders, such as government, media, NGOs, and so on, they generally act on the requirements of their stakeholders. This is because they tend to have the necessary capabilities to implement such practices. However, not all organizations will be able to integrate these into their business practices, particularly when it concerns SMEs [9]. Although SMEs are a significant driver of economic growth, wealth, and job creation [22], their lack of awareness and interest in sustainability is well acknowledged in literature [10]. Furthermore, SMEs are generally not under the attention of the media or other stakeholders and so enjoy less accountability for their activities in society. As they operate below the radar screen of regulators and the general public, they can easily escape public scrutiny. Strict regulations are sometimes considered as a possible option to make firms comply with sustainability issues.

Although legislation could drive SMEs to engage in sustainability activities, legislative pressure, or other forms of pressure, alone is not enough because of the poor surveillance of sustainability activities in developing countries. For instance, research by Baden, Harwood, and Woodward [46] found that pressure by government and big organizations to include social sustainability requirements in procurement works as an incentive for only 49% of SMEs. Moreover, because SMEs' prime concern is general competitiveness and a preference for short-term monetary benefits, it has become very complex for policymakers to set up a proper policy for SMEs to adopt sustainable practices. As a result, mere pressure from government and policymakers cannot substantially improve SSPs in SMEs. In addition, SMEs only strive to meet the requirements of their immediate partners, especially powerful ones, in the supply chain [47]; hence, any pressure that comes from other supply chain partners does not work profoundly for the implementation of SSPs.

Instead, support from stakeholders can work better to motivate SMEs to implement SSPs. Many governments and business associations across the globe have reward packages, policy support, and other schemes to encourage SMEs to implement SSPs [48]. A number of schemes, such as the "Living Business Programme" and "Caring Company Award", to improve SMEs' sustainability performance, can be found in Studer et al. [48] in the context of Hong Kong. Many of these schemes are provided because small firms do not have sufficient internal resources or the expertise to implement sustainability practices on their own, and they tend to have limited access to information concerning sustainability management [49]. Egels-Zandén [50] found that implementing SSPs is an emergent and complex process, as it involves political consideration, traceability, and trade-offs in product design. Due to having limited skills, SMEs find it hard to undertake initiatives to implement SSPs on their own. Hence, support for the implementation of and apprehension about adopting SSPs could be a big factor in this regard. In addition, policy support with direct and indirect monetary benefits could be an added factor.

In short, support from and the involvement of various stakeholders, such as governments, business associations, buyers, and other stakeholders, can boost SMEs' confidence in implementing SSPs. If the stakeholders pressurize them to implement the practices without proper support, the traditional command-and-control approach would be less likely to work. Therefore, it can be proposed that:

Proposition 2a (P2a). *When support from various stakeholders, such as governments and buyers, is available, SMEs are more motivated to implement SSPs. Mere pressure or law enforcement might not work for SMEs.*

Such support becomes more important during a disruption that has a severe impact on the operations of SMEs. SMEs are generally more vulnerable to disruption. For example, disruption has a more severe impact on the profit and other financial measures, such as return on assets and return on sales, of SMEs than those of large corporations [51,52]. With higher impacts of disruptions and a lack of resources, SMEs find it difficult to survive and maintain SSPs. For example, a recent study [53] reported that SMEs struggle to practice social sustainability during a disruption or an extended economic crisis due to a lack of liquidity. As a result, they tend to sidestep social sustainability commitments and initiatives to ensure their survival and overcome the problem of liquidity. During such a disruption, SMEs need support from their various stakeholder groups to continue their commitment towards SSPs.

Take the example of the current pandemic outbreak of acute respiratory syndrome coronavirus 2, known as COVID-19. Organizations of all sizes, including SMEs, across the world are facing problems in retaining their workforce and maintaining social sustainability. According to the International Labour Organization (ILO), around 200 million employees globally could lose their jobs due to this extraordinary outbreak [54]. While larger firms are struggling to retain their employees and perform other sustainable practices because of continued loss [55,56], it can be clearly understood how difficult it is for SMEs to retain their employees and show commitment to social sustainability. During such a crisis period, support from various stakeholder groups can serve as an incentive to practice social sustainability. The government and policymakers of a country can play a crucial role in this regard [53]. For example, Australian SMEs could access job-keeper support, an incentive provided by the government, to retain their employees [57]. Similar incentives from the government are also provided in other countries to assist SMEs in maintaining SSPs, such as retaining staff, paid leave, and other subsidies. For instance, in Cambodia, the government has contributed to the salary of the staff of SMEs who were rendered jobless because of a factory closure and confirmed their eligibility to receive 60 percent of their salary for six months [53]. The government of Sri Lanka also confirmed that employees are entitled to paid leave if factories are temporarily closed [58]. Not only the government, but other stakeholders, such as large buyers of SMEs, can also support SMEs to maintain SSPs during a disruption. Support from various stakeholder groups can give a positive impression of connectedness to SMEs. In return, they show their commitment to their employees and communities via performing SSPs. Therefore, we propose:

Proposition 2b (P2b). *SMEs require support from various stakeholders to perform their SSPs during disruption.*

4.3. Cost Implications

The implementation of various SSPs incurs costs for organizations. Depending on the types of costs incurred, the practices can be divided into a two-by-two matrix, showing operating cost and initial investment (Figure 1). Some practices require a high initial investment but incur low operating costs over time. For example, developing infrastructure for workplace safety and amenities, such as a cafeteria and restroom, needs substantial investment at the beginning [59]. However, once developed, the operating cost of these facilities is not too high. On the other hand, there are some other SSPs that incur high operating costs over time but need less initial investment. For instance, a firm that wants to adopt a decent pay structure for the staff may need to increase remunerations substantially [8]. In such a situation, firms need to pay additional remuneration every month. Similarly, if a firm joins with a local NGO to participate in a poverty reduction program, it may need to contribute to the program every month or year over a certain period.

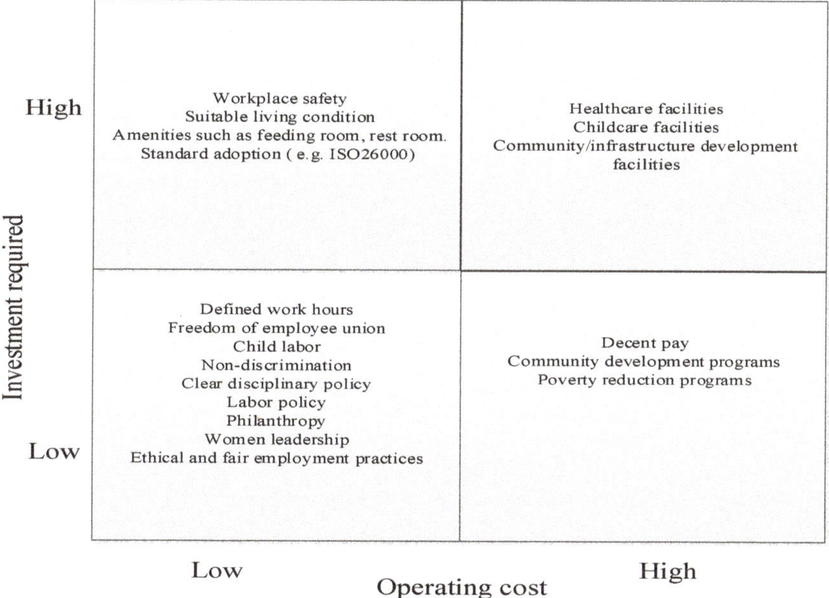

Figure 1. Social sustainability practices (SSPs) and types of cost required.

On the other hand, some practices may need both high operating costs and high initial investment. For example, if a firm wants to develop childcare facilities, first it needs to make an initial investment in developing a childcare center within the organization. Then, the firm also needs to hire one or more staff to take care of the children, which increases the operating costs for the firm. Similarly, when a firm decides to develop and run a community facility, such as a local school or health care facilities for local people, it needs to invest in the infrastructure as well as carry the running cost [11,16]. Finally, there are some SSPs that neither need high initial investment nor incur a high operating cost. Rather, these types of SSP need a proactive mindset and organizational governance structure. For example, to promote non-discrimination, a clear disciplinary policy, and women leadership, a firm just needs to value such practices.

SMEs face huge challenges in implementing SSPs that require a high initial investment. This is mainly because they have a lack of financial resources and are characterized by low capital [29]. It is not feasible for SMEs to make such a huge initial investment to adopt these SSPs. Moreover, SMEs face problems in obtaining compliance loans because of their lack of collateral. This challenge of SMEs is already noted in the literature. For example, the initial cost or investment is found to be one of the main barriers to why SMEs do not implement many SSPs in their businesses [41]. For adopting SSPs that need high initial investment, SMEs need financial support from other stakeholders. The support can be provided in various forms. For example, government or relevant SME policymakers, or even NGOs, can provide donations to SMEs to develop infrastructure to adopt SSPs. On the other hand, buyers can provide finance to SMEs to adopt SSPs, such as giving a certain amount to SMEs for adopting an SSP which will be later adjusted with the price of the products. In such a situation, SMEs would be able to develop infrastructure without obtaining a loan from a bank or other financial institution. Such financial support from the stakeholder is likely to encourage SMEs to do their part by adopting SSPs in the firm. Therefore, we propose:

Proposition 3a (P3a). *When social sustainability implementation attracts high investment, SMEs are less likely to adopt such practices on their own; hence, they require donations, finance, or other financial support from governments or other stakeholders, such as buyers or NGOs.*

When the implementation of an SSP requires a low initial investment, it is the SME's responsibility to increase the efficiency of the operation so that it can cover the increased operating cost. It is very natural that the implementation of SSPs has an impact on other enterprise functions, for example, the impact on the operating cost [37]. For example, to ensure SSPs in sourcing, firms may need to change the location of suppliers, which may incur higher transportation costs [23]. In adopting these SSPs, SMEs need to be innovative so that they can reduce the cost of the activities in their operation [60]. As an example, while SMEs implement decent pay, they can be innovative in designing shifts for the employees to maximize the use of human resources. Such an innovation in the operation can also increase productivity for the firms. SMEs should not expect that their stakeholders will provide financial assistance over the entire period needed to adopt these high operating cost-generating SSPs. Rather, they need to increase their efficiency and productivity to increase the flow of liquidity to practice these SSPs. Hence, rather than expecting financial support from the stakeholders, SMEs can seek know-how from the stakeholders so that they can efficiently implement these SSPs [61]. SMEs can also create a separate budget for covering the running cost of SSPs, as such a budget is found to be effective in implementing SSPs [62]. Based on the above discussion, we propose the following:

Proposition 3b (P3b). *When an SSP requires an increased operating cost, it is the SMEs' responsibility to increase their productivity and/or efficiency to make the SSP financially viable.*

4.4. Role of Awareness Development

Employees' and employers' awareness both play a significant role in implementing SSPs in SMEs [46]. If the entrepreneurs of SMEs have the proper knowledge and understanding of their employees' human rights, for example, the availability of proper sanitation facilities, a safe workplace considering occupational health and safety (OHS) requirements and standards, wages paid on time, and provisions for working overtime, the implementation of SSPs becomes easier. SMEs can be opportunistic, as in many countries, the legislation for SMEs is not in place or rules of laws are not strictly maintained throughout society [48]. So, it is critical that SMEs are aware of the benefits of social sustainability in the long run to the company and to society, in addition to the ethics of business and their responsibility to society. For example, Zhang and Zhang [15] found that, when SME practitioners understand their role in protecting local people or the local culture, they tend to hire more local people and develop their brands with local culture, as well as train local people. Similarly, employees' awareness can make big differences as well. If the employees are aware of their rights, it is difficult for SMEs to show opportunistic behavior. They will be forced to take proper measures to ensure the health and hygiene of the workplace, pay on time, and act according to local legislation. In this regard, internal communication regarding SSPs with employees is suggested [61]. An appropriate level of education of employees could help to increase awareness regarding their rights and responsibilities. As mentioned earlier in this section, regulations are not strictly enforced in many countries, especially in developing countries; hence, awareness among employees and society is the key to social sustainability. For example, in the least developed countries, child labor is a big problem. Children are often forced to work due to the financial constraints of the family in the least developed countries, although child labor is banned in almost all the world by law. Only the awareness of both parties can help to overcome this situation where law enforcement is lacking.

Proposition 4 (P4). *SMEs' and their employees' awareness may further encourage SMEs to adopt SSPs.*

5. Strategies to Handle the Mismatch Problem

A number of strategies can be undertaken to tackle the matching problem identified in this study. We propose some strategies (shown in Table 1) that we believe are pragmatic and useful. Depending on each SME's circumstances, any or a combination of strategies could be adopted. Building a cooperative relationship is one of the most cited strategies in supply chain research and it could be very relevant in this context. The implementation and practice of social sustainability issues are complicated [50]; hence, they need support from stakeholders. The support can be in the form of receiving know-how and technical support from a stakeholder [15]. For example, when a buyer provides their requirements for SSPs, SMEs can actively seek advice, training, and assistance from the buyer for the implementation of such practices. Moreover, stakeholders can give support via providing immediate finances for implementing the SSPs. For enhancing SMEs' capabilities to practice social sustainability, various stakeholders need to extend their support via sharing knowledge and providing technical assistance and finance. In order to obtain such support, SMEs need to build trustworthy relationships with the relevant stakeholders. SMEs need to understand that support from stakeholders can only be achieved when there exists a reciprocal relationship between them. In this regard, they need to maintain their commitment to the business partners, such as by providing superior services within their capacity and sharing relevant and timely information so that a cooperative and respectful relationship is created [44].

The next strategy that we suggest to reduce the gap is undertaking awareness development programs for creating awareness in both SME owners and employees towards SSPs. We suggest this because it has been revealed that the positive attitude and orientation of SME owners/managers is one of the most crucial factors that motivates them to implement SSPs [46]. The government and other policymakers should organize campaigns and other awareness development programs to demonstrate cases (SMEs) that successfully implement SSPs. In these programs, both the benefits of SSPs and the roles of SME practitioners in implementing such practices should be highlighted. Such understanding is likely to encourage and guide them to undertake necessary actions for implementing SSPs. SMEs also need to organize internal programs for their staff so that they become aware of their roles in performing SSPs. A culture for the timely and free flow of internal communication on social sustainability should be created to develop awareness within the organization.

Table 1. Suggested strategies for reducing the gap between stakeholders' expectations and SSPs adopted by small- and medium-sized enterprises (SMEs).

Strategy	Implications for SMEs	Implication for SMEs' Stakeholders
Building cooperative relationships	Maintain commitment; Information sharing; Provide premium customer service.	Technical support; Specific investment; Training; Knowledge sharing.
Awareness development programs towards social sustainability.	Internal communication; Internal programs; Knowledge development.	Policy support by the government; Demonstration of successful cases (SMEs that implemented SSPs)
Sustainable finance mechanism.	Participation in buyers' financial packages.	Creation/development of financial packages for SMEs; Easy access to finance from buyer; Government support in the form of loan, subsidy.
Governance structure.	Initiative to develop corporate governance structure that supports social sustainability adoption.	Compliance audit for SSPs.
Human capital development.	Development of various programs such as training, education, and skill database.	Assist SMEs in developing programs; Expert knowledge sharing by buyers.
Implementation of innovative changes.	Efficient and innovative implementation of changes.	Guide SMEs in innovative changes.
Unified standards to follow.	Liaise with buyers to come up with standard adoption requirements.	Develop a framework by involving various stakeholders and come up with a unified standard for SMEs.

In the literature, there is a notion that a sustainable supply chain finance mechanism could enhance sustainability practices and thus improve the sustainable performance of supply chains [63]. We propose that this could be an effective strategy to minimize the identified gap between expectation and ability. A sustainable finance mechanism is a platform where buyers come up with some financial packages for SMEs to easily access cash in return for discounted prices or any other agreed conditions. Financial institutions are invited to facilitate those packages. This mechanism can work in different ways; however, it increases the cash flow in the supply chain, which is absolutely critical for SMEs to operate. Understandably, SMEs have less access to capital and so always seek loans/credit from financial institutions, which is often difficult for them to get due to mortgages/bonds and other contractual arrangements. Financial packages from buyers in sustainable finance packages could help to build trust, collaboration, and eventually build up SMEs' confidence and motivate them to work towards implementing buyers' social sustainability requirements [63]. Therefore, the government and buyers should create a sustainable financial package for SMEs and ensure easy access to this package.

The next strategy that we suggest is maintaining a sound organizational governance structure that supports the proper implementation of SSPs in SMEs. While social sustainability fosters non-discrimination and career progression, these should be well articulated by the organizational governance mechanism. SME entrepreneurs/managers should ensure that organizational policies and processes for social sustainability issues are clear and communicated properly to the staff so that they are aware of their roles in implementing SSPs. In addition to developing a positive orientation, such clear policies and processes can also the build capabilities of the practitioners, as these may serve as a guide to improve the understanding of social sustainability issues [12]. Policymakers and governments can also play a role in this regard. For example, an audit program could be initiated to track whether SMEs properly maintain SSPs and guide them to improve such practices.

It is also essential to develop the human capital in SMEs to improve SSPs. Dynamic entrepreneur leadership is found to have a positive effect on SSPs in the context of Malaysian SMEs [32]. Such dynamic leadership or managerial competencies cannot be achieved without providing training to the practitioners on various issues. Therefore, SMEs need to organize various skill and knowledge development programs, such as training and education, on social sustainability for their staff. In addition, other relevant stakeholders should also help develop the human capital of SMEs. For example, buyers can provide technical support to SMEs and the government can create education programs for SME practitioners to improve their skills base on SSPs.

We also suggest that SMEs need to be innovative in changing their operations to implement SSPs. A previous study [43] showed how Metalquimia, a medium-sized family business in Spain, successfully adopted SSPs through innovation in the adoption of change. While implementing SSPs, firms may need to change their other organizational practices. For example, a firm that wants to strictly comply with maximum work hours may need to reduce the working hours of the staff. Without ensuring innovation in the operations, such as process innovation to maximize the utilization of the working hours, the cost will increase as a result of the reduced work hours of the staff. SMEs may not be able to tolerate a substantial increase in cost by adopting SSPs. The implementation of SSPs may become a burden to SMEs if they are not able to ensure innovation in the adoption of change. Therefore, SMEs need to be innovative in implementing SSPs. In this regard, various stakeholder groups, especially buyers and policymakers, should guide SMEs.

Finally, we suggest a unified social sustainability standard for SMEs to follow. The literature argues that the complexity of social sustainability standards is a major reason why SMEs do not implement SSPs. These standards are generally developed to account for international, and at least national, issues. On the other hand, SMEs act at a local level; hence, they find adopting standards to be challenging unless they are well supported by stakeholders [10]. SMEs are often required to adopt the standard certifications suggested by their stakeholders. Depending on the statutory or non-statutory obligations, stakeholders pressurize them to implement various standards, such as ISO26000, LEED, to ensure social sustainability standards are being properly followed by SMEs. There are many different types of standards required by buyers because of the local and international regulations and it is often impossible to have them all implemented due to a lack of resources. In addition, there will be a problem of fit even when they overcome the problem of the lack of resources [64]. Therefore, SMEs need a unified standard to be implemented. This will save a lot of resources, decrease the complexity of implementation, and motivate them to implement more SSPs, as the cost of implementation will be reduced greatly. Therefore, policymakers need to create a unified social sustainability standard for SMEs by involving various stakeholders. SMEs also need to provide inputs, such as requirements for buyers, so that policymakers can create a comprehensive standard to follow.

6. Contributions of the Study

This study investigates why there is a social sustainability gap in SMEs. It formulates some propositions and provides strategies to reduce the gap. By doing so, the study makes some notable contributions to the literature. The main contribution of this study is to identify the gaps in SMEs' SSPs and subsequent strategies, which will help the SME sector to be more socially sustainable. While previous reports broadly highlight the importance of social sustainability in SMEs and the difficulties and challenges in implementing SSPs in SMEs, none of them spotted the gap and provided a comprehensive understanding on this issue. Similarly, most of the academic articles that focus on SSPs in SMEs mainly discuss the importance, challenges, and barriers to implementing SSPs. For example, according to MacGregor and Fontrodona [43], the majority of the studies on SSPs in SMEs mostly focus on the development of awareness of SSPs. While enough justifications are already available in the literature as to why SMEs should implement SSPs, at this stage, it is more important to know the specifics of how SMEs can implement SSPs. This study discusses how social sustainability can be achieved in SMEs, which is a unique and noble contribution to the existing body of knowledge. Addressing such adequacy is important given that it is vital to know how SMEs can implement SSPs despite having resource and skill constraints and other challenges.

Next, the study supplements the limited research focusing on social sustainability, or sustainability in general, in SMEs [4,11,65]. Through a systematic literature review on social sustainability, Nakamba, Chan, and Sharmina [3] reported that research to date mostly focuses on large firms, and studies investigating the practices of and approaches to social sustainability in SMEs are inadequate. Finally, the study contributes to the literature on social sustainability. While environmental sustainability has been rigorously investigated in the previous literature, research on social sustainability is scarce [7].

7. Conclusions and Future Research Directions

Aiming to provide guidelines to make SMEs socially sustainable, this study has explored why there is a gap between stakeholders' expectations and the SSPs adopted by SMEs, and how the gap can be minimized. Based on a rigorous literature review, the study has revealed that both internal and external factors are responsible for this social sustainability gap in SMEs. Various internal issues, such as poor organizational orientation and culture and lack of resources, skills, and knowledge, create the gap. Moreover, external factors, such as a lack of a tailored standard for SMEs and a lack of support from stakeholders, are also responsible for creating this gap or making the gap bigger. In order to minimize the gap, we propose that stakeholder support, a supply chain finance mechanism, the improvement of SMEs' efficiency and productivity, and the development of the awareness of employees will lead SMEs to be more socially sustainable. In line with these propositions, we have also provided several strategies, outlining the implications for both SMEs and their stakeholders. The strategies include building cooperative relationships with various stakeholders, developing programs to create awareness of SSPs, creating a sustainable finance package, ensuring a supportive governance structure, developing human capital, innovation in implementing SSPs, and developing a unified social sustainability standard for SMEs.

While the study findings make notable contributions to the literature, it also has a number of limitations that can be seen as the directions for future research. For example, in designing the proposition, this study only uses the current body of literature and does not consider any particular context. Therefore, future studies could empirically investigate and validate the propositions. For validating the findings of this study, a future study could formulate and test several hypotheses grounded in appropriate theories and based on the propositions suggested in this research. Moreover, a comparative study could be undertaken to explore how the proposed strategies contribute to increasing SSPs in SMEs in developed and developing countries. Such a study might provide useful information, as SSPs and their characteristics vary between developed and developing countries [4]. Moreover, there is a difference in implementing SSPs in developing and developed countries. For example, a study by Malesios et al. [18] revealed that French SMEs are different from Indian SMEs in practicing social sustainability. Therefore, a comparative study could be useful to further refine the strategies in the context of developed and developing economies. Finally, we suggest a future study for unveiling whether, how, and to what extent the strategies proposed in this research can contribute to the achievement of the sustainable development goals of the United Nations.

Author Contributions: Conceptualization, P.C. and R.S.; Literature review and information search, P.C. and R.S.; Draft writing, P.C. and R.S.; Reviewing and editing: P.C. and R.S., Administration, P.C. All authors have read and agreed to the published version of the manuscript.

Funding: This research received no external funding.

Conflicts of Interest: The authors declare no conflict of interest.

References

1. Silvestre, B.S. Sustainable supply chain management in emerging economies: Environmental turbulence, institutional voids and sustainability trajectories. *Int. J. Prod. Econ.* **2015**, *167*, 156–169. [CrossRef]
2. Mani, V.; Gunasekaran, A.; Delgado, C. Enhancing supply chain performance through supplier social sustainability: An emerging economy perspective. *Int. J. Prod. Econ.* **2018**, *195*, 259–272. [CrossRef]

3. Nakamba, C.C.; Chan, P.W.; Sharmina, M. How does social sustainability feature in studies of supply chain management? A review and research agenda. *Supply Chain Manag. Int. J.* **2017**, *22*, 522–541. [CrossRef]

4. Mani, V.; Jabbour, C.J.C.; Mani, K.T.N. Supply chain social sustainability in small and medium manufacturing enterprises and firms' performance: Empirical evidence from an emerging Asian economy. *Int. J. Prod. Econ.* **2020**, *227*, 107656. [CrossRef]

5. European Union 2015. The Revised User Guide to the SME Definition. Available online: https://ec.europa.eu/growth/content/revised-user-guide-sme-definition-0_en (accessed on 5 May 2020).

6. Klassen, R.D.; Vereecke, A. Social issues in supply chains: Capabilities link responsibility, risk (opportunity), and performance. *Int. J. Prod. Econ.* **2012**, *140*, 103–115. [CrossRef]

7. Huq, F.A.; Stevenson, M.; Zorzini, M. Social sustainability in developing country suppliers: An exploratory study in the ready made garments industry of Bangladesh. *Int. J. Oper. Prod. Manag.* **2014**, *34*, 610–638.

8. Turyakira, P.; Venter, E.; Smith, E. The impact of corporate social responsibility factors on the competitiveness of small and medium-sized enterprises. *S. Afr. J. Econ. Manag. Sci.* **2014**, *17*, 157–172. [CrossRef]

9. Lee, C.M.J.; Che-Ha, N.; Alwi, S.F.S. Service customer orientation and social sustainability: The case of small medium enterprises. *J. Bus. Res.* **2020**. [CrossRef]

10. Johnson, M.P.; Schaltegger, S. Two decades of sustainability management tools for SMEs: How far have we come? *J. Small Bus. Manag.* **2016**, *54*, 481–505. [CrossRef]

11. Kot, S. Sustainable supply chain management in small and medium enterprises. *Sustainability* **2018**, *10*, 1143. [CrossRef]

12. Pirnea, I.C.; Olaru, M.; Moisa, C. Relationship between corporate social responsibility and social sustainability. *Econ. Transdiscipl. Cogn.* **2011**, *14*, 36–43.

13. Mani, V.; Agrawal, R.; Sharma, V. Supply chain social sustainability: A comparative case analysis in Indian manufacturing industries. *Procedia-Soc. Behav. Sci.* **2015**, *189*, 234–251. [CrossRef]

14. Wolf, J. The relationship between sustainable supply chain management, stakeholder pressure and corporate sustainability performance. *J. Bus. Ethics* **2014**, *119*, 317–328. [CrossRef]

15. Zhang, L.; Zhang, J. Perception of small tourism enterprises in Lao PDR regarding social sustainability under the influence of social network. *Tour. Manag.* **2018**, *69*, 109–120. [CrossRef]

16. Masocha, R. Social sustainability practices on small businesses in developing economies: A case of South Africa. *Sustainability* **2019**, *11*, 3257. [CrossRef]

17. Brandenburg, M.; Gruchmann, T.; Oelze, N. Sustainable supply chain management—A conceptual framework and future research perspectives. *Sustainability* **2019**, *11*, 7239. [CrossRef]

18. Malesios, C.; Skouloudis, A.; Dey, P.K.; Abdelaziz FBen Kantartzis, A.; Evangelinos, K. The impact of SME sustainability practices and performance on economic growth from a managerial perspective: Some modeling considerations and empirical analysis results. *Bus. Strategy Environ.* **2018**, *27*, 960–972. [CrossRef]

19. Moore, S.B.; Manring, S.L. Strategy development in small and medium sized enterprises for sustainability and increased value creation. *J. Clean. Prod.* **2009**, *17*, 276–282. [CrossRef]

20. Mani, V.; Gunasekaran, A. Four forces of supply chain social sustainability adoption in emerging economies. *Int. J. Prod. Econ.* **2018**, *199*, 150–161. [CrossRef]

21. Gama, A.P.M.; Geraldes, H.S.A. Credit risk assessment and the impact of the New Basel Capital Accord on small and medium-sized enterprises: An empirical analysis. *Manag. Res. Rev.* **2012**, *35*, 727–749. [CrossRef]

22. Madanchian, M.; Hussein, N.; Noordin, F.; Taherdoost, H. The impact of ethical leadership on leadership effectiveness among SMEs in Malaysia. *Procedia Manuf.* **2018**, *22*, 968–974. [CrossRef]

23. Ciliberti, F.; Pontrandolfo, P.; Scozzi, B. Investigating corporate social responsibility in supply chains: A SME perspective. *J. Clean. Prod.* **2008**, *16*, 1579–1588. [CrossRef]

24. Borga, F.; Citterio, A.; Noci, G.; Pizzurno, E. Sustainability report in small enterprises: Case studies in Italian furniture companies. *Bus. Strategy Environ.* **2009**, *18*, 162–176. [CrossRef]

25. Ammenberg, J.; Hjelm, O. Tracing business and environmental effects of environmental management systems—A study of networking small and medium-sized enterprises using a joint environmental management system. *Bus. Strategy Environ.* **2003**, *12*, 163–174. [CrossRef]

26. Lawrence, S.R.; Collins, E.; Pavlovich, K.; Arunachalam, M. Sustainability practices of SMEs: The case of NZ. *Bus. Strategy Environ.* **2006**, *15*, 242–257. [CrossRef]

27. Lee, S.Y.; Klassen, R.D. Drivers and enablers that foster environmental management capabilities in small- and medium-sized suppliers in supply chains. *Prod. Oper. Manag.* **2008**, *17*, 573–586. [CrossRef]

28. Karuppiah, K.; Sankaranarayanan, B.; Ali, S.M.; Chowdhury, P.; Paul, S.K. An integrated approach to modeling the barriers in implementing green manufacturing practices in SMEs. *J. Clean. Prod.* **2020**, *265*, 121737. [CrossRef]

29. Jorgensen, A.L.; Knudsen, J.S. Sustainable competitiveness in global value chains: How do small Danish firms behave? *Corp. Gov.* **2006**, *6*, 449–462. [CrossRef]

30. Mani, V.; Agrawal, R.; Sharma, V. Social sustainability in the supply chain: Analysis of enablers. *Manag. Res. Rev.* **2015**, *38*, 1016–1042. [CrossRef]

31. Clarke-Sather, A.R.; Hutchins, M.J.; Zhang, Q.; Gershenson, J.K.; Sutherland, J.W. Development of social, environmental, and economic indicators for a small/medium enterprise. *Int. J. Account. Inf. Manag.* **2011**, *19*, 247–266. [CrossRef]

32. Nor-Aishah, H.; Ahmad, N.H.; Thurasamy, R. Entrepreneurial leadership and sustainable performance of manufacturing SMEs in Malaysia: The contingent role of entrepreneurial bricolage. *Sustainability* **2020**, *12*, 3100. [CrossRef]

33. Burke, S.; Gaughran, W.F. Developing a framework for sustainability management in engineering SMEs. *Robot. Comput. Integr. Manuf.* **2007**, *23*, 696–703. [CrossRef]

34. Thakkar, J.; Deshmukh, A.K.S. Supply chain management in SMEs: Development of constructs and propositions. *Asia Pac. J. Mark. Logist.* **2008**, *20*, 97–131. [CrossRef]

35. Falkner, E.M.; Hiebl, M.R.W. Risk management in SMEs: A systematic review of available evidence. *J. Risk Financ.* **2015**, *16*, 122–144. [CrossRef]

36. Brammer, S.; Hoejmose, S.; Marchant, K. Environmental management in SMEs in the UK: Practices, pressures and perceived benefits. *Bus. Strategy Environ.* **2012**, *21*, 423–434. [CrossRef]

37. Kerr, I.R. Leadership strategies for sustainable SME operation. *Bus. Strategy Environ.* **2006**, *15*, 30–39. [CrossRef]

38. Wiesner, R.; Chadee, D.; Best, P. Insights into sustainability change management from an organisational learning perspective: Learning from SME sustainability champions. In Proceedings of the 10th International Research Conference on Quality, Innovation and Knowledge Management, Kuala Lumpur, Malaysia, 15–18 February 2011; pp. 268–281.

39. Marshall, D.; McCarthy, L.; McGrath, P.; Claudy, M. Going above and beyond: How sustainability culture and entrepreneurial orientation drive social sustainability supply chain practice adoption. *Supply Chain Manag.* **2015**, *20*, 434–454. [CrossRef]

40. Burgstaller, J.; Wagner, E. How do family ownership and founder management affect capital structure decisions and adjustment of SMEs? Evidence from a bank-based economy. *J. Risk Financ.* **2015**, *16*, 73–101. [CrossRef]

41. Castka, P.; Balzarova, M.A.; Bamber, C.J.; Sharp, J.M. How can SMEs effectively implement the CSR agenda? A UK case study perspective. *Corp. Soc. Responsib. Environ. Manag.* **2004**, *11*, 140–149. [CrossRef]

42. Maldonado-Erazo, C.P.; Álvarez-García, J.; del Rama, R.; Correa-Quezada, R. Corporate social responsibility and corporate performance in Romania. *Sustainability* **2020**, *12*, 2332. [CrossRef]

43. MacGregor, S.P.; Fontrodona, J. Strategic CSR for SMEs: Paradox or possibility? *Universia Bus. Rev.* **2011**, *30*, 80–94.

44. Chowdhury, P.; Lau, K.H.; Pittayachawan, S. Operational supply risk mitigation of SME and its impact on operational performance: A social capital perspective. *Int. J. Oper. Prod. Manag.* **2019**, *39*, 478–502. [CrossRef]

45. Gereffi, G.; Humphrey, J.; Sturgeon, T. The governance of global value chains. *Rev. Int. Political Econ.* **2005**, *12*, 78–104. [CrossRef]

46. Baden, D.A.; Harwood, I.A.; Woodward, D.G. The effect of buyer pressure on suppliers in SMEs to demonstrate CSR practices: An added incentive or counter productive? *Eur. Manag. J.* **2009**, *27*, 429–441. [CrossRef]

47. Ciliberti, F.; de Groot, G.; de Haan, J.; Pontrandolfo, P. Codes to coordinate supply chains: SMEs' experiences with SA8000. *Supply Chain Manag. Int. J.* **2009**, *14*, 117–127. [CrossRef]

48. Studer, S.; Tsang, S.; Welford, R.; Hills, P. SMEs and voluntary environmental initiatives: A study of stakeholders' perspectives in Hong Kong. *J. Environ. Plan. Manag.* **2008**, *51*, 285–301. [CrossRef]

49. Hillary, R. Environmental management systems and the smaller enterprise. *J. Clean. Prod.* **2004**, *12*, 561–569. [CrossRef]

50. Egels-Zandén, N. Not made in China: Integration of social sustainability into strategy at Nudie Jeans Co. *Scand. J. Manag.* **2016**, *32*, 45–51. [CrossRef]
51. Hendricks, K.B.; Singhal, V.R. Association between supply chain glitches and operating performance. *Manag. Sci.* **2005**, *51*, 695–711. [CrossRef]
52. Kaufmann, L.; Carter, C.R.; Rauer, J. The coevolution of relationship dominant logic and supply risk mitigation strategies. *J. Bus. Logist.* **2016**, *37*, 87–106. [CrossRef]
53. Majumdar, A.; Shaw, M.; Sinha, S.K. COVID-19 Debunks the Myth of Socially Sustainable Supply Chain: A Case of the Clothing Industry in South Asian Countries. *Sustain. Prod. Consum.* **2020**, *24*, 150–155. [CrossRef]
54. ILO 2020. COVID-19: Impact could Cause Equivalent of 195 Million Job Losses. Available online: https://news.un.org/en/story/2020/04/1061322 (accessed on 29 June 2020).
55. Amankwah-Amoah, J. Stepping Up and Stepping Out of COVID-19: New Challenges for Environmental Sustainability Policies in the Global Airline Industry. *J. Clean. Prod.* **2020**, *271*, 123000. [CrossRef]
56. Barneveld, K.V.; Quinlan, M.; Kriesler, P.; Junor, A.; Baum, F.; Chowdhury, A.; Junankar, P.N.; Clibborn, S.; Flanagan, F.; Wright, C.F.; et al. The COVID-19 pandemic: Lessons on building more equal and sustainable societies. *Econ. Labour Relat. Rev.* **2020**, *31*, 133–157. [CrossRef]
57. Australian Small Business and Family Enterprise Ombudsman 2020. JobKeeper to Help Small Businesses Survive Coronavirus Crisis. Available online: https://www.asbfeo.gov.au/news/news-articles/jobkeeper-help-small-businesses-survive-coronavirus-crisis (accessed on 29 June 2020).
58. Zarocostas, J. COVID-19 Crisis Triggering Huge Losses in Textile, Apparel Sector. 21 April 2020. Available online: https://wwd.com/business-news/government-trade/covid-19-crisis-triggering-huge-losses-in-textile-apparel-sector-1203566328/ (accessed on 2 August 2020).
59. Altmann, M. A supply chain design approach considering environmentally sensitive customers: The case of a German manufacturing SME. *Int. J. Prod. Res.* **2015**, *53*, 6534–6550. [CrossRef]
60. Klewitz, J.; Hansen, E.G. Sustainability-oriented innovation of SMEs: A systematic review. *J. Clean. Prod.* **2014**, *65*, 57–75. [CrossRef]
61. Henriques, J.; Catarino, J. Sustainable value and cleaner production—Research and application in 19 Portuguese SME. *J. Clean. Prod.* **2015**, *96*, 379–386. [CrossRef]
62. Burlea-Schiopoiu, A.; Mihai, L.S. An integrated framework on the sustainability of SMEs. *Sustainability* **2019**, *11*, 6206. [CrossRef]
63. Jia, F.; Zhang, T.; Chen, L. Sustainable supply chain Finance: Towards a research agenda. *J. Clean. Prod.* **2020**, *243*, 118680. [CrossRef]
64. Simpson, D.; Power, D.; Klassen, R. When one size does not fit all: A problem of fit rather than failure for voluntary management standards. *J. Bus. Ethics* **2012**, *110*, 85–95. [CrossRef]
65. Chowdhury, P.; Paul, S.K. Applications of MCDM methods in research on corporate sustainability: A systematic literature review. *Manag. Environ. Qual. Int. J.* **2020**, *31*, 385–405. [CrossRef]

 *sustainability*

Article

A Multi-Item Replenishment Problem with Carbon Cap-and-Trade under Uncertainty

Jiseong Noh [1], Jong Soo Kim [2] and Seung-June Hwang [1,*]

[1] Institute of Knowledge Services, Hanyang University, Erica, Ansan 15588, Korea; tppeon@gmail.com

[2] Department of Industrial and Management Engineering, Hanyang University, Erica, Ansan 15588, Korea; pure@hanyang.ac.kr

* Correspondence: sjh@hanyang.ac.kr

Received: 26 May 2020; Accepted: 12 June 2020; Published: 15 June 2020

Abstract: Recently, as global warming has become a major issue, many companies have increased their efforts to control carbon emissions in green supply chain management (GSCM) activities. This paper deals with the multi-item replenishment problem in GSCM, from both economic and environmental perspectives. A single buyer orders multiple items from a single supplier, and simultaneously considers carbon cap-and-trade under limited storage capacity and limited budget. In this case we can apply a can-order policy, which is a well-known multi-item replenishment policy. Depending on the market characteristics, we develop two mixed-integer programming (MIP) models based on the can-order policy. The deterministic model considers a monopoly market in which a company fully knows the market information, such that both storage capacity and budget are already determined. In contrast, the fuzzy model considers a competitive or a new market, in which case both of those resources are considered as fuzzy numbers. We performed numerical experiments to validate and assess the efficiency of the developed models. The results of the experiments showed that the proposed can-order policy performed far better than the traditional can-order policy in GSCM. In addition, we verified that the fuzzy model can cope with uncertainties better than the deterministic model in terms of total expected costs.

Keywords: green supply chain management; carbon tax and cap; can-order policy; mixed-integer programming; fuzzy constraints

---

1. Introduction

Since the 1997 Kyoto protocol, many countries and organizations have presented legislation or policies about managing carbon emissions as global warming destroys the Earth's ecosystem. Accordingly, any company can concentrate on reducing and managing its carbon emissions in a variety of areas such as supply chain management (SCM), production, contracts, inventory, and replenishment, with the result that green supply chain management (GSCM) is quickly spreading [1]. The main purpose of GSCM is to minimize supply chain costs and simultaneously reduce carbon emissions. In line with this goal, many governments have implemented carbon cap-and-trade regulation, which is well known as an effective economic-based mechanism [2]. Under such regulation, each company is allocated limited carbon emissions credits from its government, and it can buy or sell rights to emit carbon emissions with other companies in the carbon trading market [3,4]. In the European Union, the European Union Emissions Trading Scheme, which is the largest carbon trading market, has covered almost 50% of total carbon emissions [5]. Therefore, it becomes important to consider GSCM in the context of carbon cap-and-trade regulation.

Beyond governments' efforts to reduce carbon emissions, some companies have tried to manage their own carbon emissions through their supply chains. For example, the retailers Asda, Tesco, Wal-Mart, and H&M require their suppliers to reduce carbon emissions during multi-item replenishment activities [6,7]. In this way, a company considers carbon emissions simultaneously with the multi-item replenishment problem under limited resources, such as storage capacity and budget [8]. However, a company could face two realistic situations based on either known or unknown market information [9]. When a company is in a monopoly market, it knows all of the relevant market information so it can easily decide on storage capacity and budgets. In contrast, when a company is in a competitive or a new market, the needed market information is difficult to grasp. In this case, because of weak market information, neither proper storage capacity nor budgets can be estimated in a stochastic sense. It is difficult to predefine them [10]. Therefore, this paper focuses on the multi-item replenishment problem with limited resources under two market information cases, the certain and the uncertain.

Given current real-world practices, this paper considers the multi-item replenishment problem with carbon cap-and-trade under limited storage capacity and budget. This work (1) develops two mixed-integer problem (MIP) models based on a periodic can-order policy, which is a well-known multi-item replenishment policy; (2) includes carbon cap-and-trade for GSCM; and (3) covers two market information cases: certain and uncertain information. This paper makes the following contributions. First, we develop a deterministic model with carbon cap-and-trade for GSCM with certain (known) market information. In this model, both storage capacity and budget are already predefined. The deterministic model can be applied in a monopoly market case. Based on this model, we develop a fuzzy model applying fuzzy constraints. Because of uncertain market information, both storage capacity and budget are considered as fuzzy numbers. The fuzzy model can be applied in a competitive market case. Second, we suggest both MIP models based on the periodic can-order policy. Thus, each model can obtain optimal results under replenishment planning and inventory control.

The structure of this paper is as follows. A literature review is presented in Section 2. Notation, assumptions, and problem definitions are introduced in Section 3. The deterministic model and the fuzzy model are developed in Sections 4 and 5, respectively. We present numerical experiments in Section 6. Academic, managerial, and environmental insights are presented in Section 7. Finally, conclusions are presented in Section 8.

2. Literature Review

This paper is related to three elements of the relevant literature: a multi-item replenishment problem for GSCM with carbon emissions, the can-order policy, and GSCM with fuzzy constraints. The research on multi-item replenishment for GSCM with carbon emissions has been approached in various ways. Konur [11] suggested an integrated inventory-transportation model which considers a carbon cap and the emissions characteristics of trucks during transportation. Nia, Far, and Niaki [8] considered an economic order quantity model under the green vendor-managed inventory (VMI) policy, which includes limited warehouse capacity, pallets, deliveries, and greenhouse-gas emissions. Mokhtari and Rezvan [12] applied VMI policy to solve a multi-item replenishment problem in GSCM. In their model, each retailer decides on a replenishment plan based on a limited amount of total GHG emissions. Noh and Kim [1] considered a single-setup/multiple-delivery policy for a green supply chain contract under an uncertain demand situation. They proved that the cooperative contract is useful to improve the performance of GSCM. Cui et al. [13] focused on a business-to-consumer (B2C) e-business company with distribution centers, and they utilized the strategy of multi-item joint replenishment-distribution.

The theory of the can-order policy was first established by Balintfy [14]. Based on that study, Silver [15] focused on an inventory replenishment problem with Poisson demand and non-zero lead time. He established that a can-order policy performed better than an independent order policy. Liu and Yuan [16] developed a Markov model for a two-item inventory system with coordinated replenishment and a heuristic method for solving the problem. Kayiş et al. [17] presented a continuous can-order policy model with two items with Poisson demand under a semi-Markov decision process, and developed a simple enumeration algorithm to solve the problem. Tsai et al. [18] developed an association clustering method that gathers items with similar demand in a hierarchal way to evaluate the correlated demand for handling a large number of items. Kouki et al. [19] considered a continuous review can-order policy, developed as a Markov process with perishable items and zero lead time.

According to the basic theory of the can-order policy, the inventory system should be continuously reviewed. However, because the supplier has limited replenishment opportunities, Johansen and Melchiors [20] suggested a can-order policy model with a periodic review system. To simulate their idea, they suggested a new method based on Markov decision theory to obtain a near-optimal solution. Nagasawa et al. [21] presented a periodic can-order policy model that uses multi-objective programming to obtain the optimal can-order level. Most previous studies dealt with the continuous review system and focused on deriving the reorder level, can-order level, and order-up-to level based on various algorithms. In the real world, a supplier might ship items once or twice a day, so a company might not receive items as frequently as desired [20]. Although those researchers assumed that those levels cannot be fixed to certain values, a decision maker usually sets those levels according to inventory strategy and service level, respectively. Besides, they did not consider carbon cap-and-trade and limited resources, which are storage capacity and budget in the uncertain market information case.

In a competitive or a new market, it is impossible to describe the market information as a specific stochastic distribution because of this uncertainty. To handle this problem, many researchers apply the fuzzy method, a common way to handle uncertainty and non-stochastic situations [22]. Some researchers have tried to apply the fuzzy method to replenishment models. Sadeghi et al. [23] considered the economic production quantity policy under the consignment stock policy in a fuzzy demand situation, and they applied particle swarm optimization to solve their model. Nia et al. [24] presented a fuzzy resource nonlinear integer programming model that regards customer demand, storage capacity, and the budget as fuzzy numbers. Sadeghi et al. [25] focused on a nonlinear integer programming model with a trapezoidal fuzzy number for demand. However, for handling uncertain resource constraints, no previous studies have applied the fuzzy method to GSCM which focuses on a multi-item replenishment problem.

3. Notation, Assumptions, and Problem Definitions

3.1. Notation

Index:

i	item, $i = 1, \ldots, I$
t	period, $t = 1, \ldots, T$

Decision variables:

y_t	binary variable indicating the order during period t
I^i_t	inventory level of item i at the end of period t
I^{i+}_t	on-hand inventory of item i at the end of period t
I^{i-}_t	backorder level of item i at the end of period t
x^i_t	order amount for item i in period t
S^i_t	the order-up-to level of item i in period t
α^i_t	if I^i_t drops below s^i, then $\alpha^i_t = 1$, otherwise $\alpha^i_t = 0$
β^i_t	if I^i_t drops below c^i, then $\beta^i_t = 1$, otherwise $\beta^i_t = 0$
γ^i_t	if I^i_t drops below c^i and at least one item is ordered in period t, then $\gamma^i_t = 1$, otherwise $\gamma^i_t = 0$
z^i_t	if minor setup is done for item i during period t, then $z^i_t = 1$, otherwise $z^i_t = 0$
e^+_t	amount of buying carbon credit in period t
e^-_t	amount of selling carbon credit in period t

Parameters:

u_t major ordering cost in period t (\$/order)

v_t^i minor ordering cost of item i in period t (\$/order)

b_t^i per unit backorder cost of item i in period t (\$/unit)

h_t^i per period holding cost of item i in period t (\$/unit)

p carbon tax (\$/ton)

CE carbon cap for entire planning horizon

$\hat{h}_t^i$ amount of carbon emissions when a buyer holds inventory of item i in period t

$\hat{v}_t^i$ amount of carbon emissions when a buyer orders inventory of item i in period t

d_t^i demand for item i during period t

o^i volume of item i

w_t storage capacity during period t

g^i purchase price of item i

P_t amount of budget during period t

c^i can-order level of item i

s^i reorder level of item i

M big M, very big number

3.2. Assumptions

1. A single buyer orders multiple items from a single supplier and simultaneously considers carbon cap-and-trade under limited storage capacity and budget.

2. The system considers a periodic review can-order policy to obtain the order-up to level. The supplier can utilize limited transportation, so the review period is dependent on the contract period between the buyer and the supplier.

3. Both the buyer and the supplier share the demand information of the items in real time. Thus, the supplier can deliver multiple items with no lead time. Also, the demand for each item is known.

4. The reorder level and the can-order level are assumed as constant.

5. The storage capacity and budget are assumed as constant in the deterministic model and fuzzy numbers in the fuzzy model.

6. The buyer's carbon emissions occur throughout the ordering item and holding inventory. A buyer has a carbon cap and could buy or sell its own carbon credit to other company depending on the carbon emissions.

3.3. Problem Definition

In this GSCM, a single buyer orders multiple items from a single supplier, considering storage capacity, budget, and carbon emissions. Because of limited market information, both the storage capacity and the budget could have some level of uncertainty. For developing multi-item replenishment, we apply a can-order policy to GSCM. Figure 1 compares the inventory patterns with the traditional can-order policy and the proposed can-order policy for two items. By the proposed can-order policy, the order-up-to level S_t^i, where $x_t^i = S_t^i - I_t^i$, is decided individually through the planning horizon. The inventory level of both items in period t_0 explains the initial inventory level. In period t_1, only the inventory level of item 1 is lower than reorder level s^1, so item 1 is replenished up to S_1^1. In period t_2, the inventory levels of items 1 and 2 are lower than the reorder level s^1 and can-order level c^2, respectively, so both items are replenished up to S_2^1 and S_2^2. In period t_3, both items are replenished because both of their inventory levels are lower than the reorder level. In contrast, in the traditional can-order policy, when the inventory level of an item drops to or below the reorder level s^i, an order is placed and the order amount x_t^i can make the inventory level up to S^i. Thus, the order-up-to level S^i, where $x_t^i = S^i - I_t^i$, is always the same through the planning horizon.

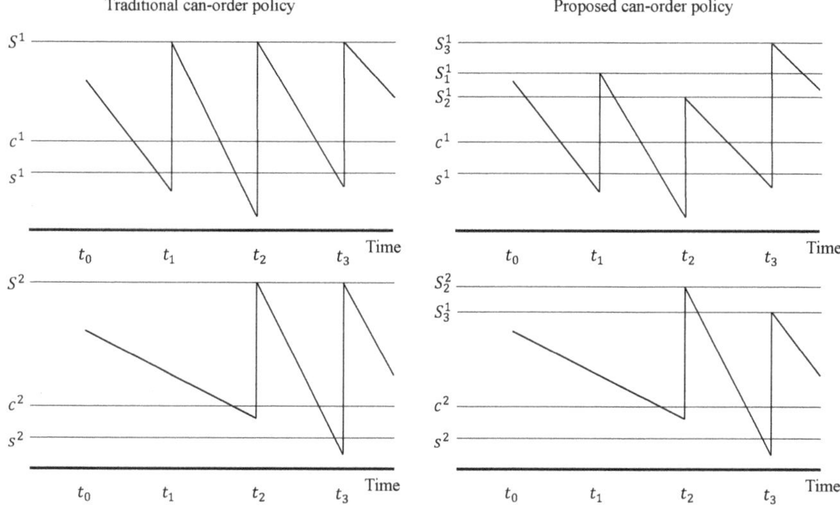

Figure 1. Inventory patterns with can-order policies for two items.

4. Deterministic Model

In this section we use MIP to develop a deterministic model for GSCM. The total cost consists of the major ordering cost, minor ordering cost, backorder cost, inventory holding cost, and carbon tax cost. The major ordering cost, incurred when an order is placed during period t, is given by:

$$\sum_{t=1}^{T} u_t y_t \tag{1}$$

where

$$y_t = 0 \text{ or } 1, t = 1, 2, \ldots, T.$$

The minor ordering cost of each item occurs when item i is ordered in period t, and it cannot be incurred unless the major ordering cost is also incurred. It is given by:

$$\sum_{t=1}^{T} \sum_{i=1}^{I} v_t^i z_t^i \tag{2}$$

The holding cost can be derived as the sum of the order-up-to level and the on-hand inventory at the end of each period. At the beginning of each period, the order-up-to level is delivered, and that amount is decreed by the amount of on-hand inventory at the end of the preceding period. Thus, the expected inventory of each item in each period is approximated as the average of those amounts.

$$\sum_{t=1}^{T} \sum_{i=1}^{I} h_t^i \left(\frac{S_t^i + I_t^{i+}}{2} \right) \tag{3}$$

The backorder cost of item i at the end of period t is:

$$\sum_{t=1}^{T} \sum_{i=1}^{I} b_t^i I_t^{i-} \tag{4}$$

A buyer has its own carbon cap. If the buyer emits less carbon than the given carbon cap, it would get rewarded by receiving money. Otherwise, the buyer pays penalties in the form of a carbon tax [26]. The buyer's carbon cap cost at period t is:

$$\sum_{t=1}^{T} p\left(e_t^+ - e_t^-\right) \tag{5}$$

Based on Equations (1)–(5), the following mathematical model can be developed:

$$Min \sum_{t=1}^{T}\sum_{i=1}^{I}\left(h_t^i\left(\frac{S_t^i + I_t^{i+}}{2}\right) + b_t^i I_t^{i-} + v_t^i z_t^i\right) + \sum_{t=1}^{T}\left(u_t y_t + p\left(e_t^+ - e_t^-\right)\right), \tag{6}$$

subject to

$$S_t^i - I_t^i = d_{t}^i, \quad i = 1,2,\ldots,I, \quad t = 1,2,\ldots,T, \tag{7}$$

$$I_t^i = I_t^{i+} - I_t^{i-}, \quad i = 1,2,\ldots,I, \quad t = 1,2,\ldots,T, \tag{8}$$

$$x_t^i \le M y_{t}, \quad i = 1,2,\ldots,I, \quad t = 1,2,\ldots,T, \tag{9}$$

$$I_{t-1}^i + M\alpha_t^i \ge s^i, \quad i = 1,2,\ldots,I, \quad t = 1,2,\ldots,T, \tag{10}$$

$$I_{t-1}^i - M\left(1 - \alpha_t^i\right) \le s^i, \quad i = 1,2,\ldots,I, \quad t = 1,2,\ldots,T, \tag{11}$$

$$S_t^i \le I_{t-1}^i + x_t^i + M\left(1 - \alpha_t^i\right), \quad i = 1,2,\ldots,I, \quad t = 1,2,\ldots,T, \tag{12}$$

$$S_t^i \ge I_{t-1}^i + x_t^i - M\left(1 - \alpha_t^i\right), \quad i = 1,2,\ldots,I, \quad t = 1,2,\ldots,T, \tag{13}$$

$$I_{t-1}^i + M\beta_t^i \ge c^i, \quad i = 1,2,\ldots,I, \quad t = 1,2,\ldots,T, \tag{14}$$

$$I_{t-1}^i - M\left(1 - \beta_t^i\right) \le c^i, \quad i = 1,2,\ldots,I, \quad t = 1,2,\ldots,T, \tag{15}$$

$$\gamma_t^i \le \beta_{t}^i, \quad i = 1,2,\ldots,I, \quad t = 1,2,\ldots,T, \tag{16}$$

$$\sum_{i=1}^{I}\alpha_t^i \le M\delta_t, \quad i = 1,2,\ldots,I, \quad t = 1,2,\ldots,T, \tag{17}$$

$$\delta_t \le \sum_{i=1}^{I}\alpha_{t}^i, \quad i = 1,2,\ldots,I, \quad t = 1,2,\ldots,T, \tag{18}$$

$$\gamma_{it} \le \delta_t, \quad i = 1,2,\ldots,I, \quad t = 1,2,\ldots,T, \tag{19}$$

$$\gamma_{it} \ge \delta_t + \beta_t^i - 1, \quad i = 1,2,\ldots,I, \quad t = 1,2,\ldots,T, \tag{20}$$

$$S_t^i \le I_{t-1}^i + x_t^i + M\left(1 - \gamma_t^i\right), \quad i = 1,2,\ldots,I, \quad t = 1,2,\ldots,T, \tag{21}$$

$$S_t^i \ge I_{t-1}^i + x_t^i + M\left(1 - \gamma_t^i\right), \quad i = 1,2,\ldots,I, \quad t = 1,2,\ldots,T, \tag{22}$$

$$x_t^i \le M\left(\alpha_t^i + \gamma_t^i\right), \quad i = 1,2,\ldots,I, \quad t = 1,2,\ldots,T, \tag{23}$$

$$S_t^i \le I_{t-1}^i + M\left(\alpha_t^i + \gamma_t^i\right), \quad i = 1,2,\ldots,I, \quad t = 1,2,\ldots,T, \tag{24}$$

$$S_t^i \ge I_{t-1}^i - M\left(\alpha_t^i + \gamma_t^i\right), \quad i = 1,2,\ldots,I, \quad t = 1,2,\ldots,T, \tag{25}$$

$$\alpha_t^i + \gamma_t^i \le M z_{t}^i, \quad i = 1,2,\ldots,I, \quad t = 1,2,\ldots,T, \tag{26}$$

$$\sum_{t=1}^{T}\sum_{i=1}^{I}\left(\hat{h}_t^i\left(\frac{S_t^i + I_t^{i+}}{2}\right) + \hat{v}_t^i z_t^i\right) + \sum_{t=1}^{T} e_t^- \le CE + \sum_{t=1}^{T} e_t^+, \tag{27}$$

$$\sum_{i=1}^{I} o^i S_t^i \leq W_t, \quad t = 1, 2, \ldots, T, \tag{28}$$

$$\sum_{i=1}^{I} g^i x_t^i \leq P_t, \quad t = 1, 2, \ldots, T, \tag{29}$$

$$y_t, \alpha_t^i, \beta_t^i, \gamma_t^i, z_t^i \in \{0, 1\}, \quad i = 1, 2, \ldots, I, \quad t = 1, 2, \ldots, T,$$

$$S_t^i, x_t^i, l_t^{i+}, l_t^{i-} \in Z_+ + \{0\}, \quad l_t^i \in Z, \quad i = 1, 2, \ldots, I, \quad t = 1, 2, \ldots, T,$$

$$e_t^+, e_t^- \in Z_+ + \{0\}, \quad t = 1, 2, \ldots, T,$$

Equation (6) presents the objective function that minimizes the total inventory holding cost, backorder cost, major ordering cost, and minor ordering cost. Equation (7) ensures that the inventory position at the end of period t is equal to the difference between the order-up-to level and demand. Equation (8) ensures that the inventory at the end of period t is equal to the difference between the on-hand inventory and the backorder level. Equation (9) presents the constraint of incurring the major ordering cost whenever at least one item is ordered. Equations (10) and (11) regulate ordering when the inventory position falls below the reorder level. If l_{t-1}^i, the initial inventory level at t, drops below s^i, then $\alpha_t^i = 1$, otherwise $\alpha_t^i = 0$. According to Equations (12) and (13), if the inventory position at the end of the previous period is below the reorder level, the item is ordered in the amount equal to the difference between the order-up-to level and the inventory position at the end of the previous period. Equations (14) and (15) regulate whether the inventory position is below the can-order level. If l_{t-1}^i, the initial inventory level at t, drops below c^i, then $\beta_t^i = 1$, otherwise $\beta_t^i = 0$. According to Equations (16)–(20), if the inventory position of at least one item is below its reorder point, the following order includes other items whose inventory positions are below their can-order level. According to Equations (21) and (22), if the inventory position at the end of the previous period is below the can-order level, that item is ordered in the amount equal to the difference between its order-up-to level and its inventory position at the end of the previous period. Equation (23) regulates the ordering of items whose inventory position is below the reorder or can-order level. Equations (24) and (25) indicate that the order-up-to level is set based on the inventory position, which is below the reorder level or the can-order level. Equation (26) indicates that any item below its reorder level or can-order level incurs the minor ordering cost. Equation (27) ensures the carbon cap-and-trade constraint. The carbon emissions are incurred from the ordering and holding activities. Based on the amount of emissions, the buyer could buy or sell the carbon credit from the participating companies [26]. Equation (28) ensures that the sum of the inventory position for each item at the beginning of each period and the order-up-to level will not exceed the storage capacity. Equation (29) ensures that the buyer cannot order items over its budget.

5. Fuzzy Model

We here develop a fuzzy model based on the deterministic model. A company that is in a competitive market or moves into a new market usually has less information about that market. In this case, to fit the information as a specific stochastic distribution is usually impossible. It is difficult to predetermine the amount of resources, such as limited storage capacity and/or budget. This uncertain market information can be handled by using the fuzzy method.

To apply the fuzzy method to our deterministic model, which is also called a crisp model, the model must first be transformed into a fuzzy model. In order to obtain a crisp value from the fuzzy model, the fuzzy model should be converted to a new crisp model through a defuzzification process. Although many researchers have developed various defuzzification methods, we used the symmetry method introduced by Zimmermann [22]. The crisp objective function with fuzzy constraints, where $A_i x \leq \widetilde{b_i}$ and $D_i x \leq C_i$ are a set of fuzzy and crisp constraints, respectively, is formulated as:

$$Min\, f(x) = c^T x, \tag{30}$$

subject to

$$A_i x \leq \widetilde{b_i}, \quad i = 1, 2, \ldots, m,$$

$$D_i x \leq C_i, \quad i = 1, 2, \ldots, m,$$

$$x \geq 0.$$

Now, using a membership function for a fuzzy set, an element x of X is mapped to a value between 0 and 1. The membership function for the fuzzy sets that represent the fuzzy constraints can be defined as:

$$\mu_i(x) = \begin{cases} 1 & if\, A_i x \leq b_i \\ \frac{b_i + p_i - A_i x}{p_i} & if\, b_i < A_i x \leq b_i + p_i \\ 0 & if\, A_i x > b_i + p_i \end{cases}, \quad i = 1, 2, \ldots, m+1. \tag{31}$$

In Equation (31), $\mu_i(x)$ could be 1 when the constraints are well satisfied, otherwise 1. The p_i is the tolerance interval which is assumed a constraint to be linearly increasing. Defining the membership function of the objective function requires solving the following two problems. The first problem is the original crisp model, and the optimal result is set as $sup_{R1} f = \left(c^T x\right)_{opt} = f_1$.

$$Min\, f(x) = c^T x, \tag{32}$$

subject to

$$A_i x \leq b_i, \quad i = 1, 2, \ldots, m,$$

$$D_i x \leq C_i, \quad i = 1, 2, \ldots, m,$$

$$x \geq 0$$

In the second problem, the fuzzy constraints in Equation (32) are changed to the crisp constraints with tolerances, and the optimal result is set as $sup_{S(\widetilde{R})} f = \left(c^T x\right)_{opt} = f_0$.

$$Min\, f(x) = c^T x, \tag{33}$$

subject to

$$A_i x \leq b_i + p, \quad i = 1, 2, \ldots, m,$$

$$D_i x \leq C_i, \quad i = 1, 2, \ldots, m,$$

$$x \geq 0.$$

In short, f_0 and f_1 are the minimum total cost with and without tolerances for resources, respectively. Based on that, the membership function of the objective function is obtained as:

$$\mu_{\widetilde{G}}(x) = \begin{cases} 1 & if\, f_1 < c^T x \\ \frac{c^T x - f_0}{f_1 - f_0} & if\, f_0 \leq c^T x \leq f_1 \\ 0 & if\, c^T x < f_0 \end{cases}. \tag{34}$$

Based on Equations (30)–(34), both the objective function and the constraints have 'symmetry' such that the crisp model by the defuzzification process is transformed.

$$Max \quad \lambda, \tag{35}$$

subject to

$$c^T x \leq \lambda\big(f_0 - f_1\big) + f_1,$$

$$A_i x \leq b_i + (1 - \lambda)p_i, \quad i = 1, 2, \ldots, m,$$

$$D_i x \leq C_i, \quad i = 1, 2, \ldots, m,$$

$$x \geq 0,$$

$$0 \leq \lambda \leq 1.$$

As mentioned earlier, we considered the storage capacity and the budget as fuzzy numbers. The following table methodically shows the fuzzy constraints of Equations (28) and (29).

$$\sum_{i=1}^{I} o^i S_t^i \leq \widetilde{W_t}, \quad t = 1, 2, \ldots, T,$$

$$\sum_{i=1}^{I} g^i x_t^i \leq \widetilde{P_t}, \quad t = 1, 2, \ldots, T.$$

Next, those constraints are converted to the defuzzification of the storage capacity and budget, where $p_{1,t}$ and $p_{2,t}$ are their respective tolerances:

$$\sum_{i=1}^{I} o^i S_t^i \leq W_t + (1 - \lambda)p_{1,t}, \quad t = 1, 2, \ldots, T, \tag{36}$$

$$\sum_{i=1}^{I} g^i x_t^i \leq P_t + (1 - \lambda)p_{2,t}, \quad t = 1, 2, \ldots, T. \tag{37}$$

Finally, the equivalent crisp model, transformed from the fuzzy model using the method of Zimmermann [22], is obtained:

$$Max \quad \lambda, \tag{38}$$

subject to

$$\sum_{t=1}^{T}\sum_{i=1}^{I}\left(h_t^i\left(\frac{S_t^i + I_t^{i+}}{2}\right) + b_t^i I_t^{i-} + v_t^i z_t^i + G^i e^i x_t^i\right) + \sum_{t=1}^{T} u_t y_t \leq \lambda\big(f_0 - f_1\big) + f_1,$$

$$\sum_{i=1}^{I} o^i S_t^i \leq W_t + (1 - \lambda)p_{1,t}, \quad t = 1, 2, \ldots, T,$$

$$\sum_{i=1}^{I} g^i x_t^i \leq P_t + (1 - \lambda)p_{2,t}, \quad t = 1, 2, \ldots, T,$$

with Equations (7)–(27)

$$y_t, \alpha_t^i, \beta_t^i, \gamma_t^i, z_t^i \in \{0, 1\}, \quad i = 1, 2, \ldots, I, \quad t = 1, 2, \ldots, T,$$

$$S_t^i, x_t^i, I_t^{i+}, I_t^{i-} \in Z_+ + \{0\}, I_t^i \in Z, \quad i = 1, 2, \ldots, I, \quad t = 1, 2, \ldots, T,$$

$$e_t^+, e_t^- \in Z_+ + \{0\}, \quad t = 1, 2, \ldots, T,$$

$$0 \leq \lambda \leq 1.$$

We denote $p_{1,t}$ and $p_{1,t}$ as the 'tolerance intervals' of W_t, and P_t, respectively.

6. Numerical Experiments

To validate our proposed two MIP models, we tested three kinds of numerical experiments. First, we conducted an efficiency test for the deterministic model by comparing it to the traditional can-order policy with predetermined values, (s^i, c^i, S^i). Second, we compared the traditional can-order policy with the proposed can-order policy. In this experiment, we determined which the policy best fit with GSCM. Third, we tested the fuzzy model with various tolerances. In all the experiments we set the very large number M at 1,000,000. To solve the MIP models, we used LINGO 17.0 software.

6.1. Efficiency Test

To show the efficiency of the proposed can-order policy, we compared it to the traditional can-order policy with predetermined values. In the traditional can-order policy, a decision maker sets the values for (s^i, c^i, S^i) based on previous data or their experience. Thus, this comparison is necessary to show whether the proposed deterministic model can reduce the total cost compared to the traditional can-order policy with predetermined values.

Table 1 presents the basic input parameters for three items in 12 periods. The major ordering cost (u_t) is \$1000/order. It is difficult to apply constraints, such as storage capacity, budget, or carbon cap-and-trade, to it in the traditional can-order policy with predetermined values, so we did not consider those constraints in this test.

Table 1. Input parameters for the basic test.

Item	1	2	3
v_t^i	$U(38, 46)$	$U(89, 91)$	$U(25, 30)$
b_t^i	$U(137, 140)$	$U(239, 288)$	$U(83, 85)$
h_t^i	$U(68, 75)$	$U(119, 144)$	$U(51, 64)$
d_t^i	$N(600, 50^2)$	$N(500, 30^2)$	$N(100, 5^2)$

Notes: $U(*,*)$: uniform distribution; $N(*,*)$: normal distribution.

Table 2 shows the initial inventory position, the reorder level, and the can-order level that we used during the test of the traditional can-order policy with predetermined values. The order-up-to level for the traditional can-order policy with predetermined values was assumed to be $S_1 = 697$, $S_2 = 590$, and $S_3 = 184$, the highest demand for each item during the 12 periods. We used Excel 2017 for calculating the traditional can-order policy with predetermined values.

Table 2. Initial inventory position, can-order level, and reorder level.

Item	I_0^i	s^i	c^i
1	0	50	80
2	0	70	100
3	0	10	20

Table 3 shows that the proposed model led to a total cost of \$757,120.50. To see the precise efficiency of the proposed model, we tested it against the traditional can-order policy with predetermined values. Otherwise, it is difficult to discover whether the order-up-to level is biased in predetermined values. We set the order-up-to level from 95% to 75% of the highest demand in 5% decrements. Table 4 shows the results of this test.

Table 3. The results of the basic test.

Traditional Can-Order Policy with Predetermined Values ($)	Proposed Model ($)
1,107,962.00	757,120.50

Table 4. The effect of the order-up-to level.

Test Number	Percent of the Highest Demand	Order-Up-to Level			Total Cost of The Traditional Can-Order Policy with Predetermined Values ($)
		S_1	S_2	S_3	
1	95%	662	561	175	928,417.00
2	90%	627	531	166	935,320.50
3	85%	592	502	157	940,683.00
4	80%	558	472	148	1,045,999.00
5	75%	523	443	138	1,155,508.00

As shown in Table 4, the set of order-up-to level $S_1 = 662$, $S_2 = 561$, and $S_3 = 175$ obtains the best result, \$928,417.00. The total cost is increased with a lower order-up-to-level because the company cannot cope with the demand. The proposed model strongly outperforms the traditional can-order policy with predetermined values. This result shows that predetermined values, which are biased by a decision maker, rarely set the order-up-to level properly, which increases the total cost.

6.2. Comparison Test between the Proposed Can-Order Policy and the Traditional can-order Policy

To determine the better policy, we compared the proposed can-order policy to the traditional can-order policy. Based on the deterministic model, we developed a new model about the traditional can-order policy, which can obtain the same order-up-to level through the planning horizon. We conducted this test using an experimental design in which we varied the number of items and periods from 6 to 10 and 10 to 20, respectively. The major ordering cost was \$1000/order and the initial inventory position was zero. The amount of carbon emissions for holding and ordering were assumed as 10% of each cost. Both carbon tax and carbon cap were assumed as \$2 and \$80,000, respectively. The storage capacity was set as 4,000 m^2 and the amount of budget was set as \$80,000. Table 5 presents each item's input data. The demands for each item were generated using a normal distribution. The reorder level was set as 99% of the safety stock, and the can-order level was assumed as 20% above the reorder level.

Table 5. Input parameters for the comparison test.

Item Number	v_t^i ($)	b_t^i ($)	h_t^i ($)	d_t^i	o^i	g^i
1	1	13	1	$N(200, 10^2)$	0.1	1
2	2	14	2	$N(400, 20^2)$	0.5	0.1
3	1	10	1	$N(300, 15^2)$	1	0.2
4	1	15	3	$N(200, 20^2)$	0.8	1
5	2	8	1	$N(500, 10^2)$	1.1	2
6	2	9	2	$N(100, 20^2)$	2	0.2
7	1	6	1	$N(60, 5^2)$	1	0.1
8	1	10	1	$N(1000, 50^2)$	0.9	0.3
9	2	11	2	$N(200, 15^2)$	1.5	2
10	1	13	1	$N(100, 5^2)$	1.3	0.1

Table 6 shows the results of the comparison test. In all cases, the proposed can-order policy outperformed the traditional can-order policy because the proposed policy produced less total cost than the traditional policy. In some of the five-period tests the total costs are negative values, which is an interesting point. Those cases illustrate that a company sells carbon caps to other companies so that it receives profits. The result proves that the proposed policy benefits the company that replenishes multi-items under carbon cap-and-trade. The proposed policy could be efficacious in the large-scale multi-item replenishment problem in GSCM. Thus, based on these discussions, the proposed model is promising for practical applications.

Table 6. The results of the comparison test.

Number of Items	Periods	Total Cost ($)	
		Traditional Can-Order Policy	**Proposed Can-Order Policy**
6	5	−5599.80	−7264.20
	10	8800.40	5471.60
	15	23,200.60	18,207.40
	20	37,600.80	30,943.20
8	5	−895.60	−4021.20
	10	18,208.80	11,957.60
	15	37,313.20	27,936.40
	20	56,417.60	43,915.20
10	5	987.80	−2448.60
	10	21,975.60	15,102.80
	15	42,963.40	32,654.20
	20	63,951.20	50,205.60

6.3. Fuzzy Model Test

We also tested the fuzzy model to demonstrate the effects of changing the tolerances. To initialize and solve the model in Equation (38), we first solved the two models in Equations (33) and (35) such that we could obtain the values of f_0 and f_1. We considered three items during 12 periods and used the parameters in Section 6.2. Table 7 shows the tolerances for the parameters of the storage capacity and the budget. The base values of the storage capacity and the budget were 1000 m^2 and $8000, respectively. Table 8 also shows the value of f_1 which is the total cost of the deterministic model, and the result of the fuzzy model.

Table 7. The various tolerances for the parameters.

Test Number	p_1	p_2
1	300	3000
2	300	4000
3	400	2000
4	400	4500
5	450	2500

Table 8. The results of the fuzzy model test.

Test Number	f_0 ($)	Fuzzy Total Cost ($)	λ	$(1-\lambda)p_1$	$(1-\lambda)p_2$	f_1 ($)
1	907,298.50	1,039,618.0	0.73	1081	8810	
2	884,448.50	1,011,266.0	0.75	1075	9000	
3	930,148.50	1,078,992.0	0.68	1128	8640	1,400,277.0
4	881,762.50	1,002,462.0	0.77	1092	9035	
5	918,723.50	1,062,622.0	0.72	1126	8700	

Table 8 shows that all the results of the fuzzy model test produced better results than the deterministic model, $1,400,277. Two factors explain this. First, the uncertain market information on storage capacity and budget might lead to large on-hand inventories to avoid large backorders. Second, the strict constraints of storage capacity and budget in the deterministic model are strongly fixed. Based on the value of λ, the decision maker can obtain the values of the storage capacity and budget, being $(1 - \lambda)p_{1,t}$ and $(1 - \lambda)p_{2,t}$, respectively. This test illustrates that the fuzzy model not only handles the uncertainty but also improves the system performance. Thus, the fuzzy model is a good option for a decision maker when the market information has uncertainty.

7. Academic, Managerial, and Environmental Insights

7.1. Academic Insights

We developed two MIP models dealing with the periodic can-order policy for GSCM with limited storage capacity, limited budget, and carbon cap-and-trade. We also developed a deterministic model under certain (known) market information, based on which we suggested a fuzzy model that considers the fuzzy numbers of storage capacity and budget. This is the first study to develop both deterministic and fuzzy models of the can-order policy under carbon cap-and-trade for GSCM. Thus, our study can be considered initial research which considers multi-item replenishment with carbon cap-and-trade for GSCM.

7.2. Managerial Insights

A company interested in GSCM could benefit from our study. The deterministic and fuzzy models developed here can help a company systematically replenish multi-items with carbon cap-and-trade regulation. The deterministic model suggests practical insights on multi-item replenishment for minimizing the total cost under limited resources and carbon cap-and-trade regulation. It can also help a company in a monopoly market make sound investment decisions. For a company that has uncertain market information, the fuzzy model can support preparation of appropriate storage capacity and budget, and also planning for cost minimization. Thus, the correct implementation of these models will give a company better decisions in managing GSCM and reducing its total cost.

7.3. Environmental Insights

This paper incorporates carbon cap-and-trade regulation into GSCM. Considering carbon cap-and-trade regulation, the companies can increase resource-use efficiently and grow together. This sustainable situation is regarded on the 2030 Agenda and the Sustainable Development Goal 9 (SDG 9), which is industry, innovation, and infrastructure. Thus, using this paper, the company can help protect environment while earning profits.

8. Conclusions

This paper presents two MIP models, a deterministic model and a fuzzy model, which address the multi-item replenishment problem with carbon cap-and-trade for GSCM under limited resources. We developed the two models based on the can-order policy, which is one of the well-known multi-item replenishment policies. Reflecting real-world situations, we considered limited storage capacity, budget, and carbon cap-and-trade regulation. The deterministic model can be used when a decision maker has solid market information, while the fuzzy model can be applied when a decision maker faces uncertain market information in a competitive or a new market. In this model, both limited storage capacity and budget are denoted as fuzzy numbers.

We carried out three experiments to test the efficiency of the two models. First we compared our deterministic model with the traditional can-order policy which already has predetermined values (s^i, c^i, S^i), and proved that our deterministic model was significantly better because it resulted in lower total costs. In the second experiment, by using our deterministic model, we compared the proposed can-order policy with the traditional can-order policy. The result showed that the proposed can-order policy outperformed the traditional can-order policy by, again, resulting in lower total costs. Finally, we quantified the effects of the fuzzy model with various tolerances. The results showed that applying fuzzy constraints is useful to make decisions under uncertain situations. We demonstrated the validity and practicality of our models in those experiments, confirming that our models can be useful for multi-item replenishment in GSCM.

There are some research limitations in this study and also some indications for possible future works. First, this paper only considered a single supplier and a single buyer. In the real world there are many suppliers, so it is an important decision to select one among several. Our current models could be applied to extend GSCM with this supplier-selection problem. Also, this paper assumed the deterministic demand. For replenishment planning, forecasting demand theory could be considered.

Author Contributions: Conceptualization, J.S.K.; Funding acquisition, S.-J.H.; Methodology, J.N.; Project administration, S.-J.H.; Writing–original draft, J.N.; Writing–review & editing, J.S.K. All authors have read and agreed to the published version of the manuscript.

Funding: This work was supported by the Ministry of Education of the Republic of Korea and the National Research Foundation of Korea (NRF-2019S1A5C2A04083153).

Conflicts of Interest: The authors declare no conflicts of interest.

References

1. Noh, J.; Kim, J.S. Cooperative green supply chain management with greenhouse gas emissions and fuzzy demand. *J. Clean. Prod.* **2019**, *208*, 1421–1435. [CrossRef]
2. Xu, J.; Chen, Y.; Bai, Q. A two-echelon sustainable supply chain coordination under cap-and-trade regulation. *J. Clean. Prod.* **2016**, *135*, 42–56. [CrossRef]
3. Cao, K.; Xu, X.; Wu, Q.; Zhang, Q. Optimal production and carbon emission reduction level under cap-and-trade and low carbon subsidy policies. *J. Clean. Prod.* **2017**, *167*, 505–513. [CrossRef]
4. Hua, G.; Cheng, T.; Wang, S. Managing carbon footprints in inventory management. *Int. J. Prod. Econ.* **2011**, *132*, 178–185. [CrossRef]
5. Hintermann, B. Allowance price drivers in the first phase of the eu ets. *J. Environ. Econ. Manag.* **2010**, *59*, 43–56. [CrossRef]
6. Dong, C.; Shen, B.; Chow, P.-S.; Yang, L.; Ng, C.T. Sustainability investment under cap-and-trade regulation. *Ann. Oper. Res.* **2016**, *240*, 509–531. [CrossRef]
7. Ramanathan, U.; Bentley, Y.; Pang, G. The role of collaboration in the uk green supply chains: An exploratory study of the perspectives of suppliers, logistics and retailers. *J. Clean. Prod.* **2014**, *70*, 231–241. [CrossRef]
8. Nia, A.R.; Far, M.H.; Niaki, S. A hybrid genetic and imperialist competitive algorithm for green vendor managed inventory of multi-item multi-constraint eoq model under shortage. *Appl. Soft Comput.* **2015**, *30*, 353–364.
9. Dimitrova, M.; Schlee, E.E. Monopoly, competition and information acquisition. *Int. J. Ind. Organ.* **2003**, *21*, 1623–1642. [CrossRef]
10. Roy, A.; Kar, S.; Maiti, M. A deteriorating multi-item inventory model with fuzzy costs and resources based on two different defuzzification techniques. *Appl. Math. Model.* **2008**, *32*, 208–223. [CrossRef]
11. Konur, D. Carbon constrained integrated inventory control and truckload transportation with heterogeneous freight trucks. *Int. J. Prod. Econ.* **2014**, *153*, 268–279. [CrossRef]
12. Mokhtari, H.; Rezvan, M.T. A single-supplier, multi-buyer, multi-product vmi production-inventory system under partial backordering. *Oper. Res.* **2020**, *20*, 37–57. [CrossRef]
13. Cui, L.; Deng, J.; Zhang, Y.; Tang, G.; Xu, M. Hybrid differential artificial bee colony algorithm for multi-item replenishment-distribution problem with stochastic lead-time and demands. *J. Clean. Prod.* **2020**, 119873. [CrossRef]
14. Balintfy, J.L. On a basic class of multi-item inventory problems. *Manag. Sci.* **1964**, *10*, 287–297. [CrossRef]
15. Silver, E.A. A control system for coordinated inventory replenishment. *Int. J. Prod. Res.* **1974**, *12*, 647–671. [CrossRef]
16. Liu, L.; Yuan, X.-M. Coordinated replenishments in inventory systems with correlated demands. *Eur. J. Oper. Res.* **2000**, *123*, 490–503. [CrossRef]
17. Kayiş, E.; Bilgic, T.; Karabulut, D. A note on the can-order policy for the two-item stochastic joint-replenishment problem. *IIE Trans.* **2008**, *40*, 84–92.

18. Tsai, C.-Y.; Tsai, C.-Y.; Huang, P.-W. An association clustering algorithm for can-order policies in the joint replenishment problem. *Int. J. Prod. Econ.* **2009**, *117*, 30–41. [CrossRef]

19. Kouki, C.; Babai, M.Z.; Jemai, Z.; Minner, S. A coordinated multi-item inventory system for perishables with random lifetime. *Int. J. Prod. Econ.* **2016**, *181*, 226–237. [CrossRef]

20. Johansen, S.; Melchiors, P. Can-order policy for the periodic-review joint replenishment problem. *J. Res. Soc.* **2003**, *54*, 283–290. [CrossRef]

21. Nagasawa, K.; Irohara, T.; Matoba, Y.; Liu, S. Applying genetic algorithm for can-order policies in the joint replenishment problem. *Ind. Eng. Manag. Syst.* **2015**, *14*, 1–10. [CrossRef]

22. Zimmermann, H.-J. *Fuzzy set Theory—And its Applications*; Springer Science & Business Media: Berlin, Germany, 2011.

23. Sadeghi, J.; Sadeghi, S.; Niaki, S.T.A. Optimizing a hybrid vendor-managed inventory and transportation problem with fuzzy demand: An improved particle swarm optimization algorithm. *Inf. Sci.* **2014**, *272*, 126–144. [CrossRef]

24. Nia, A.R.; Far, M.H.; Niaki, S. A fuzzy vendor managed inventory of multi-item economic order quantity model under shortage: An ant colony optimization algorithm. *Int. J. Prod. Econ.* **2014**, *155*, 259–271.

25. Sadeghi, J.; Mousavi, S.M.; Niaki, S. Optimizing an inventory model with fuzzy demand, backordering, and discount using a hybrid imperialist competitive algorithm. *Appl. Math. Model.* **2016**, *40*, 7318–7335. [CrossRef]

26. Benjaafar, S.; Li, Y.; Daskin, M. Carbon footprint and the management of supply chains: Insights from simple models. *IEEE Trans. Autom. Sci. Eng.* **2012**, *10*, 99–116. [CrossRef]

 *sustainability*

Article

Narrowing the Gaps: Assessment of Logistics Firms' Information Technology Flexibility for Sustainable Growth

Jeong Hugh Han [1,*], Yingli Wang [2] and Mohamed Naim [2]

[1]   Asia Pacific School of Logistics, Inha University, Incheon 22212, Korea
[2]   Logistics and Operations Management Section, Cardiff Business School, Cardiff University, Cardiff CF10 3EU, UK; wangy14@cardiff.ac.uk (Y.W.); naimmm@cardiff.ac.uk (M.N.)
*   Correspondence: hanjh@inha.ac.kr

Received: 28 April 2020; Accepted: 21 May 2020; Published: 26 May 2020

Abstract: In a supply chain management context, the effective management of Information Technology (IT) flexibility has been an issue to be resolved. However, no analytical method that calculates the required and actual level of IT flexibility dimensions has been proposed. This paper aims to provide an analytical tool that measures the required and actual levels of IT flexibility dimensions to provide the best value from a logistics firm's IT flexibility. To do so, we propose a combined Importance-Performance Analysis (IPA) and Partial Least Squared Structured Equation Modelling (PLS-SEM) method based on a multidimensional IT flexibility model. By comparing industry-level data with client firm data, our method allows for effective identification of a client logistics company's multiple IT flexibility gaps and indicates where particular management interventions are required. By proposing importance and performance as measurement scales, our research suggests an analytical tool that managers can utilize to assess IT flexibility and identify any gaps that exist between actual and required flexibility levels. This allows managers to effectively address areas that demand further attention. This approach also leads to an improved understanding of how organisations can extract the best value from their investment in IT flexibility to contribute to sustainable growth.

Keywords: flexibility; IT flexibility; importance-performance analysis; partial least squared structured equation modelling; performance gap; sustainable growth

---

1. Introduction

Information Technology (IT) flexibility is one of the most widely used concepts for identifying a firm's ability to cope with the variation generated by its business environment [1–4]. With the recognition of IT flexibility as a multidimensional concept, previous research has focussed on the identification of, and validated the dependence of firm performance on, IT flexibility dimensions [5–9]. Further, as echoed by several researchers [4,10–13] an investigation of the mismatches between the actual and required level of each flexibility dimension is required to execute efficient resource allocation to each dimension so that finite firm resources can be used effectively. However, little attention has been given to a method to improve firm performance through a flexibility requirements analysis. Specifically, there has been a lack of analytical tools that calculate the required and actual level of IT flexibility dimensions, hence impeding strategic decision-making in resource investment.

To fill this research gap, we suggest the use of Importance-Performance Analysis (IPA), combined with Partial Least Squared Structured Equation Modelling (PLS-SEM), to identify the gaps between the required and actual levels of each flexibility dimension. We particularly highlight the usefulness of this application, which indicates particular dimensions that might be under- or over-resourced. The combined use of IPA and PLS-SEM is a largely neglected method, particularly in the Technology

Management field. To the best of our knowledge, our paper is the first to apply combined IPA and PLS-SEM in the context of operational flexibility more generally, and IT flexibility specifically.

The rest of the paper is organised as follows. Section 2 offers a theoretical background for managing multiple flexibility dimensions. Section 3 discusses our methodological approach in applying IPA to examine the IT flexibility gap. This is followed by a discussion of our research findings in Section 4. We draw conclusions in Section 5 by highlighting our theoretical and practical contributions. We also acknowledge our research limitations and discuss future research directions.

2. Literature Review—Managing Multiple Dimensions of IT Flexibility

There is general agreement that IT should be flexible to help companies deal with outward uncertainties through advancing, adapting or coordinating the functionalities of the IT. IT flexibility is thought to increase the capacity for adjust to variations in internal and external business circumstances. For instance, adaptability to novel or dissimilar circumstances and scalability [8], IT investment and IT infrastructure to adapt to a changing business environment [10], support to alter business strategies [14], information system functionality, database, interface and processing capacity [3], compatibility, information sharing, modularity and capacity to handle multiple applications [15], adaptability for changing business partners and environment, reconfiguration of communication linkages and capability to redesign business process are highlighted as the principal components of IT flexibility [1].

IT flexibility is a multidimensional concept. The seminal work of Duncan [2] categorised IT flexibility dimensions into compatibility, connectivity and modularity, focusing on IT infrastructure. On the other hand, Saraf et al. [6] claimed that the value of IT flexibility depends on the ability to adapt to the different types of business requirements that emerge from different organisational levels (e.g., operational tactical and strategical levels). So, to generate value, a continuous redesign of IT infrastructure that supports incremental and revolutionary environmental changes is required. By defining IT flexibility as the manner in which a firm's IT is organised and integrated to adapt to rapid changes, Saraf et al. [6] proposed scalability, system design for new business relationships, and system design for rapid business requirement change as the primary components of IT flexibility. Gosain et al. [16] give examples of different resource uses. According to Gosain et al. [16], if there is a need to change business partners quickly then IT should be exploited to lower the switching costs. If the capability requirement is to increase the volume of interfirm business transactions, IT should enable enhanced information sharing via the standardisation of processes.

Lee and Xia's [17] findings also strongly support the multidimensional characteristics of IT flexibility. They identified that two types of IT flexibility co-exist, namely response extensiveness and response efficiency. The multidimensionality of IT flexibility is also recognised in a comprehensive review by Kumar and Stylianou [4]. They pointed out that prior research focused on information technology infrastructure flexibility and highlighted the importance of IT flexibility as both a strategic and an organizational capability. Specifically, flexibility in IT operations, IT service development and IT management that responds to changing business process and consumer requirements are viewed as IT flexibility categories. Another contribution to defining the different roles of IT flexibility dimensions was supplied by Han et al. [11]. By integrating the traditional infrastructure-focused view and value creation in their IT flexibility dimensions, they revealed that IT flexibility encapsulates transactional, operational and strategic flexibilities, and further tested the causal effect between three dimensions and firm performance. From the perspective of a technology acceptance model, Kwak et al. [18] showed that information reliability, networking capability and security of a logistics platform are the logistics platform dimensions that increase the logistics platforms' dynamic capabilities. In particular, Kwak et al. [18] highlighted the role of scalability as a part of networking capability. They argue that the scalability of the logistics platform enables network effects so increases the flexibility of the logistics platform in responding to changing business demand.

Despite previous research efforts in articulating the multiple dimensions of IT flexibility and their influence on firm performance, there have been limited discussions of how to interrogate the effectiveness of each flexibility dimension. Lee and Xia [17] and Kumar and Stylianou [4] are the notable exceptions that discuss the possible trade-offs among different dimensions of flexibility for firm performance. Kumar and Stylianou [4] argued that different IT flexibility dimensions play different roles and make synergies or conflicts in different situations requiring IT flexibility. Even though Kumar and Stylianou [4] did not provide the exact meaning of the trade-off, synergies and conflict, they implicitly argued that a set of concurrent actions that address specific needs for IT flexibility could be developed when the required and actual levels of flexibility are identified. Lee and Xia [17] argued that the gap between existing and desired IT flexibility should be identified by helping "managers attack the flexibility gap by developing theories" (p. 88). However, none of them was able to propose how such trade-offs or gap closing could be conducted.

Given the limited IT flexibility literature on assessing how well a firm performs against industry norms, we refer to the well-established OM flexibility literature that indicates the following practices are required to measure the flexibility gap. First, the desired configuration of heterogeneous dimensions of flexibility should be identified; that is, identifying the required level of flexibility needed within each dimension to yield performance benefits [19–22]. Second, an analysis of actual performance within each flexibility dimension is undertaken to identify performance mismatches between the required and observed levels of IT flexibility [20,23–26]. This process of performance measurement identifies where the flexibility levels may need to be raised or reduced [21]. Third, once a given performance gap has been identified, a management decision is required as to where to close the gaps at an acceptable cost [21,23,26,27].

The aforementioned process may provide a good guideline to identify the gaps in IT flexibility. However, the flexibility measuring process still has a number of limitations. One of the core reasons is the absence of universal criteria that encompass different flexibility dimensions. Different flexibility dimensions are not homogenous and hence require different measures [28–30]. For instance, Cousens et al. [27] propose a series of steps for increasing manufacturing flexibility and suggested six key performance indicators (KPIs) to measure volume and mix flexibility in manufacturing. However, because the KPIs are very factory-specific, such as the number of variants per key product family, one cannot use them as a total set for the measurement for other dimensions such as labour or material handling flexibility. Seebacher and Winkler [31] also measured flexibility by developing a two-dimensional framework to identify the performance and usefulness of batch production systems. Their method of evaluating manufacturing flexibility is to compute a coefficient of variation from the deviations of the manufacturing order lead times and then calculate an efficiency in manufacturing performance. Although their approach is effective for evaluating manufacturing flexibility, they acknowledge that the application of the model is restricted to discrete manufacturing due to the specific parameters adopted (i.e., coefficient of variation).

Flexibility is a relative, situation-specific concept, so a certain dimension is viewed as a more imperative dimension when a specific environmental necessity emerges [32]. This relative importance in different situations also makes the flexibility measurement difficult. For instance, when a large variety of service accessibility and speedy transition proficiency exists, they both denote the flexibility concept. When the marketplace needs an advanced level of service diversity for a specific situation or time, the flexible capability for a large variety of services would take greater value [32,33]. The biased measurements used for flexibility performance are a similar problem. For example, existing empirical studies, such as Chang [34], have only prioritized the required flexibility dimensions in environmental uncertainty, and therefore are unable to identify actual levels of flexibility dimensions. As a result, they offer limited insights into identifying the flexibility gap to be closed.

Further, previous research has also suggested the need to show the link between an increased level of flexibility and the improvement of firm performance. To show that firm performance is conditional and dependent on flexibility levels, concurrent validation of the positive impact of flexibility configurations should also be undertaken [35–40]. However, much of the performance research in flexibility-related works has focused on the justification of IT flexibility dimensions and/or their impact on firm performance and has not extended to requirements analysis [38–40]. The most relevant work regarding the decision-making process and how to fill the performance gap is probably the work of He et al. [23]. By developing the concept of "flexibility fit," they showed that the levels of required and available flexibility can be determined through a set of simulations. However, as their guidelines were limited to a single dimension of process flexibility, i.e., range, it is not clear if their approach can be applied to the multiple flexibility dimensions found within a firm. Focusing on a single dimension overlooks the fact that multidimensionality is an essential attribute of a flexibility construct, and resources within a firm need to be shared in a cost-effective manner.

Therefore, given the limitations of existing flexibility research as Table 1 presents, namely the lack of objective measurements for different dimensions of flexibility and the lack of methods to determine the correlation between the level of IT flexibility and firm performance, the opportunity to close the flexibility gaps is lacking in previous literature. Furthermore, it would be difficult to pinpoint specific areas where a proper action plan can be devised for resource allocation or adjustment. The same problem exists for IT flexibility gap measurements. There is therefore a need for an effective tool that measures both the required and observed flexibility levels exploiting comprehensive, objective criteria. Those criteria also need to be aligned with different flexibility dimensions while showing uniformity towards performance improvement.

Table 1. Key literature related to efficient IT flexibility management.

Study	Research Objectives	Key Findings/Limitation
Bamel and Bamel, 2018 [35]	To investigate the relationship of organizational resources and strategic flexibility through knowledge management process capability	Organizational resources are associated positively with strategic flexibility, and knowledge management process capability have mediating impact on these relationships/Not extended to the flexibility gap closing process
Benitez et al. 2018 [36]	To investigate how information technology infrastructure flexibility influence merger and acquisition (M&A) of firms	A flexible IT infrastructure facilitates business flexibility in capturing M&A opportunities and increasing post-M&A IT integration capability/Not extended to the flexibility gap closing process
Benitez et al. 2018 [37]	To capture the positive relationships between IT infrastructure capability and business flexibility.	IT-enabled business flexibility supports firms to develop the operational proficiency to capture the new business opportunities and increase their performance/Not extended to the flexibility gap closing process
Boyle, 2006 [19]	To develop a research framework that provides best management practices in implementing manufacturing flexibility.	Measurement of required flexibility and processing of achieving required flexibility process is proposed/No empirical research is presented
Cousens et al. 2009 [27]	To design a process that define the key activities of a strategic manufacturing plan for the improved manufacturing flexibility	A change management process for flexibility performance improvement is identified/Focusing on factory-specific flexibility so one cannot use them as a total set for measurement for IT flexibility dimensions
Chaudhuri et al. 2018 [41]	To examine the impact of internal integration, external integration and supply chain risk management on manufacturing flexibility.	Internal integration and supply chain risk management have a direct influence on manufacturing flexibility/Not extended to the flexibility gap closing process
Gao et al. 2020 [38]	To investigate how IT business spanning capability interacts with IT flexibility and IT integration, which influence organizational agility.	IT flexibility and IT integration are positively inter-related with organizational agility/Not extended to the flexibility gap closing process
He et al. 2012 [23]	To guide process flexibility investment by establishing a flexibility fit index	'Flexibility fit' is acquired by quantifying the required process flexibility/Flexibility fit is limited to a single specific dimension of process flexibility (i.e., range)
Hou, 2019 [39]	To investigate the mediating role of supply chain capabilities on the inter-relationships between IT infrastructure flexibility, integration and firm performance.	IT infrastructure integration and flexibility indirectly and positively influence organizational performance with the mediating role of supply chain capability/Not extended to the flexibility gap closing process
Irfan et al. 2019 [40]	To analyse the influence of IT capabilities on supply chain capabilities and organizational agility.	IT infrastructure and IT assimilation affect information integration and operational coordination, and these capabilities also positively influence organizational agility/Not extended to the flexibility gap closing process

Table 1. *Cont.*

Study	Research Objectives	Key Findings/Limitation
Kemmoe et al. 2014 [42]	To evaluate production systems by measuring excess demand that can be satisfied with the systems	A model accommodate unexpected peaks in demand in production capacity is developed/Focusing on factory-specific flexibility so one cannot use them as a total set for measurement for IT flexibility dimensions
Kumar and Stylianou, 2014 [4]	To supply an IT flexibility dedicated management process framework	A framework for identifying flexibility categories, types of flexibility needed, understanding synergies and trade-offs between different flexibility types is developed/No empirical research is presented
Lee, 2012 [43]	To develop a theoretical model that explains how firms achieve business agility from their deployment and utilization of IT.	Theoretical development on IT exploitation and IT exploration is achieved/No empirical research is presented.
Merschmann and Thonemann, 2011 [44]	To highlight the relationship between environmental uncertainty, supply chain flexibility and firm performance	Proved that the firm performance is conditional and dependent on flexibility levels and configurations/Not extended to the flexibility gap closing process
Seebacher and Winkler, 2015 [31]	To evaluate supply chain flexibility by capturing the performance and efficiency of batch production systems.	A supply chain's flexibility that satisfies its delivery dates and its operational costs in the case of changing environment is identified/The application of the model is restricted to manufacturing process

3. Methodology

3.1. A Combined IPA and PLS-SEM Method: Background

This study proposes the combined use of an IPA matrix and PLS-SEM to identify the flexibility gap. While the IPA matrix concept has its origins in marketing [45], it has since been applied in a number of different industry management settings [46–50]. IPA allows a company to detect which attributes of its product or service ought to be improved. Its main structure is a four-dimensional grid based on the importance and performance level of the identified attributes. For instance, for a particular product attribute, importance ratings could be obtained, from "extremely important" to "not important," and, similarly, performance ratings could be obtained, from "excellent" to "poor." Attributes can then be classified according to their relative importance and performance ratings by mapping the scores in a four-dimensional plot. Its introduction to operations management was through Slack [49], who modified the classic 2 x 2 importance-performance grid into alternative zones allowing a more constant evolution in inferred priorities. The matrix was later extended by Tontini and Silveira [50]. By incorporating the Kano Model for the arrangement of service features, namely basic, performance and excitement attributes, they developed a way to identify gaps between expected satisfaction, current and average market satisfaction.

A notable aspect of Slack's research [49] is that, based on focus group discussions with company personnel, he configured a zoning representation, as shown in Figure 1. In this representation, companies must determine how well the performance aligns with the line AB, which represents the "best fit" with respect to the performance level. Anything below the line requires improvement—or, in extreme cases, as defined by curve CD, urgent action. Anything above the line may be deemed appropriate or, if above the curve EF, questionably excessive. As indicated by Slack [49], the AB line does not provide a clear cutoff point where being over (or under) the line definitely indicates being over- (or under-) resourced. Nonetheless, it helps to reveal the potential gap and signals that organisations may need to investigate that are the potential causes of the performance gap.

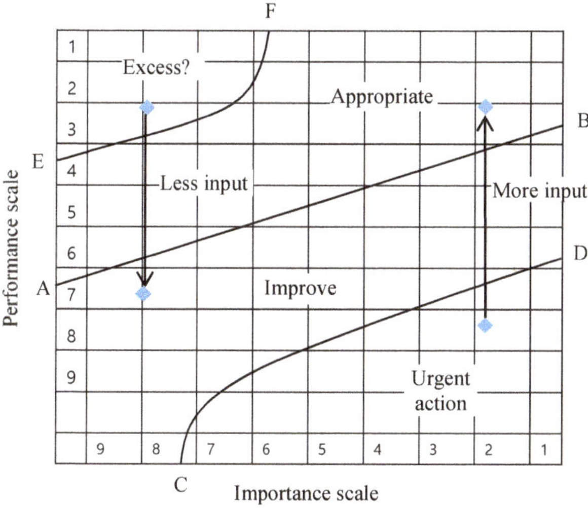

Figure 1. Zoning representation of the IPA matrix (Source: Adapted from Slack 1994).

Recent developments of the IPA matrix have combined the analysis with PLS-SEM applications by Hair et at al. [46] and Ringle and Sarstedt [51]. The process of identifying the flexibility performance gap is elaborated as follows. On the one hand, the importance level on the x-axis of a given matrix denotes the valuation of the direct, indirect and overall relationships between latent constructs. This is computed with the inner and outer coefficients. The coefficients range from 0 to 1.0. On the other hand, the performance level is rescaled to 0 to 100 on the y-axis of a given matrix, according to the average scores of the latent construct values. In addition, the scores for the importance level and performance level of each variables are united in a matrix with a bootstrapping method that is employed to measure the significance level of the indicators' importance scores.

3.2. Application of the Method to IT Flexibility Dimensions

In this analysis, one can interpret the importance level as the required flexibility level and the performance as the actual flexibility level. Ideally, the most important dimension will show the highest performance score. If the actual performance does not meet the required level, the IPA matrix identifies the performance gap (i.e., the mismatch). Moreover, with this tool we can expand our analysis to the indicator level, thus identifying specific areas that may contribute to the under- or overperforming flexibility dimensions, which may then require rectifying actions [46,51,52].

This method resolves the aforementioned issues in flexibility measurement. First, by providing a universal, all-encompassing measurement, namely performance, different dimensions can be measured in a consistent manner. Second, by providing the two types of measurement, i.e., the importance and performance, this method measures the required and actual flexibility levels simultaneously. Further, as PLS-SEM structural model analysis can demonstrate the impact of independent variables (flexibility dimensions in this case) on dependent variables (firm performance in this case) [53,54], it validates the correlations between flexibility dimensions and firm performance concurrently while testing the uniformity of the dimensions towards firm performance.

Applying IPA with PLS-SEM in our research, we undertook the following steps. First, the results of the IT flexibility model analysis with PLS-SEM with industry-level data are incorporated into an IPA matrix. This step is meant to validate our method in a generalized industry setting, and also to determine whether the general performance levels are consistent with the line AB in Figure 1. If the performance level of each dimension is appropriate when compared to its importance level, that is, consistent with AB, then this tends to indicate that the resources are fairly distributed, and no resource reallocation is required. Second, a case firm that needed IT flexibility improvement was selected, and its data were analysed with the same method to determine if the performance levels are consistent with the line AB in Figure 1. We also compared the importance and performance levels of the case firm to the general industry results. If the performance levels show different distribution patterns compared to the industry norm, with performance gaps, this strongly indicates that there may be under- or over-resourcing from the case firm. Third, where the case firm's data are not consistent with the line AB, a further analysis was conducted to identify which indicators may require more or fewer resources.

To apply IPA combined with the PLS-SEM method in closing IT flexibility gaps, a model that meets IT flexibility-specific requirements needs to be employed. We applied our method to an existing IT flexibility model proposed by Han et al. [11]. By incorporating exploitive purposes of IT use and an explorative view [55,56], Han et al. [11] classified multiple dimensions of IT flexibility such as transactional IT flexibility, operational IT flexibility and strategic IT flexibility. Transactional (TR) IT flexibility refers to a capability to utilise advances in IT infrastructure. Operational (OP) flexibility is the ability to use IT for information distribution and process enhancement. Strategic (STR) flexibility is the ability to use IT expertise to generate novel, future-oriented operations together with supply chain partners [11]. Such a classification enables us to measure the gap in IT flexibility given that each dimension has different roles. We required a prevalidated model that shows the uniformity of each firm's IT flexibility performance. Based on the theory that IT interacts with intermediate business practices [57], Han et al. [11] showed that IT flexibility is created when the integration of supply chain operations inside the firm and with external business partners is ensured. Thus, Han et al.'s model is well placed to assess the firm performance. One can refer to Han et al. [11] for detailed model development discussions and the hypothesis of the IT flexibility research model. Measurement indicators of the three different dimensions of IT flexibility and the hypotheses are presented in Appendix A.

3.3. Data Collection

Due to the different IT flexibility dimensions at different organizational levels, the current research requires every respondent to acquire adequate interfirm and functional understanding and experiences at all areas and levels. Our rationale is that senior executives would have a more integrative and strategical perspective, but they might not automatically possess in-depth understanding or experience regarding present working systems. On the other hand, more junior member of staff may be well acquainted with certain IT systems, due to their close engagement with the systems, but may lack a holistic view. Further, the respondents should be capable of assessing the promoting roles of IT flexibility for intra-/interfirm process improvement and the enhancement of firm performance. Respondent validation ensures that the acquired data are reliable and credible for our empirical analysis. Such complex qualifications inevitably reduce the availability of suitable respondents. Because of the aforementioned constraint, this study opted for nonprobability sampling.

Specifically, a mixture of purposive data collection and convenience data collection method was chosen for the current research. The purposive data collection method was selected because it employs the experiences and skills of the researcher to gain well-informed respondents [58,59]. With this data collection method, researchers stipulate the features of a population of interest and attempt to find people who have those features. In the convenience data collection method, the informants are asked to contact new informants, as a focal contact point, who satisfy specific requirements and are willing to take part in specific research [58]. Our survey was organized into five parts (1–5). In order to confirm that the informants satisfy the inclusion principles and measure their ability to assess the dimensions of IT flexibility, a supplementary check was conducted within part 1. It also inquired about the technologies that the informants' firms use to check if the informants are familiar with different IT use patterns, based on the recommendation of Kumar et al. [60]. Further, only the responses from respondents successfully answering the full questionnaire were used for empirical analysis. This data collection method is in line with prior IT flexibility research. By employing a nonprobability data collection method, Gosain et al. [16] collected dependable information from both senior-level respondents and junior-level employees. This is to integrate the insights from the workers associated in daily operations. Rai et al. [61] used similar data collection method as high-level managers are not responsible for repetitive problem settlement and the principal responsibility of the executives is more tactical in nature than the operations conducted by junior-level workers. Parts 2-4 supply questions measuring the level of different dimensions of IT flexibility and other variables (Appendix A). The questions in Part 5 in our survey questionnaire supply general information on the respondents.

For the industry data, the professional network at the authors' university was accessed, which contains professionals who worked at the university for several years on shared research work and information assimilation projects, as well as graduates likely to be familiar with the current research subject. Furthermore, professionals were encouraged to distribute the survey to their co-workers to invite them to take part in the current study. A firm-level dataset was collected within a case company, hereafter known as MultiLogistics. MultiLogistics is a multinational logistics service provider that provides a diverse range of logistics services, such as warehousing, transportation, custom clearance and freight forwarding. Acting as an intermediary, it provides order fulfilment services for a large number of customers, such as telecommunication manufacturers and fashion retailers. It also works with shipping lines and freight transport companies for supply chain execution. The nature of the business indicates that it transmits a large quantity of information and has exhaustive information interchange activities with its customers and business partners. Therefore, IT is critical to the successful execution of MultiLogistics' operations. Under intensifying marketing pressures, whereby logistics services are increasingly seen as a commodity, senior executives feel IT flexibility is an important enabler to allow the company to respond and adapt to a changing environment quickly and to remain competitive. Therefore, they feel the need to assess their current IT flexibility and compare it with industry practices to see whether the company is underperforming in certain areas, and to identify opportunities for improvement. We distributed the survey to 62 key informants from over 20 operating units of MultiLogistics globally. In doing so, an international director of the MultiLogistics was actively involved in the data collection. He distributed the survey questionnaire to potentially suitable respondents. Follow-up emails were sent two weeks and four weeks after the initial questionnaire distribution to encourage the respondents to participate in the survey. We followed up our quantitative analysis with qualitative interviews with key respondents from the case company in order to understand the significance of the results from the quantitative stage.

Regarding the industry data (n = 128), the analysis of the questionnaire showed that the questionnaires were answered by professionals from production and manufacturing (22%, n = 29), warehousing and inventory management service (28%, n = 36), integrated transport service providers (30%, n = 38), logistics and transport service brokers like 3PL firms (16%, n = 20) and others (4%, n = 5). The sample included vice presidents or higher position professionals (4%, n = 5), directors or vice directors (16%, n = 21), managers or assistant managers (42%, n = 54), supervisors (12%, n = 15), operators and clerks (24%, n = 31) and others (2%, n = 2), so the survey obtained information covering different areas and levels of an interorganizational business, namely transactional, operational and strategic operations. A majority of informants were from supply chain-related areas (13%, n = 17) and the transport and logistics field (71%, n = 90). The sample also included marketing position informants (8%, n = 10), IT personnel (2%, n = 3), CEOs (2%, n = 3) and others (4%, n = 5). The company's age and number of staff were also captured. To preserve the case firm's anonymity, we will not disclose detailed background information on MultiLogistics. We were able to attain 35 returns. Our sample size seems to be relatively small. However, this study falls into the category of exploratory research. So, a 10% significance level was thought to be theoretically adequate [46,62]. With a minimum R^2 of 0.25–0.50, the required sample size was 34–53 [46] (p. 38). Bearing in mind that the R^2 from the client firm's model analysis was 0.261–0.735, the acquired sample size of 35 satisfies the recommended criteria.

4. Data Analysis

We investigated the industry-level data by employing PLS-SEM, and then extended the investigation to an IPA matrix method by using SmartPLS 3.0. The scores we calculated serve as the foundation for our investigation.

4.1. Industry-Level Analysis

In Table 2, each score of importance and performance level of the different IT flexibility dimensions using the industry data is provided. This is produced using the method described in Section 3. TR IT flexibility showed the highest importance score (0.369). OP IT flexibility had the second-highest score (0.201). STR IT flexibility had the lowest score (0.186) in importance among the three flexibility dimensions. Such an output strongly indicates that TR flexibility's score in performance (i.e., desired performance) should be the highest among the three dimensions. In fact, TR flexibility's actual score in performance was 26.276. This was the highest score, while OP IT flexibility's performance score is the second highest (23.835). STR flexibility's performance came in third (20.459). The output of the structural model analysis is summarised in Appendix C. In this analysis, the direct influence of STR flexibility on firm performance and indirect influence of TR and OP flexibility on firm performance via PIC is captured. Therefore, correlations between IT flexibility dimensions and firm performance are demonstrated. A validity test for the measurement models is as reported in Han et al. [11].

Table 2. Importance–performance level analysis of industry: construct scores.

Constructs (Dimensions)	Importance	Performances
TR IT flexibility	0.369	26.276
OP IT flexibility	0.201	23.835
STR IT flexibility	0.186	20.459

Source: authors.

From Table 2, we observe that the actual performance levels of different IT flexibility dimension are suitable to their relative levels of importance. Specifically, at the industry level, the resource distribution for the three different IT flexibility dimensions follows a line of "best fit," as given in Figure 1. This is also demonstrated by presenting a trend line among the three different flexibility levels that establish the AB line in Figure 2.

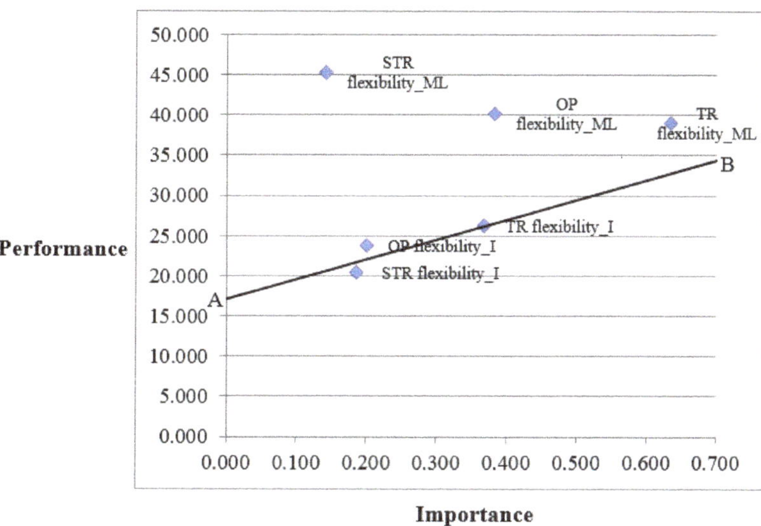

Figure 2. Importance–performance analysis of MultiLogistics: construct matrix. Source: authors. (Note: flexibility_I stands for industry and flexibility_ML is for MultiLogistics.)

4.2. Firm-Level Analysis

Table 3 shows that MultiLogistics' importance scores are in line with the findings from the industry-level data investigation. However, its performance scores in the three dimensions are inconsistent with the industry results. Notably, TR flexibility's performance score (i.e., desired performance) should be the highest among the three dimensions, as this dimension has been deemed the most important among the three. However, in the case of MultiLogistics, TR IT flexibility showed an actual performance score of 39.013 (the lowest score). In terms of OP flexibility's importance, it had the second-highest score (40.184). With regard to the STR IT flexibility's performance, it came in first (45.363). This indicates that MultiLogistics performed best at the STR dimension, followed by OP, then TR. The detailed PLS-SEM analysis results for MultiLogistics are provided in Appendix D.

Table 3. Importance–performance analysis of MultiLogistics: Construct scores.

Construct (Dimension)	Importance	Performance
TR IT flexibility	0.635	39.013
OP IT flexibility	0.384	40.184
STR IT flexibility	0.142	45.363

Source: authors.

Both the industry and MultiLogistics scores were united in a plot, given in Figure 2. As can be identified from Figure 2, MultiLogistics' perception of the order of importance of the three IT flexibility dimensions conforms with the industry-level analysis. However, its performance for the three dimensions is greater than the general industry measure. In particular, MultiLogistics' STR flexibility performance far exceeds the industry norm. This indicates that MultiLogistics may have invested excessively and hence overperformed in this dimension. If the company invested unnecessarily in resources in an effort to build STR flexibility, it might not get the rewards it expected. Alternatively, resources could have been better allocated to other areas, which would have a bigger influence on firm performance. What exactly could have contributed to this potential overperformance? The next section offers further insights via an indicator-level analysis.

4.3. Indicator-Level Analysis—Development of a Resource Allocation Action Plan

To identify which indicators of STR flexibility may be consuming resources that could be reallocated to other indicators, the performance of indicators is analysed. Table 4 provides the performance scores of each indicator for both MultiLogistics and the industry as a whole, as well as the score difference between MultiLogistics and the industry. We examine the three STR indicators in particular.

Table 4. The relative performance analysis for MultiLogistics versus the industry: indicator scores.

Dimensions	Indicator	Performance Score		
		MultiLogistics	Industry	Difference
TR flexibility	HW	40.952	23.177	17.775
	SW	37.619	21.654	15.966
	NW	28.095	19.271	8.824
	ACC	34.706	29.134	5.572
	LINK	45.455	28.042	17.412
	INTP	42.857	38.320	4.537
OP flexibility	QLT	37.255	20.604	16.651
	VIS	39.524	24.147	15.377
	SPD	40.476	22.572	17.904
	STMR	46.667	24.800	21.867
	OPT	36.667	26.640	10.026
STR flexibility	**PTN1**	**40.000**	**8.889**	**31.111**
	PTN2	**48.095**	**30.577**	**17.518**
	OFF	**47.059**	**23.228**	**23.830**

Source: authors.

All three indicators of STR show high performance scores, as expected. However, if one examines the score difference between MultiLogistics and the industry as a whole, there is evidence that some indicators have consumed extensive resources. First, the actual performance of PTN 1 (the ability of the company to establish and adjust information connections with existing supply chain partners) is considerably higher than the industry norm with a gap of 31.111—the largest gap among all 14 flexibility indicators. Moreover, the observed performance of OFF (the ability of the company to use ICT in offering novel products and services to their customers) (47.059) also shows a relatively high score difference (23.830) when compared to the industry score (23.228). The performance of PTN2 (the ability of the company to establish and adjust information linkages with new supply chain partners) is of less concern as it is more in line with the other indicators from the TR and OP dimensions.

Such potential overcommitment on PTN1 and OFF could be appropriate due to MultiLogistics' mission to be a truly customer-centric company. Given the increasing market pressure, and the fact that logistics and freight forwarding services are increasingly being perceived as commodities, the company has seen its profit margin being squeezed to a single digit in some regions. Therefore, the company has invested heavily in IT, including implementing a popular commercial transportation management system (TMS) globally, and an enterprise resource planning (ERP) customer relationship management suite to manage its air and freight transactions. According to its annual report, it expected that the flexibility brought about by such investments would provide better visibility to existing and new customers, simplify and speed up information flows, and streamline financial transactions across all its divisions. A lack of information on how the rest of the industry has performed could have led to a potential overcommitment. However, the seeming overinvestment could also be a deliberate act from the company in order to outperform its competitors and retain its leading position in the marketplace.

In order to find the underlying reasons why there is a large gap in PTN1 and OFF between the case company and the industry norm, we conducted a follow-up study. Interviews were conducted with four senior staff from the company: global innovation manager, IT manager, country fulfilment manager and a senior supply chain executive from one of the company's biggest clients, which, in order to retain anonymity, we call TelCo. With each interviewee, we first talked through the rationale of our study and the data analysis results. We then pointed out the gap we identified in the strategic dimension and asked our interviewees what might have contributed to the gap.

The interviews with participants from the case company largely confirmed our initial speculation that the "customer-centric" strategy drives the company's investment decisions in strategic IT flexibility. Investments have been made to streamline internal information integration (e.g., investment on Transport Management Systems and Enterprise Resource Planning) for efficiency and productivity gains. "This is the area that we can control," commented the IT manager. Areas that the case company has less control over but nonetheless must be committed to are building interorganisational information links and improving communications with various clients, particularly with large clients. The biggest challenge is that those large clients tend to have different in-house information systems, and the case company often has to build a dedicated information link with each of the clients, rather than a standard and cost-effective interface with all. The bespoke information connectivity demands a heavy resource commitment and contributes largely to the gap identified in factor PTN1.

The interviewee from TelCo explained why bespoke connectivity is needed in order for them to work with MultiLogistics: "We are a large global manufacturing company and have multiple factories in Europe. MultiLogistics is in charge of our UK order fulfilment process. This means the company needs to be able to interact with a number of our in-house systems. For inbound logistics, they also need to interact with each factory's ERP system to manage and receive goods coming into the UK. We have a central inventory management system that they need to access in order to gain visibility to stock levels. This means their WIS [warehouse information system] needs to integrate with our WIS. For outbound logistics, every time we issue them our customer's PO [purchasing order] and a packing list. They will then have to pick and configure the parts needed for that PO and send us a picking list via EDI [Electronic Data Interchange] link to our CRM [Customer Relationship Management] system."

As to the gap identified in OFF, the global innovation manager from the case company explained that, different from other logistics companies that tend to outsource many of their IT functions, MultiLogistics manages most of its IT in-house and sees technology as the core of its ability to adapt to changing demands in industry. Unlike asset-heavy logistics companies like DHL, the company's core competitiveness lies in its dynamic capability to continuously innovate and provide value-adding services to their clients. "So rather [than] just managing stocks for our clients, we actually work with our clients to reduce their overall inventories. We'll get paid less [for] warehousing, but we could then implement [a] VMI [vendor-managed inventory] type of exercise and take over our client's replenishment function," commented the innovation manager. Other interviewees also commented that a number of explorative initiatives took place in MultiLogistics—for example, introducing manufacturing services and 3D printing. Hence it seems that "overcommitment" under the factor of OFF is a deliberate act that the company undertook to differentiate itself from its competitors. They did acknowledge that they had limited knowledge about how the rest of industry performed, and therefore our tool helped them to gain a clear sense of their competitive position in the marketplace.

By translating the PLS-SEM analysis output to the IPA matrix, the current research has assessed the performance gap in different IT flexibility dimensions. The assessment was based on their relative levels of importance to firm performance. In the case of MultiLogistics, our analysis shows that STR flexibility may overperform, given that its importance does not deviate much from the industry standard. A further indicator-level analysis offers additional insight as to which factors contribute to the possible excessive performance of STR flexibility and where downscaling might be possible. Our follow-up study reveals the complex coercive and competitive forces that drove the case company's deliberate action to commit considerable resources.

5. Conclusions

5.1. Theoretical Contribution

Previous research has stressed the importance of effective IT flexibility management through a requirements analysis that mitigates deficiencies or excesses in dissimilar flexibility dimensions. Yet there has been a lack of analytical tools that evaluate the required and actual level of IT flexibility dimensions. To resolve this issue, we propose a combined method of IPA and PLS-SEM. Our method shows that prioritization among multiple dimensions of IT flexibility is made by employing the two universal measurements, namely importance and performance. Furthermore, the distribution of firm resources to the most important dimensions is advised.

Application of this method to a client company's data (Multilogistics in this study) also visualizes how this client firm can distribute its resources through the prioritization of different IT flexibility dimensions. In doing so, we proposed an action plan to distribute finite resources to different IT flexibility dimensions in a well-organized and efficient method. The method of PLS-SEM combined with the IPA matrix revealed the most important variables that contribute to the highest level of performance. To the best of the authors' knowledge, there has been no study that has proposed such an analytical tool that assesses multiple dimensions of IT flexibility. The unique contribution of this study is, therefore, an improved understanding of how to get the greatest value out of an investment in IT resources by revealing areas where particular management attention and subsequent rectification may be necessary. By proposing a combined method of IPA and PLS-SEM, we provide a novel insight about the mechanisms that firms can utilize to control their IT flexibility levels in rapidly changing business environments.

The IPA-PLS-SEM tool, however, does not offer detailed explanations as to why companies overcommit in certain areas. There may well be positive and legitimate reasons why companies acted in this way—for example, to maintain a competitive advantage by managing clients who require a certain IT flexibility, as illustrated in the case of MultiLogistics. Or equally, the areas identified could be "blind spots," or mismatches between required and committed investment, that senior executives were not aware of. Those areas could then be targeted for further improvement, or resources could be redeployed elsewhere to deliver improved value to the businesses. Furthermore, without this tool it is not possible to detect the IT flexibility gaps at an indicator level. Consequently, it will be difficult for organisations to develop a strategy for better resource exploitation. Once organisations are confident that the flexibility gaps identified via the use of the IPA—PLS-SEM tool are due to misinvestment in certain areas, action can be undertaken to rectify the situation. Certain resources could be transferred from one area to another in order to achieve the desired IT flexibility outcomes. The concept of resource mobility is well established in the literature, and reallocating resources from one dimension to another often incurs lower penalties when compared with additional investment.

5.2. Practical Contribution

Our research also suggests a measurement tool that managers can utilize to assess IT flexibility and identify any gaps that exist between current and desired flexibility levels. This output encourages logistics managers to consider the importance of IT flexibility in a more integrative and clear manner. It also helps logistics managers to decide how to coordinate their IT flexibility dimensions efficiently to deal with upcoming changes with finite firm resources. The current literature on quantifiable flexibility measurements has been accused of not being relevant to real-life industry settings. However, in this study, we propose that the IPA tool can be applied directly to industrial practice. Both supply chain and IT managers can use the IPA tool to understand what their competitive priorities should be when examining the performance of their IT flexibility, and identify whether some dimensions may be over- or underperforming. This will enable managers to review their IT flexibility capabilities and make more informed choices about where to best concentrate their resources, ultimately increasing managers' abilities to control and manipulate organizational factors to increase firm performance.

5.3. Limitations and Future Research

Even though the we did our best to gain many samples for applying the IPA and PLS-SEM combined method to the client firm, the application was performed with a comparatively small sample size ($n = 35$). This is because we required respondents to be key informants who are skilled and well-informed about inclusive IT use, interorganizational process connections and firm performance attributes. Moreover, acquiring 100 individuals from a single firm was a difficult task as the sample pool in a specific company is narrower than in an overall industry sector. If we acknowledge the exploratory characteristics of this study, the sample size is statistically acceptable [46]. However, even though our sample size satisfies the recommended criteria, future research should seek to collect more data to enhance the validity of both industry- and firm-level data.

Although IPA is particularly useful for indicating areas that may either under- or overperform, it does not offer an explanation as to why this happens. As the development of IT flexibility is specific to the operational context, environmental (e.g., industrial advances and market circumstances), organisational (e.g., firm scale, tactics and monetary condition) and technological (e.g., IT proficiency and human resources, architecture and IT merchants) issues can affect the required IT flexibility level. Further investigation is thus needed to understand whether a company is indeed over- or underperforming in a specific dimension. As the IPA matrix assumes linear relationships between importance and performance, this study does not address lines CD and EF (shown in Figure 1). Considerations of possible nonlinear relationships, such as the one proposed by Tontini and Silveira [50], could complement our approach. Future research could also attempt to apply this tool to other, non-IT types of flexibility.

Author Contributions: Conceptualization, J.H.H.; methodology, J.H.H.; Y.W. and M.N.; formal analysis, J.H.H. and Y.W.; writing—original draft preparation, J.H.H. and Y.W.; writing—review and editing, Y.W. and M.N.; visualization, J.H.H.; supervision, Y.W. and M.N.; project administration, Y.W. All authors have read and agreed to the published version of the manuscript.

Funding: This research received no external funding.

Conflicts of Interest: The authors declare no conflict of interest.

Appendix A

Table A1. Multiple dimensions of IT flexibility.

Dimensions	Subdimensions	Indicators (Abbreviations)	Explanations
Transactional Flexibility	IT Infrastructure	Hardware (HW)	We can successfully transact with external firms by using our advanced hardware (e.g., computers, field devices, sensors, meters, servers, etc.)
		Software (SW)	We can successfully transact with external firms by using our advanced software and applications (e.g., logistics portals, email systems, etc.)
		Networks (NW)	We can successfully transact with external firms by using our advanced network (e.g., internet, LAN, telephone, text)
	Connectivity	Access (ACC)	We can effectively access our IT network properly and securely to communicate with external firms (e.g., network access anytime anywhere)
		Linkages (LINK)	We can access a wide range of external firms through our IT network (e.g., number of external firms we can access through our portal
		Interoperability (INTP)	We can effectively transact with our external firms through standardized information format (e.g., Excel, PDF, HTML, EDI)
Operational Flexibility	Information sharing	Quality (QLT)	We can share accurate and timely information
		Visibility (VIS)	We can gain good visibility of supply chain processes
		Speed (SPD)	We can complete transactions rapidly
	Process improvement	Streamlining (STMR)	We can integrate and automate supply chain processes
		Optimisation (OPT)	We can optimise the supply chain processes with external firm
Strategic Flexibility	Partnering	Partnering1 (PTN1)	We can easily build and alter our information linkages to our existing supply chain partners providers
		Partnering 2 (PTN2)	We can easily build and alter our information linkages to new supply chain partners
	Offering	Offering (OFF)	We are actively exploring innovative ways of using ICT in offering new products or services to customers
Process integration capability (PIC)		PIC 1	We have a capability to integrate sourcing, transport, service process and other areas internally
		PIC 2	We have a capability to integrate sourcing, transport, service process and other areas with suppliers
		PIC 3	We have a capability to integrate sourcing, transport, service process and other areas with customers

Table A1. *Cont.*

Dimensions	Subdimensions	Indicators (Abbreviations)	Explanations
Firm performance (FP)		Cost (COST)	Transaction costs for your supply chain operations is reduced
		Service (SRV)	Level of service provided to customer is improved
		Speed (SPD_P)	Speed of supply chain operations is improved
		Quality (QLT_P)	Quality of service to customers is improved
		Value (Value)	Value creation in the supply chain is improved

Source: adapted from Han et al. [11].

Appendix B

Table A2. Hypotheses for IT flexibility research model.

Types	Hypotheses
Hierarchical structure of IT flexibility	Transactional IT flexibility positively affects Operational IT flexibility. Transactional IT flexibility positively affects Strategic IT flexibility. Operational IT flexibility positively affects Strategic IT flexibility.
Indirect impact of IT flexibility dimensions on firm performance	Transactional IT flexibility positively affects Process Integration Capability. Operational IT flexibility positively affects Process Integration Capability. Strategic IT flexibility positively affects Process Integration Capability.
Direct impact of IT flexibility dimensions on firm performance	Transactional IT flexibility positively affects firm performance. Operational IT flexibility positively affects firm performance. Strategic IT flexibility positively affects firm performance.
Impact of mediator on firm performance	Process Integration Capability positively affects firm performance.

Source: adapted from Han et al. [11].

Appendix C

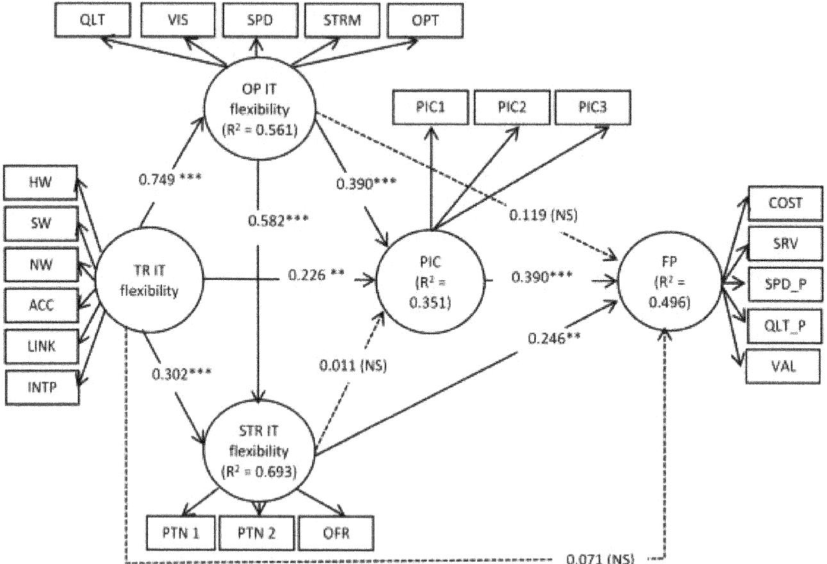

Figure A1. PLS-SEM test results for industry-level data (** $p < 0.05$, *** $p < 0.01$, NS: Non-Significant).

Appendix D

Table A3. PLS-SEM test results for MultiLogistics. Summary of validity test results of the measurement model.

Latent Variables	Number of Indicators	Internal Consistency Reliability		Convergent Validity	Indicator Reliability
		Composite Reliability	Cronbach's Alpha	AVE	Factor Loadings
TR IT flexibility	6	0.918	0.894	0.655	0.619 to 0.898
OP IT flexibility	5	0.940	0.920	0.758	0.831 to 0.898
STR IT flexibility	3	0.919	0.868	0.792	0.854 to 0.945
Process integration capability	3	0.911	0.856	0.773	0.869 to 0.888
Firm performance	5	0.954	0.940	0.807	0.848 to 0.953

Table A4. PLS-SEM test results for MultiLogistics. Fornell-Larcker criterion.

Latent Variables	Process Integration Capability	Firm Performance	Operational Flexibility	Strategic Flexibility	Transactional Flexibility
Process integration capability	**0.879**				
Firm performance	0.422	**0.898**			
Operational flexibility	0.498	0.757	**0.871**		
Strategic flexibility	0.361	0.715	0.739	**0.890**	
Transactional flexibility	0.401	0.689	0.692	0.805	**0.809**

Table A5. PLS-SEM test results for MultiLogistics. Cross-loading analysis.

	TR Flexibility	OP Flexibility	STR Flexibility	Process Integration Capability	Firm Performance
HW	**0.898**	0.715	0.759	0.446	0.588
SW	**0.889**	0.603	0.785	0.208	0.584
NW	**0.817**	0.331	0.544	0.168	0.349
ACC	**0.724**	0.404	0.514	0.341	0.638
LINK	**0.870**	0.762	0.874	0.474	0.678
INTP	**0.619**	0.312	0.287	0.173	0.397
QLT	0.550	**0.854**	0.556	0.404	0.700
VIS	0.633	**0.897**	0.711	0.448	0.671
SPD	0.565	**0.872**	0.677	0.543	0.732
STMR	0.678	**0.898**	0.742	0.376	0.603
OPT	0.582	**0.831**	0.508	0.389	0.584
PTN1	0.588	0.710	**0.851**	0.227	0.576
PTN2	0.810	0.635	**0.945**	0.337	0.661
OFF	0.783	0.640	**0.871**	0.385	0.665
PIC1	0.428	0.439	0.412	**0.869**	0.430
PIC2	0.215	0.314	0.178	**0.888**	0.224
PIC3	0.363	0.509	0.309	**0.880**	0.401
COST	0.590	0.751	0.680	0.351	**0.848**
SVC	0.694	0.663	0.637	0.399	**0.913**
SPD_P	0.589	0.684	0.628	0.304	**0.917**
QLT_P	0.685	0.673	0.685	0.429	**0.953**
VAL	0.523	0.619	0.570	0.415	**0.856**

Table A6. Effects and variance explained.

Effects on Endogenous Variable with Hypotheses	Path Coefficient β (t-Value)	Variance Explained (R^2)
Effects on OP flexibility	-	0.478
H1a: TR $\rightarrow$ OP	0.692 *** (7.718)	-
Effects on STR flexibility	-	0.735
H1b: TR $\rightarrow$ STR	0.600 *** (5.418)	-
H1c: OP $\rightarrow$ STR	0.324 *** (3.020)	-
Effects on PIC	-	0.261
H2a: TR $\rightarrow$ PIC	0.203 (0.659, NS)	-
H2b: OP $\rightarrow$ PIC	0.474 ** (2.123)	-
H2c: STR $\rightarrow$ PIC	-0.157 (0.397, NS)	-
Effects on FP	-	0.639
H3a: TR $\rightarrow$ FP	0.179 (0.921 NS)	-
H3b: OP $\rightarrow$ FP	0.446 ** (2.224)	-
H3c: STR $\rightarrow$ FP	0.220 (0.971, NS)	-
H4: PIC $\rightarrow$ FP	0.049 (0.320, NS)	-

Note: *** $p < 0.01$; ** $p < 0.05$; * $p < 0.1$ (all two-tailed).

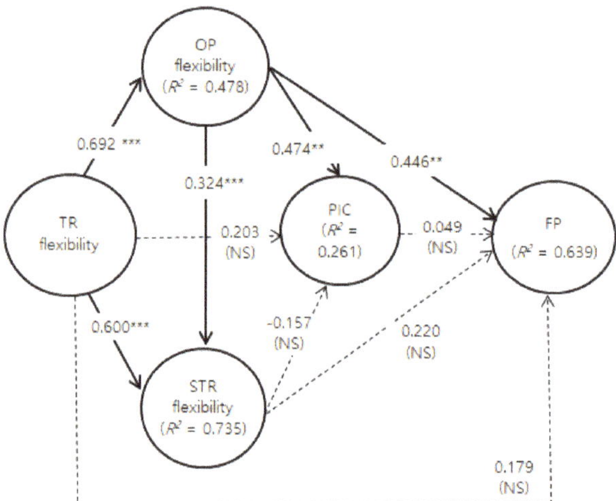

Figure A2. Result of path analysis—Company A. (** $p < 0.05$, *** $p < 0.01$, NS: Non-Significant).

References

1. Duncan, N.B. Capturing Flexibility of Information Technology Infrastructure: A Study of Resource Characteristics and their Measure. *J. Manag. Inf. Syst.* **1995**, *12*, 37–57. [CrossRef]
2. Gebauer, J.; Schober, F. Information system flexibility and the cost efficiency of business processes. *J. Assoc. Inf. Syst.* **2006**, *7*, 122–147. [CrossRef]
3. Bush, A.; Tiwana, A.; Rai, A. Complementarities between product design modularity and IT infrastructure flexibility in IT-enabled supply chains. *Eng. Manag. IEEE Trans.* **2010**, *57*, 240–254. [CrossRef]
4. Kumar, R.L.; Stylianou, A.C. A process model for analyzing and managing flexibility in information systems. *Eur. J. Inf. Syst.* **2014**, *23*, 151–184. [CrossRef]
5. Ngai, E.; Chau, D.; Chan, T. Information technology, operational, and management competencies for supply chain agility: Findings from case studies. *J. Strateg. Inf. Syst.* **2011**, *20*, 232–249. [CrossRef]
6. Saraf, N.; Langdon, C.S.; Gosain, S. IS Application Capabilities and Relational Value in Interfirm Partnerships. *Inf. Syst. Res.* **2007**, *18*, 320–339. [CrossRef]
7. Tafti, A.; Mithas, S.; Krishnan, M.S. The effect of information technology-enabled flexibility on formation and market value of alliances. *Manag. Sci.* **2013**, *59*, 207–225. [CrossRef]
8. Tallon, P.P.; Pinsonneault, A. Competing perspectives on the link between strategic information technology alignment and organizational agility: Insights from a mediation model. *MIS Q.* **2011**, *35*, 463–484. [CrossRef]
9. Zhang, J.; Li, H.; Ziegelmayer, J.L. Resource or capability? A dissection of SMEs' IT infrastructure flexibility and its relationship with IT responsiveness. *J. Comput. Inf. Syst.* **2009**, *50*, 46–53.
10. Nelson, K.M.; Ghods, M. Measuring technology flexibility. *Eur. J. Inf. Syst.* **1998**, *7*, 232–240. [CrossRef]
11. Shi, D.; Daniels, R. A survey of manufacturing flexibility: Implications for e-business flexibility. *IBM Syst. J.* **2003**, *42*, 414–427. [CrossRef]
12. Byrd, T.A.; Turner, D.E. Measuring the Flexibility of Information Technology Infrastructure: Exploratory Analysis of a Construct. *J. Manag. Inf. Syst.* **2000**, *17*, 167–208.
13. Han, J.H.; Wang, Y.; Naim, M. Reconceptualization of information technology flexibility for supply chain management: An empirical study. *Int. J. Prod. Econ.* **2017**, *187*, 196–215. [CrossRef]
14. Armstrong, C.; Sambamurthy, V. Information technology assimilation in firms: The influence of senior leadership and IT infrastructures. *Inf. Syst. Res.* **1999**, *10*, 304–327. [CrossRef]
15. Bhatt, G.; Emdad, A.; Roberts, N.; Grover, V. Building and leveraging information in dynamic environments: The role of IT infrastructure flexibility as enabler of organizational responsiveness and competitive. *Inf. Manag.* **2010**, *47*, 341–349. [CrossRef]

16. Gosain, S.; Malhotra, A.; El Sawy, O.A. Coordinating for Flexibility in e-Business Supply Chains. *J. Manag. Inf. Syst.* **2004**, *21*, 7–45. [CrossRef]
17. Lee, G.; Xia, W. The ability of information systems development project teams to respond to business and technology changes: A study of flexibility measures. *Eur. J. Inf. Syst.* **2005**, *14*, 75–92. [CrossRef]
18. Kwak, S.-Y.; Cho, W.-S.; Seok, G.-A.; Yoo, S.-G. Intention to Use Sustainable Green Logistics Platforms. *Sustainability* **2020**, *12*, 3502. [CrossRef]
19. Boyle, T.A. Towards best management practices for implementing manufacturing flexibility. *J. Manuf. Technol. Manag.* **2006**, *17*, 6–21. [CrossRef]
20. Fayezi, S.; Zutshi, A.; O'Loughlin, A. Developing an analytical framework to assess the uncertainty and flexibility mismatches across the supply chain. *Bus. Process Manag. J.* **2014**, *20*, 362–391. [CrossRef]
21. Gerwin, D. Manufacturing Flexibility: A Strategic Perspective. *Manag. Sci.* **1993**, *39*, 395–410. [CrossRef]
22. Jordan, W.C.; Graves, S.C. Principles on the Benefits of Manufacturing Process Flexibility. *Manag. Sci.* **1995**, *41*, 577–594. [CrossRef]
23. Suarez, F.F.; Cusumano, M.A.; Fine, C.H. *Flexibility and Performance: A Literature Critique and Strategic Framework*; Sloan School, Massachusetts Institute of Technology: Cambridge, MA, USA, 1991.
24. Kumar, V.; Fantazy, K.A.; Kumar, U.; Boyle, T.A. Implementation and management framework for supply chain flexibility. *J. Enterp. Inf. Manag.* **2006**, *19*, 303–319. [CrossRef]
25. He, P.; Xu, X.; Hua, Z. A new method for guiding process flexibility investment: Flexibility fit index. *Int. J. Prod. Res.* **2012**, *50*, 3718–3737. [CrossRef]
26. Jain, A.; Jain, P.K.; Chan, F.T.; Singh, S. A review on manufacturing flexibility. *Int. J. Prod. Res.* **2013**, *51*, 5946–5970. [CrossRef]
27. Cousens, A.; Szwejczewski, M.; Sweeney, M. A process for managing manufacturing flexibility. *Int. J. Oper. Prod. Manag.* **2009**, *29*, 357–385. [CrossRef]
28. Gerwin, D. An Agenda for Research on the Flexibility of Manufacturing Processes. *Int. J. Oper. Prod. Manag.* **1987**, *7*, 38–49. [CrossRef]
29. Sethi, A.; Sethi, S. Flexibility in manufacturing: A survey. *Int. J. Flex. Manuf. Syst.* **1990**, *2*, 289–328. [CrossRef]
30. Upton, D. The management of manufacturing flexibility. *Calif. Manag. Rev.* **1994**, *36*, 72–89. [CrossRef]
31. Seebacher, G.; Winkler, H. A capability approach to evaluate supply chain flexibility. *Int. J. Prod. Econ.* **2015**, *167*, 177–186. [CrossRef]
32. Stevenson, M.; Spring, M. Flexibility from a supply chain perspective: Definition and review. *Int. J. Oper. Prod. Manag.* **2007**, *27*, 685–713. [CrossRef]
33. Upton, D. What really makes factories flexible? *Harv. Bus. Rev.* **1995**, *73*, 74–84.
34. Chang, A.-Y. Prioritising the types of manufacturing flexibility in an uncertain environment. *Int. J. Prod. Res.* **2012**, *50*, 2133–2149. [CrossRef]
35. Lee, O.K. IT-Enabled Organizational Transformations to Achieve Business Agility. *Rev. Bus. Inf. Syst.* **2012**, *16*, 43. [CrossRef]
36. Hou, C.K. The effects of IT infrastructure integration and flexibility on supply chain capabilities and organizational performance: An empirical study of the electronics industry in Taiwan. *Inf. Dev.* **2019**, 1–27. [CrossRef]
37. Gao, P.; Zhang, J.; Gong, Y.; Li, H. Effects of technical IT capabilities on organizational agility: The moderating role of IT business spanning capability. *Ind. Manag. Data Syst.* **2020**, *120*, 941–961. [CrossRef]
38. Irfan, M.; Wang, M.; Akhtar, N. Impact of IT capabilities on supply chain capabilities and organizational agility: A dynamic capability view. *Oper. Manag. Res.* **2019**, *12*, 113–128. [CrossRef]
39. Benitez, J.; Ray, G.; Henseler, J. Impact of Information Technology Infrastructure Flexibility on Mergers and Acquisitions. *MIS Q.* **2018**, *42*, 25–43. [CrossRef]
40. Bamel, U.K.; Bamel, N. Organizational resources, KM process capability and strategic flexibility: A dynamic resource-capability perspective. *J. Knowl. Manag.* **2018**, *22*, 1555–1572. [CrossRef]
41. Benitez, J.; Llorens, J.; Braojos, J. How information technology influences opportunity exploration and exploitation firm's capabilities. *Inf. Manag.* **2018**, *55*, 508–523. [CrossRef]
42. Chaudhuri, A.; Boer, H.; Taran, Y. Supply chain integration, risk management and manufacturing flexibility. *Int. J. Oper. Prod. Manag.* **2018**, *38*, 690–712. [CrossRef]

43. Kemmoe, S.; Pernot, P.-A.; Tchernev, N. Model for flexibility evaluation in manufacturing network strategic planning. *Int. J. Prod. Res.* **2014**, *52*, 4396–4411. [CrossRef]
44. Martilla, J.A.; James, J.C. Importance-Performance Analysis. *J. Mark.* **1977**, *41*, 77. [CrossRef]
45. Slack, N. The Importance-Performance Matrix as a Determinant of Improvement Priority. *Int. J. Oper. Prod. Manag.* **1994**, *14*, 59–75. [CrossRef]
46. Tontini, G.; Silveira, A. Identification of satisfaction attributes using competitive analysis of the improvement gap. *Int. J. Oper. Prod. Manag.* **2007**, *27*, 482–500. [CrossRef]
47. Pezeshki, V.; Mousavi, A.; Grant, S. Importance-performance analysis of service attributes and its impact on decision making in the mobile telecommunication industry. *Meas. Bus. Excell.* **2009**, *13*, 82–92. [CrossRef]
48. Hair, J.F.; Ringle, C.M.; Sarstedt, M. Partial Least Squares Structural Equation Modeling: Rigorous Applications, Better Results and Higher Acceptance. *Long Range Plan.* **2013**, *46*, 1–12. [CrossRef]
49. Lai, I.K.W.; Hitchcock, M. Importance-performance analysis in tourism: A framework for researchers. *Tour. Manag.* **2015**, *48*, 242–267. [CrossRef]
50. Rigdon, E.; Ringle, C. Assessing heterogeneity in customer satisfaction studies: Across industry similarities and within industry differences. *Adv. Int. Mark.* **2011**, *22*, 169–194.
51. Ringle, C.M.; Sarstedt, M. Gain more insight from your PLS-SEM results. *Ind. Manag. Data Syst.* **2016**, *116*, 1865–1886. [CrossRef]
52. Hair, J.F., Jr.; Hult, G.T.M.; Ringle, C.; Sarstedt, M. *A Primer on Partial Least Squares Structural Equation Modeling (PLS-SEM)*, 1st ed.; SAGE Publication Ltd.: Thousand Oaks, CA, USA, 2013.
53. Rigdon, E.E. Choosing PLS path modeling as analytical method in European management research: A realist perspective. *Eur. Manag. J.* **2016**, *34*, 598–605. [CrossRef]
54. Hair, J.F.; Hult, G.T.M.; Ringle, C.; Sarstedt, M. *A Primer on Partial Least Squares Structural Equation Modeling (PLS-SEM)*, 2nd ed.; SAGE Publications: Thousand Oaks, CA, USA, 2017.
55. Schmitz, K.W.; Teng, J.T.C.; Webb, K.J. Capturing the Complexity of Malleable IT Use: Adaptive Structuration Theory for Individuals Availability: In stock. *MIS Q.* **2016**, *40*, 663–686. [CrossRef]
56. Subramani, M. How do suppliers benefit from information technology use in supply chain relationships? *MIS Q.* **2004**, *28*, 45–73. [CrossRef]
57. Melville, N.; Kraemer, K.; Gurbaxani, V. Review: Information technology and organizational performance: An integrative model of IT business value. *MIS Q.* **2004**, *28*, 283–322. [CrossRef]
58. Bryman, A.; Bell, E. *Business Research Methods*; Oxford University Press: New York, NY, USA, 2012.
59. Saunders, M.N.K.; Lewis, P.; Thornhill, A. *Research Methods for Business Students*, 6th ed.; Pearson Education Limited: Harlow, UK, 2012.
60. Kumar, N.; Stern, L.; Anderson, J. Conducting interorganizational research using key informants. *Acad. Manag. J.* **1993**, *36*, 1633–1651.
61. Rai, A.; Pavlou, P.A.; Im, G.; Du, S. Interfirm IT Capability Profiles and Communications for Cocreating Relational Value: Evidence from the Logistics Industry. *MIS Q.* **2012**, *36*, 233–262. [CrossRef]
62. Reinartz, W.; Haenlein, M.; Henseler, J. An empirical comparison of the efficacy of covariance-based and variance-based SEM. *Int. J. Res. Mark.* **2009**, *26*, 332–344. [CrossRef]

sustainability

MDPI

Article

Sustainability of Investment Projects with Energy Efficiency and Non-Energy Efficiency Costs: Case Examples of Public Buildings

Mališa Đukić [1,*] and Margareta Zidar [2]

[1] Belgrade Banking Academy, Faculty of Banking, Insurance and Finance, Union University, 11000 Belgrade, Serbia

[2] Department for Energy Efficiency, Energy Institute Hrvoje Požar, HR-10000 Zagreb, Croatia; mzidar@eihp.hr

* Correspondence: djukicmalisa@hotmail.com

Abstract: According to the European Commission Energy Union strategy from 2015, some of the main objectives are to improve energy efficiency, reduce dependence on energy imports, cut emissions, and drive jobs and growth. Achieving the objectives of the Energy Union requires significant financing, particularly for investments in energy efficiency. Serbia and Croatia included the objectives of the Energy Union in their national strategies and have implemented various investment projects in this area. This paper focuses on the sustainability of energy efficiency projects for public buildings which include not only energy efficiency investment cost but also non-energy efficiency investments. By applying the European Commission methodology for cost-benefit analysis, we assessed the sustainability of several projects in Serbia and Croatia. The sustainability assessment is done by quantifying energy savings, greenhouse gas emission reductions and the social and economic benefits that are related to non-energy efficiency project components. The values of economic performance indicators imply that society would be better off with projects that would contribute to achieving not only the targets set in national energy strategies but also to creating broader social benefits.

Keywords: investment projects; financial analysis; economic analysis; energy efficiency; public buildings; sustainability

Citation: Đukić, M.; Zidar, M. Sustainability of Investment Projects with Energy Efficiency and Non-Energy Efficiency Costs: Case Examples of Public Buildings. *Sustainability* **2021**, *13*, 5837. https://doi.org/10.3390/su13115837

Academic Editors: Golam Kabir, Sanjoy Kumar Paul, Syed Mithun Ali and Marc A. Rosen

Received: 30 March 2021
Accepted: 18 May 2021
Published: 22 May 2021

1. Introduction

The buildings sector in the EU contributes to approximately 36% of CO_2 emissions and 40% of final energy consumption, while 75% of the building stock is energy inefficient and 35% of buildings are over 50 years old [1]. Due to the Energy Efficiency Directive (2012/27/EU and 2018/2002) (EED), the annual rate of refurbishment should amount to 3% of the total surface area of buildings occupied by the central government [2]. The Directive on Energy Performance of Buildings (2010/31/EU and 2018/844) (EPBD) requires decarbonizing the national building stocks by 2050 by establishing national long-term renovation strategies [3]. In addition to these energy efficiency related directives, the Directive 2018/2001 on the Promotion of the use of Energy from Renewable Sources (REDII) calls for more renewable-based heating and cooling in buildings [4]. Current annual renovation rates of the building stock across Member States ranges from 0.4% to 1.2%, and to reach the targeted 3%, many barriers to renovation need to be overcome [5].

In this paper, we focus on analyzing the sustainability of investing in several public buildings in Croatia and Serbia. The objective of the study is to assess the technical, financial, and economic feasibility of investment projects in three public buildings by calculating corresponding performance indicators. This paper addresses the following questions: What is the energy consumption in each case? What are the technical options and what are their implications on the energy performance of buildings? Do the options contribute to the decarbonization of the building stock? What are the investment, operation

---

and maintenance costs and energy savings per every option? Which option offers the most cost-effectiveness? Is the proposed project financially viable? What are the energy efficiency and non-energy efficiency benefits in every case? Do the energy efficiency benefits justify the costs? If not, how can the economic viability of the cases presented in this paper be assessed?

In Croatia, from 2014 until 2020, the energy refurbishment of buildings was co-financed by the European Structural and Investment funds with 60% for energy efficiency measures and 85% for design documentation, energy audit and energy certificate costs [6]. To qualify for the co-financing, at least 50% of heating energy savings was required and usually this was achieved by the thermal insulation of the building envelope and the installation of energy efficient windows. The market was highly motivated to use grants for financing the energy refurbishment costs. There were 758 projects of public buildings and the total co-financing amount was EUR 211.81 million. In residential buildings, i.e., multiapartment buildings and family houses, which included a total of 16,000 households, there were 584 projects and the total co-financing amount was EUR 100 million [7]. Even though this is a relatively high number of projects and financial resources invested, an annual refurbishment rate of 0.7% of the total building stock was achieved and 1.35 million m^2 of the building surface area were refurbished. According to the recently published National building stock refurbishment strategy, a gradual increase in the annual refurbishment rate is targeted, starting from 1% in 2021 and reaching 4% in 2050, and expecting to refurbish a total of 104 million m^2 and to achieve 2260.9 $ktCO_2$ savings [8]. Decarbonization of the building stock will include additional energy efficiency measures, such as modernizing the building systems, using on-site renewable energy, and developing neighborhood renovation projects instead of single building projects. In Serbia, according to the Energy sector development strategy until 2025, the energy refurbishment of buildings has been set out as a priority activity.

To assess the sustainability of the investment projects in several public buildings in Serbia and Croatia, we took into consideration both energy efficiency and non-energy efficiency costs and benefits. Economic performance indicators imply that energy efficiency benefits may not be sufficient to justify the project costs. The inclusion of non-energy efficiency benefits improves the results of the economic analysis, making the projects desirable from a socioeconomic perspective.

The paper is organized as follows. Section 2 provides a literature review. Methods on energy audits, certificates of buildings and the cost-benefit analyses of investment projects are introduced in Section 3. Case examples of a building in Croatia and two buildings in Serbia are elaborated in Section 4. Results and discussion on energy cost savings, option analysis and cost-benefit analyses are presented in Section 5. Finally, conclusions are presented in Section 6.

2. Literature Review

Improving energy efficiency has been a way to increase the productivity and sustainability of society, primarily through the delivery of energy savings. The impact of energy efficiency measures can be an important contributor to economic growth and social development [9]. To investigate the complex and controversial relationship between energy consumption and GDP, a study concluded that there is mixed evidence on the direction of causality between energy consumption and GDP [10]. Job creation is an important socioeconomic effect associated with the development of energy efficiency technologies. The approaches to estimate job creation effects of the renewable energy and energy efficiency sectors, as well as the existing estimations of job creation figures by technology and capacity, were summarized in the literature [11,12]. Kerr et al. considered how different benefits have been used within the overall rationale for energy efficient retrofit policies in different contexts [13]. The identification of multiple benefits may not imply multiplied policy support, and instead it is more likely that different rationales will have relevance at different times, for different audiences [13].

Bleyl et al. analyzed the economic and financial implications for renovating an office building to the "Passive House" standard and concluded that the Life Cycle Cost & Benefit Analysis cash flow model can be used not only for deep energy retrofit business case analysis, project structuring, and financial engineering, but also for policy design [14]. Gelatioto et al. simulated a public historic building and some permitted retrofit actions were applied to analyze the effectiveness of national measures in four different climatic zones [15]. Energy efficiency retrofitting of existing buildings is a key program for improving building energy efficiency in northern regions of China. A methodological framework to conduct an economic cost-benefit analysis (CBA) for energy efficiency retrofit projects was applied and the research found that the retrofitting of existing buildings generally lacks attractiveness to investors from an economic perspective [16]. Energy saving retrofits of residential and public buildings positively contributed to economic growth, employment, and protection of the environment in a post-transition country [17].

Novikova et al. analyzed the costs and benefits of the different thermal efficiency retrofits as well as the impact of user behavior, implying that energy usage in the public building sector could increase and mitigate savings resulting from energy-saving measures, making saved energy costs invisible [18]. By reducing energy consumption, other benefits can be achieved such as reduced government subsidies and improved health due to less air pollution and a better indoor climate [19]. Filippidou and Jimenez Navaro [1] have used CBA to identify the cost effective and cost optimal solutions for current and future EU building stock. Both energy efficiency and sustainable heating and cooling should be considered when planning the decarbonization of the building sector, thus avoiding lock-in effects in terms of investments in less than cost-optimal energy efficiency renovations. Cost-optimal solutions require moderate thermal insulation in southern Europe, and efficient heating and cooling technologies should be prioritized. On the other hand, in central Europe, deep energy efficiency improvements are cost-optimal and should be combined with efficient heating and cooling technologies.

The renovation of large scale buildings is still a difficult task to be accomplished. The most prevalent reasons for this are, among others, large investments and lack of awareness of the potential benefits and of the skills required [20,21]. Papers published so far evaluate the co-benefits (or multiple benefits) of a portfolio of energy efficiency projects by quantifying the effects of such projects on selected macroeconomic variables [9–11,17–19,22]. In our paper, we try to contribute to the literature by estimating the economic performance of energy efficiency investments in public buildings on a case-by-case basis.

3. Methods

The European Commission methodology is applied to assess the technical, financial and economic feasibility of proposed investments in three public buildings [23]. Figure 1 presents the appraisal steps.

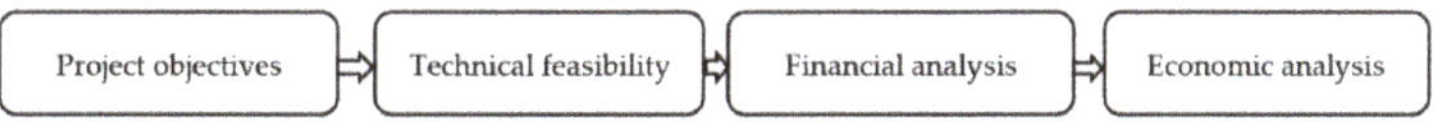

Figure 1. Project appraisal methodology steps (Source: Adapted from Sartori et al. [23]).

Project objectives are elaborated in Section 4. The methodology on technical feasibility with option analysis as well as the financial and economic analysis is presented in Sections 3.1 and 3.2, respectively.

3.1. Energy Audits and Certificates of Buildings

Energy savings in buildings are determined in an energy audit. There are several standards for conducting energy audits, and the most used ones are ISO 50002:2014 and EN 16247-1:2012, which promote similar basic steps targeted to identify current energy performance of a building and building systems and the energy performance improvement measures. The energy audit process has two implementation steps, (1) Data collection & Site Visit, and (2) Analysis & Reporting.

The data collection includes discussion with the building owner and user to define the energy audit purpose and scope, to acquire data on the building and the energy bills. Next, a site visit is undertaken to record findings on the current condition of the building's elements and systems, and to identify opportunities for energy savings [24].

Savings are achieved by considering various energy efficiency measures that improve energy performance of building components and systems. Baseline energy consumption, costs and CO_2 emissions are determined from the energy bills. Feasibility of energy efficiency measures is determined by technoeconomic analysis of the investment costs and the saving potential presented in energy, CO_2 emissions and costs (kWh/y, tCO_2/y, EUR/y, respectively). Additional financial indicators as simple payback period (SPB), discounted payback period (DPB), internal rate of return (IRR) and investment net present value (NPV) are also prepared to describe the effectiveness of the energy efficiency measures. All findings on the current energy performance, energy efficiency measures and the improved energy performance of a building are described in a report, which is presented to the building owner and user to make the decision on the implementation of energy efficiency measures.

The EPBD has defined a general framework for the calculation of the energy performance of buildings where the heat energy demands for heating and cooling, along with energy efficiency levels of the building systems, are assessed. The energy performance certificate is a document where the unique energy performance indicator is presented, grading the building's energy performance in energy classes, from A+ as the most efficient level, to G as the lowest efficiency level.

For existing buildings, which are usually graded in low energy efficiency classes from D to G, there is a high energy saving potential. Deep energy refurbishment measures, such as thermal insulation of the building envelope and on-site renewable energy-based technologies, achieve energy savings of up to 80% and the highest energy classes. Major refurbishment is usually focused on thermal insulation measures and saving up to 50% more energy and energy classes B or C are reached. Minor refurbishment includes improvements of the existing building systems while savings can range up to 30% and the energy class usually improves one level up.

In the option analysis, cost effectiveness rather than cost efficiency should be targeted. Implementing lower levels of energy performance may lead to locking-in energy consumption to unsustainable patterns due to long-term lack of proper maintenance, which remain unchanged for several decades until the next renovation cycle [25]. As future renovation rates are expected to remain close to the current level, it is highly important to make sure that the best available energy efficiency measures are included [26]. In our paper, the selection of the preferred options is based on cost effectiveness, which allows for high energy savings.

3.2. CBA Methodology

CBA is defined as an activity that enables calculating and comparing costs and benefits of an investment project. The impact of financial, economic, social and other factors should be considered in order to assess the financial and economic viability of projects [27]. When the value of benefits exceeds the costs incurred, the project is feasible. When an activity undertaken by an individual or a company affects other individuals or companies who do not pay for that activity or are not paid, the effect can be either positive or negative; therefore, there are positive or negative externalities [28]. Sáez and Requena emphasize that CBA includes our own intergenerational ethics, expressing what our present generation is willing to pass on to future generations and not the future generations' preferences [29]. They propose the use of the common Social Discount Rate for market goods and a lower discount rate (environmental discount rate) for non-market goods in the same CBA exercise as a way to include a certain level of intergenerational equity [29].

In this paper, CBA is performed in line with the European Commission Guide to cost-benefit analysis of investment projects, hereafter the EC CBA Guide [23]. CBA includes financial analysis and economic analysis. The financial analysis is made on behalf of the owner of the infrastructure. The methodology used is the Discounted Cash Flow (DCF) method. Only cash inflows and outflows are considered in the analysis, i.e., depreciation, reserves, price and technical contingencies and other accounting items which do not correspond to actual flows are disregarded. The reference period is 20 years. The analysis is carried out in constant (real) prices, net of value added tax (VAT), both on costs and revenues.

The incremental approach is applied, comparing the with-project scenario and the without-project scenario. The with-project scenario assumes the implementation of the proposed investment and operations within the reference period. The without-project scenario is a counterfactual scenario and indicates what would happen in the absence of the project.

It is assumed that the energy price will be increased according the electricity price forecasts in the South East Europe Electricity Road Map Country Report [30]. Projected electricity price growth rates are applied to estimated electricity cost. Fuel scenarios developed for the Ten-Year Network Development Plan (TYNDP) 2018 from the European Network of Transmission System Operators (ENTSOs) were considered with regards to gas price [31]. Projected gas price growth rates are applied to estimated heating cost.

The economic analysis is made on behalf of the whole society and appraises the project's contribution to the economic welfare of the region or a country. Investment projects often have impacts that have no direct market values such as the impacts on the environment. These effects can be monetized through different valuation techniques. The key objective of the economic analysis is to prove that the present value of the project's economic benefits exceeds the present value of its economic costs, which means that the project has a positive net contribution to society. This is expressed as a positive Economic Net Present Value of the net cash flow, a Benefit/Cost (B/C) ratio higher than 1,0 or a project's economic rate of return (ERR) exceeding the social discount rate. The economic analysis is based on an incremental approach, comparing economic cost and benefits of the project. Constant prices are applied and a social discount rate of 5% is used. As no particular salary distortions are foreseen, the conversion factor of 1 is used. The following five steps are applied: (1) conversion from market to accounting prices, (2) monetization of non-market impacts, (3) inclusion of additional indirect effects, if relevant, and (4) discounting of the estimated costs and benefits and calculation of the economic performance indicators [23].

In this paper, we assessed energy efficiency and non-energy efficiency benefits. Energy efficiency benefits included O&M cost savings, avoided CO_2 emissions, avoided cost of airborne pollutants and enhanced security of supply. The non-energy efficiency benefits are specific for every building and may comprise education-related benefits, avoided healthcare costs and avoided cost of child care, if applicable.

The recommended values of European Commission (EC) for the shadow price of CO_2 are used in the economic analysis and are available in the European Commission Directorate-General for Climate Action paper on Climate Change and Major Projects [32]. Changes in the emissions of airborne pollutants are also monetized (PM, NOx, SOx) with unit damage values from the NEEDS study [33]. If the with-project temperature will be different from the current temperature, the project would create incremental comfort benefits. The project is expected to generate security-of-supply benefits at a shadow price used by the European Investment Bank (EIB) and the EC for power generated from a combined cycle gas turbine (CCGT) [23].

When applicable, education related benefits are assigned. Such benefits would result from improved quality of health care services and an increase in the number of survivors. Value of a Statistical Life (VSL) and income elasticities are elaborated by Viscusi and Masterman [34].

4. Case Examples of Public Buildings

4.1. Building 1—Hospital

Building 1 is a clinical hospital built in 1988 and is located in Zagreb, Croatia. The building is graded in energy class F > 220%. Electricity accounts for 48% of total operational costs, natural gas for 19% and water for 33%. Although the investment maintenance has been implemented over the years, the building systems have reached the end of life and many damages can be observed on the building envelope. Leakage of the flat roofs is occurring continuously due to the large roof surface area, damaged stormwater outlets and damaged skylights. Building systems lack all functionalities that would enable stable and efficient supply of energy, as part of the damaged equipment like heat recovery and humidification systems were not replaced due to high investment costs. The window opening mechanisms, window sealing and prefabricated façade panels with thermal insulation layers are damaged, causing increased heat losses. The water supply and hydrant network lack proper system pressure due to frequent pipe bursting while all water fixtures are damaged with continuous leakage. Investment not only in energy efficiency improvement but also in other major building properties is needed, but the respective refurbishment program finds only energy efficiency measures and water saving measures eligible for co-financing.

Energy efficiency measures proposed for this building include reducing heat demand by improving the thermal insulation of the external envelope, the full replacement of the centralized air-conditioning and mechanical ventilation systems and the replacement of steam boilers with hot water boilers. A new lighting system and automation and control for all building systems with integration in the central management and surveillance system are suggested. In addition, the revitalization of laundry and kitchen facilities, refurbishment of the external hydrant network, introducing new efficient water saving fixtures, refurbishment of the entire indoor supply pipeline, photovoltaic system and solar thermal collectors are included. Option analysis is performed for this building indicating various energy efficiency levels of the energy efficiency measures proposed. Option A presents the minimum improvements of the building envelope and building systems. Option B aims at achieving the overall highest energy efficiency level. In addition to the improvements of option B, option C includes a higher building envelope performance. The building will be graded in energy class C < 100% following the implementation of the proposed investment.

4.2. Building 2—Emergency Health Care Center

Building 2 was built in 1975 as a medical facility in Belgrade, Serbia. The building is undergoing significant surface expansion to improve substandard working conditions and to introduce new medical services. The building is graded in energy class E > 150%. Electricity accounts for 30% of total operational costs, heat energy for 67% and water for 3%. The overall poor conditions due to lack of adequate maintenance on all installations and the building structure can be observed, such as leakage of the flat roofs and façade damages. The existing building systems have substandard energy performance levels, and additional power capacity, proper cabling and illumination level of the lighting system, replacement of damaged sanitary furniture, improvement of indoor air quality and indoor comfort both in summer and winter period are required.

The energy efficiency measures proposed for this building include the reduction of heat demand by improving the thermal insulation of the external envelope. The supply of heat from the existing district heating system will remain, but the heat distribution and transfer system will be completely replaced. Moreover, solar thermal collectors, a new centralized cooling system, a ventilation system with heat recovery for high priority areas and building energy management system will be introduced. The sustainability measures include the reinforcement of the existing loadbearing structure, fire protection systems and installations, new water and sewage installations, the complete replacement of the electrical installations, as well as back-up energy generators and reliable and high-capacity telecommunication installations. Auxiliary services (kitchen, restaurant, laundry, car wash), accessibility infrastructure for disabled persons, a video surveillance system and traffic and landscape infrastructure are provided. Option A includes the minimum energy efficiency improvement of all building elements and systems and is in line with national energy efficiency requirements, while option B introduces solar thermal system additionally. Option C incorporates the mechanical ventilation systems with heat recovery in high priority areas. The building will be graded in energy class B < 50% in the with-project scenario. CBA is performed for the preferred option.

4.3. Building 3—Kindergarten Complex

Building 3 is a kindergarten complex with administration offices and a kitchen facility in Belgrade, Serbia. The building was built in 1975 and is graded in energy class E > 200%. Electricity accounts for 37% of total operational costs, heat energy for 58% and water for 5%. The building was thermally insulated in 2009 but with the minimum performance of the façade and windows. The indoor thermal comfort in heating season is low due to insufficient heating capacity of the central light heating oil boiler plant. Individual electric heating units are used for the additional heating of rooms. The power capacity is not sufficient to cover all electric equipment installed.

Energy efficiency measures proposed for this building include reducing heat demand by improving the thermal insulation of the external envelope. Modernization of building systems introduces heat pumps for heating and cooling, a mechanical ventilation system with heat recovery, solar thermal collectors, a new lighting system and a building energy management system. Option A includes the minimum energy efficiency improvement of all building elements and systems and is in line with national energy efficiency requirements. Option B introduces the solar thermal system and heat pump, while option C incorporates the mechanical ventilation and heat recovery. The building will be graded in energy class B < 50% following the implementation of the proposed investment. The with-project scenario of CBA is based on the preferred option.

5. Results and Discussion

5.1. Building 1—Hospital

The investment cost in various options is from 432 to 500 EUR/m^2 while the operational cost savings range from 21 up to 39 EUR/m^2 (Table 1). The simple payback period is from 10 to 21 years. Option B is the most cost-effective one with the highest operational cost saving, and with an annual electricity saving of 24%, heat saving of 67%, water saving of 42% and CO_2 saving of 57%. Renovation costs in hospitals in Albania for building envelopes and heating systems were assessed in the literature [18]. Due to the difference in investment objectives, their findings are not comparable to the results of option analysis.

For Option B, the NPV amounts EUR −406,840 and the IRR is 3.73%. With 40% co-financing, the NPV is EUR 8.8 million and the IRR is 12.53%. The results suggest that an energy efficient renovation investment, if not supported by government grants, may not be financially viable. This is in line with previous findings that government support could significantly improve the financial viability for investors [17,19].

Table 1. Investment cost and savings for various options for Building 1.

Building 1	Option A	Option B	Option C
Investment [EUR/m^2]	432	480	500
Heat saving [kWh/m^2]	325	590	598
Electricity saving [kWh/m^2]	29	37	40
Water saving [m^3/m^2]	4	4	4
Energy class	D	C	C
CO_2 saving [tCO$_2$/m^2]	107	145	146
Operational cost saving [EUR/m^2]	21	46	39
Simple payback period [years]	21	10	13

Source: Authors' calculations.

5.2. Building 2—Emergency Health Care Center

The investment cost ranges from 67 up to 94 EUR/m^2 while the operational cost saving varies from 5 up to 14 EUR/m^2 (Table 2). Total energy savings vary from 41% up to 49%. Option B has the shortest simple payback period, however, option C is the most cost-effective as it provides higher indoor comfort.

Table 2. Investment cost and savings for various options for Building 2.

Building 2	Option A	Option B	Option C
Investment [EUR/m^2]	149	157	192
Heat saving [kWh/m^2]	99	131	147
Electricity saving [kWh/m^2]	35	36	24
Energy class	C	B	B
CO_2 saving [tCO$_2$/m^2]	0.011	0.017	0.015
Operational cost saving [EUR/m^2]	6	8	9
Simple payback period [years]	24	19	23

Source: Authors' calculations.

The investment cost is estimated to be EUR 7.3 million and includes energy efficiency (EE) components and non-EE components with 54% and 46% share, respectively (Table 3). EE components comprise architecture, thermotechnical and electrical items, while non-EE components contain architectural, structural, sanitary, telecommunication, fire detection and alarm system, vertical transport, demolition works, smoke extraction, sprinkler system and landscaping items.

Table 3. Investment cost of Building 2.

Item	EUR	%
EE components	3,950,000	54%
Non-EE components	3,330,000	46%
Total investment cost	7,280,000	100%

Source: Authors' calculations.

Operation and maintenance (O&M) costs include electricity, heating and maintenance costs of both EE and non-EE components (Table 4). Due to the energy cost savings that would result from the project implementation, the incremental operation and maintenance costs are negative from the first year of operations in the amount of EUR −5.804. The energy costs are adjusted to the forecasted changes of energy prices as elaborated in the methodology section of the paper.

Table 4. Incremental O&M cost in the first year of project operations.

Item	EUR
Electricity cost	−29.647
Heating cost	−16.077
Maintenance cost	39.920
Total incremental O&M costs	−5.804

Source: Authors' calculations.

The economic costs include the investment cost, replacement cost and residual value (Table 5). As the weighted average life time of investment cost items is equal to the reference period of 20 years, the replacement costs and residual value are set to EUR 0. Since the project generates energy savings, incremental O&M costs are presented as economic benefits. The net present value of economic costs is EUR 6.9 million.

Table 5. CBA results for Emergency Health Center.

Item	Net Present Value	%
Investment cost	6,933,333	100.0%
Replacement cost	0	0.0%
Residual value	0	0.0%
Total economic costs	6,933,333	100.0%
Avoided CO_2 emissions	56,950	0.7%
O&M cost savings	157,678	2.0%
Avoided cost of airborne pollutants	85,441	1.1%
Enhanced security of supply	13,936	0.2%
Education benefits	7,674,430	96.1%
Total economic benefits	7,988,436	100.0%
Economic net present value	1,055,103	

Source: Authors' calculations.

The O&M cost savings are a major energy efficiency related benefit with the net present value of EUR 157 k and a 2.0% share of total benefits. The avoided cost of airborne pollutants follows in importance with a 1.1% share (EUR 85 k). The avoided CO_2 emissions and the enhanced security of supply account for 0.9% of the total. The project implementation would not generate incremental revenues. Benefits from education with a net present value of EUR 7.98 million are considered as a non-EE benefit and are a result of improved quality of health care services. The economic net present value is EUR 1.0 million while the benefit/cost ratio is 1.15. Such economic performance indicators imply that the society would be better off with the project. The energy efficiency benefits are not sufficient to justify the EE investment cost. The inclusion of non-EE costs and benefits in the analysis provides the basis for the comprehensive assessment of the project's feasibility. The assessment of co-benefits or multiple project benefits of various projects, which include the impact on GDP, employment, labor income, rental income and building sales price, is presented in the literature [11,14,16–18]. Since these benefits are not included or applicable to the cases in this paper, the results of the CBA are not comparable to the indicators presented in the literature.

A sensitivity analysis was then performed to identify the effect of the choice of a discount rate on the weighing of costs and benefits. If the social discount rate would increase, the net present value would decrease as the costs come up-front and benefits come later. A relevant element of the sensitivity analysis is a value that a variable (social discount rate) would take in order for the net present value of the project to become zero or negative.

If a rate of 7% is used, the benefit/cost ratio is 0.99% and the net present value is negative (Table 6). Such results would suggest that the project is not economically viable only due to the increase in the social discount rate from 5% to 7%. In the case of the

Emergency Health Center, an additional increase in the discount rate (7% or higher) would further deteriorate the results of the cost-benefit analysis.

Table 6. Results of economic analysis under different social discount rates for the Emergency Health'Center.

Social Discount Rate	5%	7%
Benefit/cost ratio	1.15	0.99
Economic net present value	EUR 1.0 million	EUR −72,013

Source: Authors' calculations.

5.3. Building 3—Kindergarten Complex

The investment cost in various options is from 203 up to 494 EUR/m^2 while the operational cost savings is from 19 up to 26 EUR/m^2 (Table 7). The total energy savings vary from 50% to 80% in different options. Option C achieves the highest savings of operational costs and CO_2 emissions and is the selected (preferred) option. The renovation costs of the building envelope and heating systems in kindergartens in Albania were assessed in the literature [18]. Since the investment objectives are not the same, their findings are not comparable to the results of option analysis.

Table 7. Investment cost and savings for various options for Building 3.

Building 3	Option A	Option B	Option C
Investment [EUR/m^2]	203	321	494
Heat saving [kWh/m^2]	114	190	302
Electricity saving [kWh/m^2]	19	1.76	15
Energy class	C	B	B
CO_2 saving [tCO$_2$/m^2]	0.06	58	92
Operational cost saving [EUR/m^2]	19	21	26
Simple payback period [years]	10	15	13

Source: Authors' calculations.

The project investment cost is estimated to be EUR 2.0 million and includes energy efficiency (EE) components and non-EE components with an 82% and 18% share, respectively (Table 8). The EE components comprise the building envelope, mechanical systems, solar thermal system, lighting system, architecture and mechanical items, while non-EE components include electrical, firefighting, interior works, structural, demolition works and sanitary items.

Table 8. Investment cost of kindergarten complex.

Item	EUR	%
EE components	1,650,000	82%
Non-EE components	370,000	18%
Total investment cost	2,020,000	100%

Source: Authors' calculations.

The project generates significant energy savings with the incremental operation and maintenance costs of EUR −45 k in the first year of operations (Table 9). The savings are attributed to the heating cost reduction. Energy costs are adjusted over the reference period as stated in the methodology section of the paper.

Table 9. Incremental O&M cost in the first year of project operations.

Item	EUR
Electricity cost	1768
Heating cost	−66,481
Maintenance cost	19,652
Total incremental O&M cost	−45,061

Source: Authors' calculations.

The net present value of economic costs is EUR 1.9 million (Table 10). Replacement costs and residual value are set to EUR 0 as the average life time of cost components correspond to the reference period.

Table 10. CBA results for the kindergarten complex.

Item	Net Present Value	%
Investment cost	1,942,308	100.0%
Replacement cost	0	0.0%
Residual value	0	0.0%
Total economic costs	1,942,308	100.0%
Comfort benefit	1,604,852	57.1%
Avoided CO_2 emissions	292,240	10.4%
Avoided cost of airborne pollutants	246,206	8.8%
Enhanced security of supply	7182	0.3%
Avoided cost of child care	544,230	19.4%
Avoided health care costs of influenza	116,529	4.1%
Total economic benefits	2,811,240	100.0%
Economic net present value	868,932	

Source: Authors' calculations.

The with-the-project temperature will be higher by 2 degrees Celsius than the current temperature, resulting in a comfort benefit. This benefit is monetized on the basis of the hypothetical additional energy savings associated with a hypothetical higher energy consumption in the without-project scenario that would have been needed to reach the temperature of the with-project scenario [23]. The comfort benefit is a major energy efficiency and project benefit, with the net present value of EUR 1.6 million and a 57.1% share of total benefits. The avoided CO_2 emissions and cost of airborne pollutants follow in importance, with EUR 292 k and EUR 246 k, respectively. The enhanced security of supply is estimated to EUR 7.2 k. Due to the improved internal air quality and temperature, a conservative decrease in absence rate by 1% is applied [35]. The benefit is quantified on the basis of avoided costs of private child care during illness and the associated health costs of EUR 544 k and EUR 116 k, respectively. The economic net present value is EUR 0.87 million while the benefit/cost ratio is 1.45. The benefits on society justify the opportunity cost of the investment. In this case, EE benefits would be sufficient to justify the EE investment cost. The inclusion of non-EE costs and benefits enables the assessment of the project's impact on society as a whole. The results are not evaluated against the results in previous studies [11,14,16–18] since the structure of the project benefits is not comparable.

A sensitivity analysis was performed to quantify the effect of increasing the social discount rate. A relevant element of the sensitivity analysis is a value that a variable (social discount rate) would take in order for the net present value of the project to become zero or negative.

If a rate of 8% is used, the benefit/cast ratio is 0.97% and the net present value is negative (Table 11). Such results would suggest that the society would not be better off with the project, only due to the increase in the social discount rate. In the case of the kindergarten complex, an additional increase in the discount rate (8% or higher) would further deteriorate the results of the cost-benefit analysis.

Table 11. Results of economic analysis under different social discount rates for the kindergarten complex.

Social Discount Rate	5%	8%
Benefit/cost ratio	1.45	0.97
Economic net present value	EUR 0.87 million	EUR −56,111

Source: Authors' calculations.

6. Conclusions

According to the Energy Efficiency Directive (2012/27/EU and 2018/2002), the annual rate of refurbishment should amount to 3% of the total surface area of buildings occupied by the central government. Decarbonization of the national building stocks by 2050 is required by the Directive on Energy Performance of Buildings (2010/31/EU and 2018/844). More renewable-based heating and cooling in buildings is called by the Directive 2018/2001 for the promotion of the use of energy from renewable sources. The paper presents results of option analyses and the CBA of investment projects in three public buildings with energy efficiency and non-energy efficiency costs and benefits in Serbia and Croatia.

- The results of a case study on a hospital building indicate a negative net present value, implying that the proposed investment is not financially viable. From the owners' point of view, such a result negatively affects the attractiveness of the project. This is in line with previous findings that government support could significantly improve the financial viability of energy efficiency projects [17,19].
- The case studies on an emergency health care center and a kindergarten complex suggest that energy efficiency benefits are not sufficient to justify the energy related costs. This does not mean that the project is not convenient for the society. Such a result requires the calculation of economic return to determine if the society would be better off with the project. Inclusion of not only EE costs and benefits, but also non-EE costs and benefits, provides the basis for the comprehensive assessment of the project's feasibility and its impact for the entire society. The non-energy benefits are project specific and consist of health benefits, education benefits and the avoided costs of child care.
- Following the inclusion of both EE and non-EE costs and benefits in CBA of emergency health care center and kindergarten complex projects, the economic net present value is positive and the benefit/cost ratio is greater than 1 in both cases. Such results suggest that the projects are economically viable. Co-benefits or multiple project benefits of various projects, which include the impact on GDP, employment, labor income, rental income and building sales price, are assessed in the literature [11,14,16–18]. Since such benefits are not applicable to these specific projects, the results of the CBA are not comparable to the results presented in the literature.
- In the case of the emergency health care center, the greatest benefit is associated with education as it would improve the quality of health care services. The O&M cost savings follow in importance due to the expected energy savings. In the case of the kindergarten complex, a major benefit is the comfort benefit as a consequence of a higher indoor temperature. The improved air quality is expected to result in avoided cost of child care and avoided health care costs. The reduction of CO_2 emissions and airborne pollutants would generate significant benefits as well.
- A factor that affects the conclusion of cost-benefit analysis is the choice of a discount rate. A sensitivity analysis was performed to present the CBA results with different social discount rates for the hospital building and the kindergarten complex. If the social discount rate would increase, the net present value would decrease as the costs come up-front and benefits come later. If a rate of 7% or higher is used, the benefit/cost ratio would be less than 1 and the net present value would be negative in both cases. Such results suggest that the conclusions of CBA would change and that projects would not be economically viable only due to the increase in the social discount rate from 5% to 7%.

The next step will be the development of an energy efficiency investment project prioritization tool. The aim of the tool is to identify projects of highest saving potential under a specific budget constraint and/or specific retrofit rate. This tool is valuable for public and any other building stock managing company as it allows the development of implementation plans for energy refurbishment of buildings, provided the building inventory is available. Indicators for investment size, energy and CO_2 saving and investment efficiency are defined for different refurbishment scenarios and then evaluated by multicriteria analysis.

6.1. Contribution

This research establishes a framework for performing option analyses and cost-benefit analyses for investments in specific public buildings (hospital, emergency health care center, kindergarten complex) which include not only energy efficiency but also non-energy efficiency costs by using empirical data. It also empirically examines the feasibility of a range of options in order to assess the technical, economic and environmental convenience of a project. The proposed option analyses and cost-benefit analyses, as well as the methods for calculating energy and non-energy benefits, provide useful cases to conduct CBA of energy efficiency projects with non-energy costs and benefits of similar characteristics.

The results are derived from case studies in Serbia and Croatia under similar conditions including the building type, climate zone and building systems. Option analyses and CBA offer a better insight to investors for their decisions. To promote energy efficiency investments with non-energy costs and benefits, the policy makers will be better informed of the possible policy directions.

6.2. Policy Implications

The following policy implications are drawn from the case studies:

- The proposed investments are not financially viable. Therefore, government policy instruments such as government subsidies need to be in place as they could significantly improve the financial feasibility for the owner of the infrastructure and investors.
- The financial and economic analysis results satisfy the requirements of applying for EU grant financing. Given the EU objective of decarbonizing the national building stocks by 2050, securing EU funds for similar projects would be justified.
- The choice of social discount rate significantly affects the conclusions of the cost-benefit analyses. The discount rate should be set at the national level to reflect the country's social view of how future benefits and costs are to be valued against present ones.
- Establishing building inventory, investigating the saving potential and preparing high quality projects able to reach the savings in real use and the use of renewable energy is the way to achieve the decarbonization of the building stock. The key is the preparation of portfolio of projects, flexibly organized to be easily adapted to a range of priorities and available financing mechanisms.

Author Contributions: Conceptualization, M.Đ. and M.Z.; methodology, M.Đ. and M.Z.; validation, M.Đ. and M.Z.; formal analysis, M.Đ. and M.Z.; investigation, M.Đ. and M.Z.; resources, M.Đ. and M.Z.; data curation, M.Đ. and M.Z.; writing—original draft preparation, M.Đ. and M.Z.; writing— review and editing, M.Đ.; visualization, M.Đ. and M.Z.; supervision, M.Đ.; project administration, M.Đ. and M.Z.; funding acquisition, M.Z. All authors have read and agreed to the published version of the manuscript.

Funding: This research is being funded by "SmartCity.Energy & Environment" project KK.01.2.2.03.0004 Competence centre for Smart Cities, implemented in Croatia.

Institutional Review Board Statement: Not applicable.

Informed Consent Statement: Not applicable.

Conflicts of Interest: The authors declare no conflict of interest.

References

1. Filippidou, F.; Jimenez Navarro, J.P. *Achieving the Cost-Effective Energy Transformation of Europe's Buildings*; EUR 29906 EN; Publications Office of the European Union: Luxembourg, 2019; ISBN 978-92-76-12394-1. [CrossRef]
2. Directive (EU) 2018/2002 of the European Parliament and of the Council of 11 December 2018 amending Directive 2012/27/EU on energy efficiency. 2018. OJ L 328, 21.12.2018. Available online: https://eur-lex.europa.eu/legal-content/EN/TXT/?uri=uriserv%3AOJ.L_.2018.328.01.0210.01.ENG (accessed on 22 May 2021).
3. Directive (EU) 2018/844 of the European Parliament and of the Council of 30 May 2018 amending Directive 2010/31/EU on the energy performance of buildings and Directive 2012/27/EU on energy efficiency 2018. OJ L 156, 19.6.2018. Available online: https://eur-lex.europa.eu/legal-content/EN/TXT/PDF/?uri=CELEX:32018L0844&from=IT (accessed on 22 May 2021).
4. Directive (EU) 2018/2001 of the European Parliament and of the Council of 11 December 2018 on the promotion of the use of energy from renewable sources 2018. OJ L 328, 21.12.2018. Available online: https://eur-lex.europa.eu/legal-content/EN/TXT/PDF/?uri=CELEX:32018L2001&from=fr (accessed on 22 May 2021).
5. European Parliament. *Report on Maximising the Energy Efficiency Potential of the EU Building Stock*; European Parliament: Brussels, Belgium, 2020. Available online: https://www.europarl.europa.eu/doceo/document/A-9-2020-0134_EN.pdf (accessed on 20 March 2021).
6. Ministry of Construction, Physical Planning and State Assets of Croatia. *Use of European Structural and Investment Funds, Operative Program Cohesion and Competitiveness*; Ministry of Construction, Physical Planning and State Assets of Croatia: Zagreb, Croatia, 2014.
7. Ministry of Construction, Physical Planning and State Assets of Croatia. *Energy Refurbishment Program od Multiresidential Buildings*; Ministry of Construction, Physical Planning and State Assets of Croatia: Zagreb, Croatia. Available online: https://mgipu.gov.hr/o-ministarstvu-15/djelokrug/energetska-ucinkovitost-u-zgradarstvu/energetska-obnova-zgrada-8321/energetska-obnova-visestambenih-zgrada-8323/8323 (accessed on 14 February 2021).
8. Ministry of Construction, Physical Planning and State Assets of Croatia. Long-Term Strategy of Energy Refurbishment of National Building Stock Until 2050, 2020, Official Gazette of the Republic og Croatia No 140/2020. Available online: https://mgipu.gov.hr/UserDocsImages//dokumenti/EnergetskaUcinkovitost//Dugorocna-strategija-2050-sazetak_EN.docx (accessed on 14 February 2021).
9. Ryan, L.; Campbell, N. *Spreading the Net: The Multiple Benefits of Energy Efficiency Improvements*; IEA Energy Papers; OECD Publishing: Paris, France, 2012. [CrossRef]
10. Saldivia, M.; Kristjanpoller, W.; Olson, J.E. Energy consumption and GDP revisited: A new panel data approach with wavelet decomposition. *Appl. Energy* **2020**, *272*, 115207. [CrossRef]
11. Sooriyaarachchi, T.M.; Tsai, I.T.; El Khatib, S.; Farid, A.M.; Mezher, T. Job creation potentials and skill requirements in, PV, CSP, wind, water-to-energy and energy efficiency value chains. *Renew. Sustain. Energy Rev.* **2015**, *52*, 653–668. [CrossRef]
12. Dell'Anna, F. Green jobs and energy efficiency as strategies for economic growth and the reduction of environmental impacts. *Energy Policy* **2021**, *149*, 112031. [CrossRef]
13. Kerr, N.; Gouldson, A.; Barrett, J. The rationale for energy efficiency policy: Assessing the recognition of the multiple benefits of energy efficiency retrofit policy. *Energy Policy* **2017**, *106*, 212–221. [CrossRef]
14. Bleyl, J.W.; Bareit, M.; Casas, M.A.; Chatterjee, S.; Coolen, J.; Hulshoff, A.; Lohse, R.; Mitchell, S.; Robertson, M.; Ürge-Vorsatz, D. Office building deep energy retrofit: Life cycle cost benefit analyses using cash flow analysis and multiple benefits on project level. *Energy Effic.* **2019**, *12*, 261–279. [CrossRef]
15. Galatioto, A.; Ricciu, R.; Salem, T.; Kinab, E. Energy and economic analysis on retrofit actions for Italian public historic buildings. *Energy* **2019**, *176*, 58–66. [CrossRef]
16. Liu, Y.; Liu, T.; Ye, S.; Liu, Y. Cost-benefit analysis for Energy Efficiency Retrofit of existing buildings: A case study in China. *J. Clean. Prod.* **2018**, *177*, 493–506. [CrossRef]
17. Mikulić, D.; Bakarić, I.R.; Slijepčević, S. The economic impact of energy saving retrofits of residential and public buildings in Croatia. *Energy Policy* **2016**, *96*, 630–644. [CrossRef]
18. Novikova, A.; Szalay, Z.; Horváth, M.; Becker, J.; Simaku, G.; Csoknyai, T. Assessment of energy-saving potential, associated costs and co-benefits of public buildings in Albania. *Energy Effic.* **2020**, *13*, 1387–1407. [CrossRef]
19. Copenhagen Economics. *Multiple Benefits of Investing in Energy Efficient Renovation of Buildings*; CE: Copenhagen, Denmark, 2012.
20. Filippidou, F.; Nieboer, N.; Visscher, H. Energy efficiency measures implemented in the Dutch non-profit housing sector. *Energy Build.* **2016**, *132*, 107–116. [CrossRef]
21. Filippidou, F.; Nieboer, N.; Visscher, H. Are we moving fast enough? The energy renovation rate of the Dutch non-profit housing using the national energy labelling database. *Energy Policy* **2017**, *109*, 488–498. [CrossRef]
22. Rosenow, J.; Platt, R.; Demurtas, A. Fiscal impacts of energy efficiency programmes—The example of solid wall insulation investment in the UK. *Energy Policy* **2014**, *74*, 610–620. [CrossRef]
23. Sartori, D.; Catalano, G.; Genco, M.; Pancotti, C.; Sirtori, E.; Vignetti, S.; Del Bo, C. *Guide to Cost-Benefit Analysis of Investment Projects. Economic Appraisal Tool for Cohesion Policy 2014–2020*; European Commission: Brussels, Belgium, 2014.
24. Thollander, P.; Karlsson, M.; Rohdin, P.; Wollin, J. *Introduction to Industrial Energy Efficiency: Energy Auditing, Energy Management, and Policy Issues*; Academic press: Cambridge, MA, USA, 2020.

25. Korytárová, K.; Knapko, I.; Šoltésová, K. Energy savings potential for space heating in public buildings in Slovakia. In Proceedings of the European Council for an Energy Efficient Economy Summer Study 2017, Hyères, France, 29 May–3 June 2017; pp. 1393–1399.

26. Sandberg, N.H.; Sartori, I.; Heidrich, O.; Dawson, R.; Dascalaki, E.; Dimitriou, S.; Vimm-r, T.; Filippidou, F.; Stegnar, G.; Šijanec Zavrl, M.; et al. Dynamic building stock modelling: Application to 11 European countries to support the energy efficiency and retrofit ambitions of the EU. *Energy Build.* **2016**, *132*, 26–38. [CrossRef]

27. Pearce, D.W. *Cost-Benefit Analysis*; Macmillan International Higher Education: London, UK, 2016.

28. Stiglitz, J.E.; Rosengard, J.K. *Economics of the Public Sector: Fourth International Student Edition*; WW Norton & Company: New York, NY, USA, 2015.

29. Sáez, C.A.; Requena, J.C. Reconciling sustainability and discounting in Cost–Benefit Analysis: A methodological proposal. *Ecol. Econ.* **2007**, *60*, 712–725. [CrossRef]

30. László, S.; Mezősi, A.; Pató, Z. South East Europe Electricitiy Roadmap: Country Report: Serbia; SEERMAP—South East European Roadmap Project ISBN 978-615-80813-5-1. 2017. Available online: https://rekk.hu/analysis-details/238/south-east-europe-electricity-roadmap---seermap (accessed on 15 February 2021).

31. *Ten Year Network Development Plan*; ENTSO: Brussels, Belgium, 2018.

32. *Climate Change and Major Projects*; European Commission, Directorate-General for Climate Action: Brussels, Belgium, 2016.

33. NEEDS (New Energy Externalities Developments for Sustainability). External Costs from Emerging Electricity Generation Technologies, Project co-funded by the European Commission. 2009. Available online: https://cordis.europa.eu/project/id/502687 (accessed on 14 February 2021).

34. Viscusi, W.K.; Masterman, C. Income Elasticities and Global Values of a Statistical Life. *J. Benefit-Cost Anal.* **2017**, *8*, 1–25. [CrossRef]

35. American Society of Heating, Refrigerating and Air-Conditioning Engineers. ASHRAE Handbook—Fundamentals. 2017. Available online: https://www.ashrae.org/technical-resources/ashrae-handbook (accessed on 15 March 2021).

Article

Benchmarking of Water, Energy, and Carbon Flows in Academic Buildings: A Fuzzy Clustering Approach

Abdulaziz Alghamdi [1], Guangji Hu [1], Husnain Haider [2], Kasun Hewage [1] and Rehan Sadiq [1,*]

[1] School of Engineering, University of British Columbia (Okanagan), 3333 University Way, Kelowna, BC V1V 1V7, Canada; aalghamd@mail.ubc.ca (A.A.); guangji.hu@ubc.ca (G.H.); kasun.hewage@ubc.ca (K.H.)

[2] Department of Civil Engineering, College of Engineering, Qassim University, Buraydah, Qassim 51452, Saudi Arabia; husnain@qec.edu.sa

* Correspondence: rehan.sadiq@ubc.ca

Received: 23 April 2020; Accepted: 21 May 2020; Published: 29 May 2020

Abstract: In Canada, higher educational institutions (HEIs) are responsible for a significant portion of energy consumption and anthropogenic greenhouse gas (GHG) emissions. Improving the environmental performance of HEIs is an important step to achieve nationwide impact reduction. Academic buildings are among the largest infrastructure units in HEIs. Therefore, it is crucial to improve the environmental performance of academic buildings during their operations. Identifying critical academic buildings posing high impacts calls for methodologies that can holistically assess the environmental performance of buildings with respect to water and energy consumption, and GHG emission. This study proposes a fuzzy clustering approach to classify academic buildings in an HEI and benchmark their environmental performance in terms of water, energy, and carbon flows. To account for the fuzzy uncertainties in partitioning, the fuzzy c-means algorithm is employed to classify the buildings based on water, energy, and carbon flow indicators. The application of the developed methodology is demonstrated by a case study of 71 academic buildings in the University of British Columbia, Canada. The assessed buildings are grouped into three clusters representing different levels of performances with different degrees of membership. The environmental performance of each cluster is then benchmarked. Based on the results, the environmental performances of academic buildings are holistically determined, and the building clusters associated with low environmental performances are identified for potential improvements. The subsequent benchmark will allow HEIs to compare the impacts of academic building operations and set realistic targets for impact reduction.

Keywords: higher educational institutions; academic buildings; environmental performance; performance benchmarking; fuzzy clustering analysis; metabolic flows

---

1. Introduction

The United Nations Paris Agreement set ambitious goals for 191 countries to reduce the anthropogenic greenhouse gases (GHG) emissions linked to climate change, in an aim to curb the global temperature increase by 1.5 °C above the pre-industrial levels [1]. Canada is among the nations that ratified the agreement and committed to reducing their emissions, by 2030, to a level that is 30% lower than level reported in 2005 [2]. In Canada, the building sector is the third largest GHG emitting source, and is responsible for 12% of the total emission [2]. Moreover, the building sector consumes 20–40% of the total produced energy in developed countries; for instance, in the US and EU this sector is the third largest in terms of energy usage [3]. Buildings are believed to be responsible for more than one-third of the GHG emissions and 32% of the total energy consumption [4]. A building emits GHG during different phases, but the largest portion of the GHG emissions is generally associated with the operational phase, i.e., about 80–94% of the entire lifecycle emissions of a building [5–7].

The educational sector stays among the largest public sectors of many countries around the world. For instance, being the largest public sector in China, it consumes 40% of the total energy supplied to the public sector [8]. Many large universities operate on a scale similar to a small city [9–13]. In Canada, higher educational institutions (HEIs) generally use 60% of the electricity allocated to the educational sector, equivalent to that consumed by a small city of 430,000 households [14]. In British Columbia (BC), Canada, the educational sector alone emitted 309,222 ton of CO_2e as of 2018, accounting for 41.25% of the total public sector emissions in the province; specifically, HEIs in the province are responsible for 19% of the entire emissions from the public sector alone [15]. Moreover, in the United States, educational buildings use approximately 6% of the entire public water usage in the country [16]. HEIs are reported to have a larger impact than any other organizations or institutions in the educational sector. It is reported that HEIs use 3–4.9 more times of energy than schools [17]. In China, the energy and water consumed by university/college students are four and two times higher than the average consumptions [18]. In Norway, the emission per student at a university is significantly higher than the national average per citizen [19]. Many studies highlighted the significance and challenges of HEIs buildings in the overall environmental impact reduction in different countries, such as Australia [20], Spain [21], China [8,22], the UK [23], Norway [19], Saudi Arabia [24], and Canada [25]. These studies also pointed out challenges that universities are facing to achieve their goals, e.g., 43% of the HEIs in Canada fell behind their energy baseline targets [26].

The National Science Foundation defines an engineering system as "a combination of components that work in synergy to collectively perform a useful function." [27] Infrastructures in HEIs like academic buildings can be considered as engineering systems because they require many different components, such as energy, water, lighting, and air circulation systems, in order to function properly. The operation of academic buildings is associated with significant amounts of water, energy, and carbon (WEC) flows. The current sustainability assessment tools aggregate the indicators of an HEI 's performance regarding the social, economic, and environmental sustainability to a final score; however, this aggregated score may not be very useful to the improvement of the HEI's environmental performance from an infrastructure management perspective. Thus, an assessment tool focusing on WEC flows is essential for enhancing the environmental performance of infrastructure in HEIs. To this end, this study aims to develop a methodology for environmental performance benchmarking of academic buildings through the lens of infrastructure management. Thus, the environmental performance in this paper is evaluated by assessing WEC flows in academic buildings [25].

Significant WEC flows in HEIs make them a pivotal point of attraction for countries to meet the international emission reduction commitments and sustainable development targets [28]. Figure 1 shows a conceptual road map to improve the sustainability performance of HEIs. Attesting to declarations is the first step that HEIs acted upon to deal with the plethora of attempts to define, raise awareness, and communicate sustainable issues on campuses. These non-statutory declarations cover a wide range of topics in sustainability (pedagogical and operational), and they impact the sustainability in HEIs in three distinctive ways: (i) They help shape an instrumental argument of the surrounding role of a university in relation to sustainable development [29,30]; (ii) these declarations help formulate national legislations around resource utilization and highlight goals towards reducing the adverse impacts of HEIs; (iii) they pave the road towards the development of tools that help rank, assess, and communicate the progress of sustainability in HEIs [25,31]. As of 2011, there were 31 declarations in the context of education, and 1400 universities have signed them [31]. However, the number of universities that have signed these declarations is small comparing to the total number of HEIs worldwide. Furthermore, the declarations primarily focused on raising the awareness of sustainability in HEIs but did not provide any mechanism to assess sustainability performance [32].

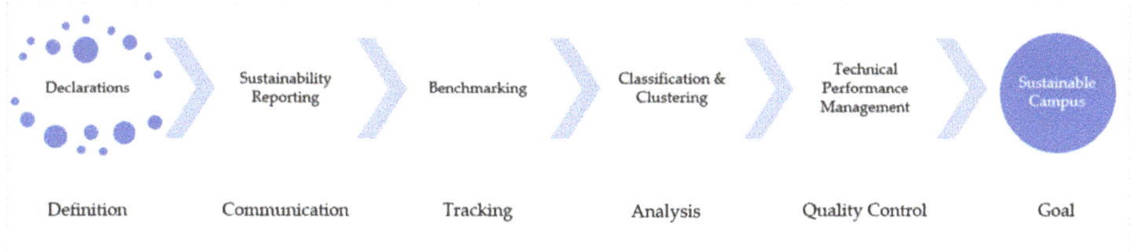

Figure 1. Roadmap to a sustainable campus [28].

HEIs began reporting their sustainability performances through reporting systems in early 2006. The first sector-specific reporting system was the Graphical Assessment of Sustainability in Universities (GASU) [33]. Many reporting systems were created afterwards, and the most used one is the sustainability tracking, assessment, and rating system (STARS) [34]. STARS evaluates an HEI's performance in five categories: engagement, planning and administration, academics, innovation, and operations. Moreover, within those categories are subcategories and criteria of measurement. STARS assesses the HEI in terms of its performance in 247 areas and then provides with one of the overall performance ranks: platinum, gold, silver, or bronze [34]. STARS was firstly used to assess the HEIs in North America, and later its application was promulgated across the world. The system launched its first reporting version in 2010 with 149 HEIs and, to date, there are nearly 1000 registered institutions. Out of the registered HEIs, over 600 reports from 40 different countries. The number of reporting HEIs is still relatively small in comparison to the overall number of HEIs [35]. This can be due to certain challenges faced by the HEIs such, e.g., complexity of the sustainability assessment methods and the limitations in resources to complete the assessment within the allocated timeframe [34]. Another limitation of these reporting systems is related to the weighting structure used [35]. As these systems cover a wide array of areas, the direct impact on climate change (i.e., GHG emissions) can be underestimated by assigning higher weights to other socio-economic parameters [36]. One of such an example is the case of the University of Alberta: The university reported an increase of nearly 34% in emissions and at the same time received a Gold ranking [37].

HEIs face several challenges in reporting their sustainability performance, such as lacks of (i) interpretation of sustainability specifically in the area of climate change [38], (ii) guides or mechanisms to provide systematic roadmaps to a sustainable campus [39], and (iii) baseline values to create a cross-institutional performance comparison [40]. To overcome these limitations, Martin and Samels [41] proposed benchmarks as a means to establish a mechanism for disseminating key information, establishing best practices, and set baseline values for the industry [41]. The current methods consider a singularity approach in benchmarking buildings, i.e., building type. Such methods may come up with misleading outcomes because of their inability to consider opposing or multiple features of a building [42].

Benchmarking is a widely used tool to compare the performance of a building or a set of buildings to those of a larger pool of similar buildings under similar pressures (e.g., GHG emissions). There are two approaches of benchmarking: The top-down approach and the bottom-up approach, and the selection of a suitable approach depends on the purpose of assessment, type of data, and the level of information available. The top-down approach is suitable for evaluating the overall building performance, such as the total energy usage intensity (EUI) [43], while the bottom-up approach builds on the aggregated values of each inner component of the building at each zone, e.g., the summation of the total heating, ventilation and air condition (HVAC), and lighting [44]. Benchmarking consists of three stages: Planning (to define the objectives and scope), analysis (to identify performance gaps), and finally an integration step to continuously and systematically implement the findings [45]. There are several methods used to benchmark buildings performance depending on the data available and the degree of benchmarking to be completed: white-box, black-box, and grey-box. The first refers to data generated from simulations, the second is referred to when statistical approaches are used, and the final is a combination of the first two [46]. A number of studies conducted on educational buildings can be found in the recent literature [20,21,47,48].

Benchmarking for buildings should be performed in buildings with similar functions and characteristics. For instance, a residential building should not be compared with a commercial building due to the differences in internal factors (e.g., demand, scope, operational hours) and external factors (e.g., climatic conditions). Therefore, data need to be collected in ways that meet the definition, scope, and strategy of the benchmarking process. However, due to the large set of data collected, issues of misleading information may arise. For example, whether or not to include water used for irrigation as part of the total water usage could lead to a significant difference in water usage benchmarking of buildings because this part of water usage is heavily influenced by climate factors, area of the landscape, and the vegetation species. To minimize the uncertainty caused by dissimilar data, use of multi-dimensional features instead of one parameter/indicator is encouraged in performance benchmarking [42].

With a continuous and rigorous data collection, the need to compare the performances of buildings based on similarities in function, size, and climatic conditions becomes prominent. The act of measuring the performances of a group (i.e., cluster) of buildings, sharing similar features and characteristics, and compare those to other building groups has emerged as a new benchmarking approach [42,49,50]. Many studies used classification methods as a means to understand energy consumption patterns in HEIs. For example, Khoshbakht et al. [20] classified 80 HEI buildings in Griffith University, Australia into six classes—office, administration, library, research, teaching, and mixed buildings—based on the major activities that are carried out within those buildings. For instance, if 40% of the area in a building is allocated to laboratories, then the building will be classified into the research type. In another study, Chihib et al. [21] classified 33 buildings of the University of Almeria into six classes, i.e., research, administration, teaching, library, sports facilities, and restaurants, and compared their performances over a time span of 8 years using independent climate variables and other dependent variables like occupancy [21]. Both studies found that buildings classified into the research type (i.e., laboratory-intensive) use a higher ratio of energy than other building types. Tan et al. [22] analyzed Tongji University in China and broke down the energy consumption for student dorms, research buildings, classrooms, office, libraries, and others, and the results showed that dorms account for 29% of the total energy consumption [8,22].

Studies also highlighted that the research buildings equipped with many laboratories consume significantly higher amounts of energy than other research buildings [20,21]. This agrees with the findings reported by Mills et al. [51] that laboratorial buildings are 4–5 times more energy-intensive than commercial and institutional (non-laboratory) buildings. Another study by Federspiel et al. [52] reported similar findings. One reason for the high energy consumption could be that the air-exchange rate uses more energy in laboratory-intensive buildings than that traditional buildings [52]. Furthermore, natural science and engineering buildings are equipped with more laboratories than buildings designed for economic, law, and art sciences, and thus, they are associated with higher energy consumptions [53]. Federspiel et al. [52] applied a model-based benchmarking methodology on an academic building at the University of California, Berkeley campus and calculated the total building energy consumption based on the minimum amount of energy required to fulfill a set of functions in compliance with code-compliant environmental controls, then used the calculated energy consumption and compared it with the actual reported data to assess the efficiencies of the cooling equipment and identified the inefficient mechanical cooling designs. The results can help identify potentials for reducing energy consumption when devising a laboratory-intensive research building.

Finally, benchmarking is a technical performance tool used in liaison with a broader management strategy to help the leading organizations to improve their performance through identification of best in class, communication of performances, and to improve resource utilization systematically and dynamically [38,54]. However, benchmarking alone does not propose a set of solutions for an organization—benchmarking is a means to an end, not the final destination [55–58]. A number of studies focused on different approaches to underpin the cause and effect of inefficient energy use in universities. Some of the studies used classification approaches to determine the characteristics of buildings in terms of their intended uses (e.g., library, office, and laboratory) [20–22]. Others attempted to pinpoint the behavioral aspect by using stochastic approaches to define the influence of occupancy [48,57,58], and finally, a macro study analyzed the sector performance in terms of their ability to achieve their commitments to reduce GHG emissions [25].

To establish a benchmark for academic buildings, several issues must be considered. Firstly, the challenge of appropriate classification of buildings: Since only the performance of similar buildings can be compared, and there is no publicly accessible database to assist the classification, a methodology that uses unsupervised learning to derive the hidden patterns of building performance data is needed. Secondly, conventional crisp clustering approaches, such as the *k*-means clustering draw hard boundaries between the classified groups, which may bring uncertainties to the clustering results. For example, two buildings with similar performance may be classified into two different groups because of the hard boundary created by the crisp clustering. Fuzzy logic can address this issue by introducing the concept of "partial truth". Thirdly, several studies have used fuzzy logic to help benchmark a building's performance; however, the performance was benchmarked based on single-dimensional data (e.g., energy consumption or carbon emission) [42,59]. Limited attention has been placed to performance benchmarking by considering a multitude of a building's characteristics (e.g., WEC flows). To address these issues, this study proposes an unsupervised fuzzy clustering analysis to reduce the uncertainty of the building classification results generated based on multi-dimensional data.

Fuzzy clustering analysis is used in the literature to limit the uncertainties that may arise from a large number of data and parameters used. Many studies applied fuzzy approaches to performance benchmarking of different systems; for example, Chung [60] applied fuzzy linear regression analysis to develop a benchmarking method for commercial buildings, Iliadis et al. [61] applied fuzzy c-means algorithms to determine the risk factors in a Greek forest, Krajnc et al. [62] applied fuzzy logic to compare performances between two plants, Santamouris [59] applied fuzzy clustering techniques to 320 schools in Greece to assess energy and environmental performance in school buildings. Kouloumpis and Azapagic [63] used fuzzy evaluation for life cycle-integrated sustainability assessment as a tool to evaluate five different sources of energy and identified the most sustainable sources of energy to help decision and policy makers. Haider et al. [64] used a fuzzy synthetic evaluation technique to develop a sustainability index for small-sized urban neighborhoods. However, limited studies have used fuzzy clustering analysis to classify academic buildings based on their WEC flows.

The objectives of this paper are to provide a review of the steps and studies taken historically to define, attain, and measure environmental performance in HEIs; to propose fuzzy clustering analysis-based framework for HEIs to benchmark the performance of academic buildings by holistically considering energy and water consumption and carbon emission. The developed framework is applied to a university in Canada, and based on the benchmarking results, potentials for environmental performance improvement in the university are recommended. The developed framework can aid decision-makers in setting, and achieving, environmental goals and targets in the context of HEIs.

2. Methodology

2.1. General Framework

The proposed fuzzy clustering-based framework is outlined in Figure 2. The framework provides a local classification of different buildings within a university according to their similarities in resource consumption and GHG emissions, and then holistically benchmarks the environmental performance of individual buildings, among others within the same class. The first step is to select the performance indicators needed to be benchmarked. In the case of this study, energy and water consumption data were collected, and subsequently, the carbon emission was derived from the energy consumption data collected. These parameters relevant to water, energy and carbon are referred to as metabolic flows (MF). The MF are used as performance indicators because they are the most important aspects influencing the environmental impacts within the context of HEIs. The second step is to normalize MF data using a common factor, such as climatic factors, area, and a number of users. After normalization, MF data can be converted to several performance indices, such as water usage intensity (WUI), energy usage intensity (EUI), and carbon usage intensity (CUI).

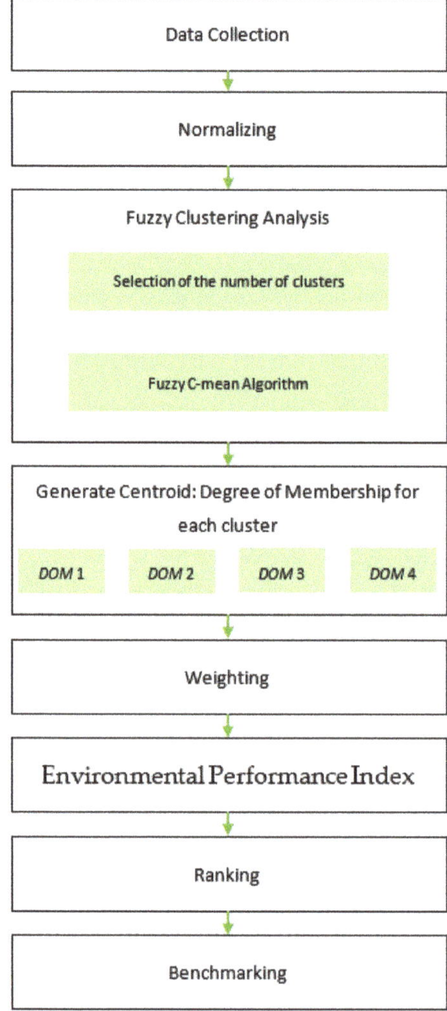

Figure 2. The framework of fuzzy clustering-based environment performance benchmarking of academic buildings.

The derived indices are used as inputs to the fuzzy clustering analysis using the fuzzy clustering algorithm. The optimal number of clusters is determined by using the elbow method. This method is used to evaluate the reduction of within-group variance that is brought by increasing the number of clusters. The reduction of within-group variation as a function of the increase of cluster number is graphically represented as an "elbow", where the "elbow point" is often selected as the optimal number of clusters [42,65]. After the fuzzy clustering analysis, each building assessed will be assigned a degree of membership (DOM) to the clusters formed. The general environmental performance of each cluster, in terms of WUI, EUI, and CUI, will be evaluated and ranked. A weighting method may be needed if the performance indices do not follow the same direction. Finally, an environmental performance index is generated for each building based on its DOMs to different clusters using a weighted aggregation method. The environmental performance of individual buildings can be ranked according to their environmental performance indices.

In this study, the framework is applied at the building level (top-down benchmarking) based on the available data and the type of study conducted [66]. However, the framework can be generalized further to include a bottom-up approach of benchmarking by aggregating data of building components (e.g., HVAC components and lighting type). By including this sort of data for each level of the building, a better understanding of these characteristics will be gained, and better information will be provided for the management to act upon (i.e., by determining inefficient components). The bottom-up approach is shown in Figure 3. As more data becomes available (through simulations and or measurements), better information can be gained which will help improve the quality of the data and subsequently improve the assessment results by pinpointing inefficiencies in a building (e.g., inefficient cooling method).

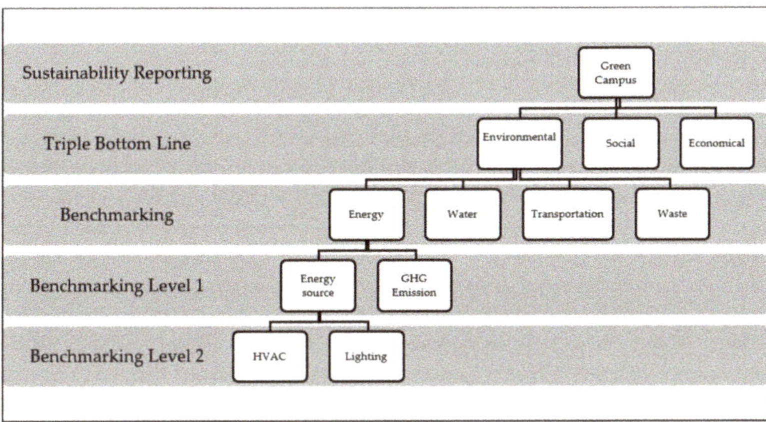

Figure 3. Scalability of the framework.

2.2. Case Study

The University of British Columbia (UBC) is investigated as a case study. UBC is one of the largest and highly ranked universities in Canada, and is home to 64,798 students and a total of 16,891 faculty and staff. UBC consists of two main campuses: The Vancouver campus (UBCV) and the Okanagan campus (UBCO) in two different yet close geographical regions. The first is in Vancouver where is a moderate oceanic climate, while the other campus in Kelowna is roughly 280 km inland towards the east. Kelowna is characterized as a subcontinental hemiboreal climate that is generally drier than Vancouver with warmer summers and colder winters. The annual precipitation in Vancouver is about 3.25 times more than Kelowna on average [67]. There are over 160 buildings in both the campuses, including classroom, laboratory, library, administration, residential, and recreational buildings. Due to the lack of continuous data recording for more than half of the buildings, only 71 buildings are included in the case study.

The university uses energy from several sources to operate buildings, such as electricity supplied by the service providers, natural gas, and biomass-derived energy. Close to 95% of BC's electricity is generated from renewable sources, with hydro being the most dominant and important source [68]. There are other small energy sources, such as geothermal and biodiesel, and limited quantities of solar energy panels on the residential buildings. UBCV reports energy by source for general usage as electricity and for the source of energy used for heating as well, in addition to water usage in the Skyspark portal [66]. The portal provides the energy and water consumption data for most of the buildings, in days, week, months and years. In addition to the types of buildings (e.g., percentage of classrooms, laboratories, libraries, and offices), and the areas of buildings are also recorded in the Skyspark portal. The distributions of heating energy by sources used in this study are presented in Table 1.

Table 1. Distribution of energy source for building heating in University of British Columbia Vancouver campus (UBCV).

Energy Source	Percent
Renewable Gas	10%
Biomass	25%
Natural Gas	65%

The present study includes 11 buildings in UBCO. The sustainability office on campus provided the information of source wise energy distribution. Electricity from the main grid and natural gas are the two primary energy sources at UBCO. The geothermal plant consumes natural gas to increase the efficiencies of heating and cooling.

To derive the carbon emissions for each building, it is important to know the energy by source for calculating the carbon equivalent of each building. Energy supplying facilities, such as the central heating plants, for both campuses were not included in this study, due to their energy exporting nature.

Based on the available data, a total of 71 buildings at both campuses were evaluated. MF of water and energy consumption data from 1st April 2018 to 31st March 2019 were obtained for the benchmarking. Carbon equivalents were calculated using the method recommended in the BC Best Practices Methodology for Quantifying GHG Emissions [69]. The sources of emission factors are calculated from the relevant literature from the utility companies on the emission factors for the energy source supplied [70]. Table 2 summarizes the emission factors for different energy sources to estimate carbon emission. Table 3 shows the reported energy and water consumptions, and the calculated GHG emissions for Alumni Centre in UBCV as an example. By using the BC Best Practices Methodology for Quantifying GHG Emissions, the CO_2e emissions were found to be 38.4 kg for Alumni Centre for both heating energy and electricity used.

Table 2. Carbon emission factors by energy source. UBCO, University of British Columbia Okanagan campus.

Source	kg CO_2e/GJ	UBCV kg CO_2e/kWh	UBCO kg CO_2e/kWh
Renewable Gas	0.29	0.0010	0.0010
Biomass	0	0.0000176	0.0000176
Natural Gas	49.87	0.1795	0.1795
Electricity	0	0.0107	0.0026

Table 3. Metabolic flow data of Alumni Centre, UBCV.

Source	Value
Electricity (kWh)	713,572
Heating (kWh)	229,117
Water (m^3)	3968
Natural gas (m^3)	2025
GHG (kg, CO_2e)	38

The overall goal is to reduce the MFs in the buildings without compromising the operational efficiencies. Part of the MFs is related to water usage; the usage of water, throughout a building's lifecycle, uses energy and emits GHGs. The upstream emissions associated with energy or water (production, transportation, and recycling) are neglected in this study because they do not fall within a university's control or sphere of influence. The WEC data of all investigated buildings are summarized in Appendix A. Normalization of the WEC data was carried out by using the building floor area as the normalizing factor.

2.3. Fuzzy Clustering Analysis

The fuzzy clustering analysis conducted in this paper is used in Hu et al. [71]. Before the application of fuzzy clustering analysis, the MF data were normalized based on the building area. The resultant WUI, EUI, and CUI have values on a scale ranging from 0 to 1. The c-means fuzzy clustering algorithm (FCA) was used to group the index values based on their performance on the three factors. One of the FCA characteristics is that it offers each data point in a dataset a DOM to every cluster formed, indicating that each data point belongs to different clusters with a different level of association. The DOM is a unique feature that distinguishes FCA from other crisp clustering algorithms, such as the k-means clustering and hierarchical clustering. This feature offers FCA great flexibility in benchmarking of buildings because it can address fuzzy uncertainties (i.e., the concept of partial truth) and is suitable for grouping data points with weakly defined boundaries [61]. A widely accepted fuzzy clustering algorithm is fuzzy c-means. For a dataset $x = (x_1, x_2, \ldots, x_n)$ comprising n data points, the fuzzy c-means algorithm classifies the data points into predefined p pclusters based on measured similarities among the data points. Each cluster has a center $e_j (j \in [1, p])$, and the Euclidean distance d_{ij} between a data point x_i and e_j can be calculated as:

$$d_{ij} = \|x_i - e_j\| \tag{1}$$

In this study, each additive was considered as a data point defined by three-dimensional values (a, b, c) the metabolic flows (i.e., WUI, EUI, and CUI). Thus, the Euclidean distance between data point $x_i(a_i, b_i, c_i)$ and $e_j(a_j, b_j, c_j)$ in a three-dimensional space was calculated as:

$$d_{ij} = \sqrt{(a_i - a_j)^2 + (b_i - b_j)^2 + (c_i - c_j)^2} \tag{2}$$

At the beginning of fuzzy c-means, random centers (usually with a value of zero) are selected for the clusters. Based on the derived d_{ij}, a DOM ($\mu(x_i)$) can be calculated as a measure of the similarity between a data point x_i and the jth cluster:

$$\mu_j(x_i) = \frac{(1/d_{ij})^{2/(m-1)}}{\sum_{k=1}^{p} (1/d_{ik})^{2/(m-1)}} \tag{3}$$

where m is a fuzzification parameter to determine the degree of fuzziness between different clusters. A higher value of m will lead to higher fuzziness between clusters. Commonly m takes values between 1.25 and 2 [72,73]. In this study, m value was set at 2. The parameter d_{ik} is the Euclidean distance between $x_i x_i$ and the center of the kth cluster. The new centers of clusters can be calculated as:

$$e_j = \frac{\sum_i [\mu_j(x_i)]^m x_i}{\sum_i [\mu_j(x_i)]^m} \tag{4}$$

Based on the new e_j, $\mu_j(x_i)$ will be updated. The iteration will continue until the minimum objective function J is achieved:

$$J = \sum_{i=1}^{n} \sum_{j=1}^{p} [\mu_j(x_i)]^m d_{ij}^2; p \le n \tag{5}$$

A total of three clusters were formed based on the available factors. The FCA process was carried out using the statistical computing software R ^{TM} (version 1.0.136). Based on the value ranges of the three factors, the benchmark characteristics of each cluster can be interpreted. After the FCA, three DOMs (μ_1, μ_2, μ_3) can be generated for each value to show the degrees of similarity between the benchmark characteristics and the number of clusters. An environmental performance index (EPI) can be calculated for each value based on the DOMs using Equation (6):

$$EPI = \sum_{j=1}^{p} \mu_j \, w_j \tag{6}$$

where w_j is the specific quality-ordered weight of cluster j, and p is the total number of clusters. The values of w_j were determined by the characteristics of different clusters. For example, w_j values (i.e., the DOMs) can be assigned to the number of clusters ordered from the lowest to the highest benchmark, respectively. The values of w_j were assigned subjectively, and they can be modified to accommodate different levels of the benchmark [74].

To determine the useful number of clusters, an elbow analysis is conducted to calculate within-cluster sum of squares or validity index (VI) as a function of the number of clusters. VI is calculated using the equation adapted from [42]:

$$VI = \left[\frac{\sum_{i=1,\cdots,k} \sum_{j=1,\cdots,d} \sum_{q=1}^{n_{ij}} (x_q - \bar{x}_j)^2}{\sum_{i=1,\cdots,k} \sum_{j=1,\cdots,d} (n_{ij} - 1)} \right]^{1/2} \tag{7}$$

where, $\bar{x}_j$ is the mean of data values of j dimension and n_{ij} is the number of data values of j dimension that belongs to cluster i.

3. Results

3.1. Benchmarking

Boxplots of the normalized MF are shown in Figure 4. The mean values are also identified within the boxplots. All variables are not normally distributed, where CUI is more skewed than WUI and EUI. Thus, assuming the mean to be the reference point (benchmark), misleading information could be generated about the distribution of the data.

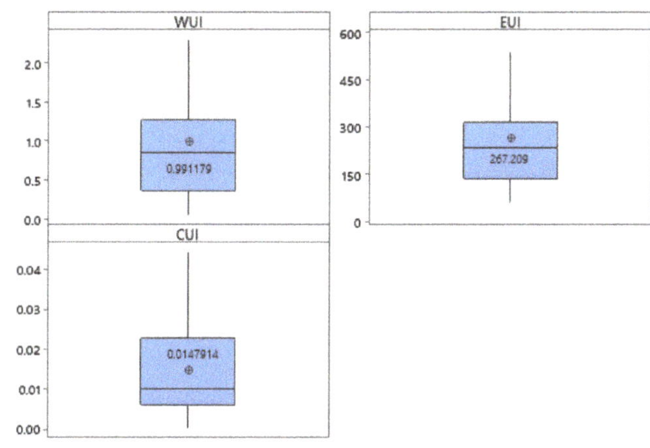

Figure 4. Boxplots of the normalized metabolic flows (MF).

An alternative approach to use the mean as a benchmark for each normalized flow, a cumulative distribution function (CDF) was applied, and the 50th and 75th percentile ranges were used as the benchmarks. This will allow the nature of the distribution in the data to be better reflected and subsequently, give more meaning to the benchmarks. Table 4 shows the suggested benchmarks, the 50th and 75th percentiles, and the number of buildings above and below the benchmark levels. Buildings above the 75th percentiles are considered the buildings with low environmental performance, and buildings below the 50th percentile are considered as the high-environmental performance buildings. A graphical presentation for each normalized MF is presented in Figure 5a–c. By assuming the 75th percentile as a benchmark, all points (buildings) above would be identified with low environmental performance and require management attention.

Table 4. Cumulative distribution of functional benchmarking.

Parameter	Percentile	Value	No. of Buildings Above	No. of Buildings Below
WUI	50th Percentile	0.84584	35	35
	75th Percentile	1.262188	18	53
EUI	50th Percentile	235.05355	35	35
	75th Percentile	315.3586	18	53
CUI	50th Percentile	0.01029	35	35
	75th Percentile	0.02273	18	35

Academic buildings were considered a subgroup of buildings as per the commercial and institutional building surveys in Canada. This raises two limitations when interpreting similar CDF benchmarking results. First, they cannot be compared against similar (academic) buildings in national reports. Second, it is difficult to conclude which set of buildings has satisfactory environmental performance in comparison with the overall population if the performance of the entire population is not satisfactory [52]. Comparing the best performers to a known high-environmental performance building may help determine whether the performance of the best performers is satisfactory or not. A clustering approach is used to resolve these two limitations.

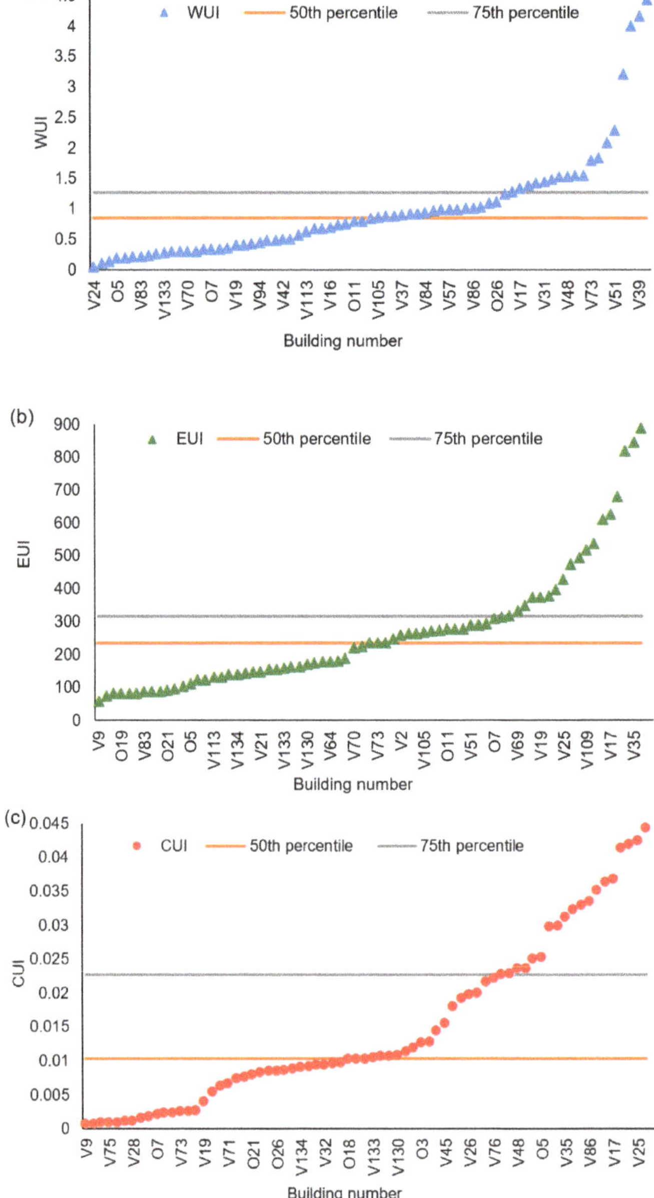

Figure 5. The 50th and 75th percentile benchmarks for (**a**) water usage intensity (WUI), (**b**) energy usage intensity (EUI), and (**c**) carbon usage intensity (CUI).

3.2. Fuzzy Clustering

The suitable number of clusters is derived from Equation (7), and the variation of the within cluster sum of squares (WCSS) as a function of the number of clusters is shown in Figure 6. The sum of squares is the sum of the squared distance between each member of the cluster and its centroid. As can be seen in Figure 6, three clusters are selected to optimally represent this case study.

The cluster centers are shown in Table 5. The lower value of the index, the better performance of the cluster. Based on the centroids for the MF, cluster 2 is identified as the group with the highest environmental performance, followed by cluster 1 and cluster 3.

Table 5. Cluster centroids.

Cluster	WUI	EUI	CUI
1	1.008317	308.9435	0.016878
2	0.796999	130.3005	0.008255
3	1.603258	714.9797	0.034582

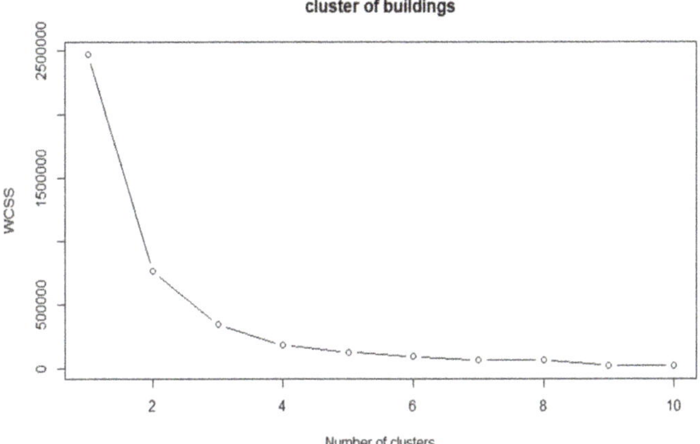

Figure 6. Within cluster sum of squares (WCSS) plot to select the number of clusters.

The detailed fuzzy clustering analysis results are shown in Appendix A. As shown in the appendix, the green shaded buildings are the buildings with a high environmental performance (cluster 2), and the red shaded are the buildings with low performance (cluster 3), while the orange being the moderately performed buildings (cluster 1). In addition to the building name, cluster number and the codes (which refers to the location of the building) of buildings are also listed in the appendix. Some buildings have information about the areas for specific functions, namely, classroom, laboratory, library, and office. For example, the given data from the portal shows that the Alumni Building has 18% of the area used as offices, 0% for libraries, 0% for laboratories, and 0% for classrooms. Similar information for the other buildings is also provided in Appendix A.

A building may serve multiple functions, e.g., one building may contain a library, classrooms, and administration offices. Therefore, to negate the intertwined effect of multiple areas within a building, a building is assumed to be a specific type of building (i.e., lab building) based on the function that most of the area is used for. For example, the Asian Centre building has 0% of area for classrooms, 3% of area for laboratories, 42% of area for a library and 14% of area for offices. Because most of its area serves as a library, it is considered as a library building. Figure 7 presents a plot of three clusters; each point in the figure represents a building and its proximity to the centroids of each cluster is presented by the distance.

To examine the variability in the data and set the benchmarks for the data in total and each cluster, a statistical summary of the obtained data is presented in Table 6. Cluster 1 has 30 buildings, cluster 2 has 33 buildings, and cluster 3 has eight buildings. The variability (IQR: Inter Quartile Range) in each cluster is a more accurate presentation than the overall buildings in combination. For example, The

WUI values for cluster 1, 2, and 3 are 0.384, 0.970, and 2.620, respectively, and the value for all of the buildings is 0.917. This shows that the cluster analysis generates different water usage assessment results for different clusters, and the different results cannot be revealed by the average water use value generated by assessing all buildings together. Cluster 3 has a larger data variance in terms of the three indices, due to the larger number of outliers.

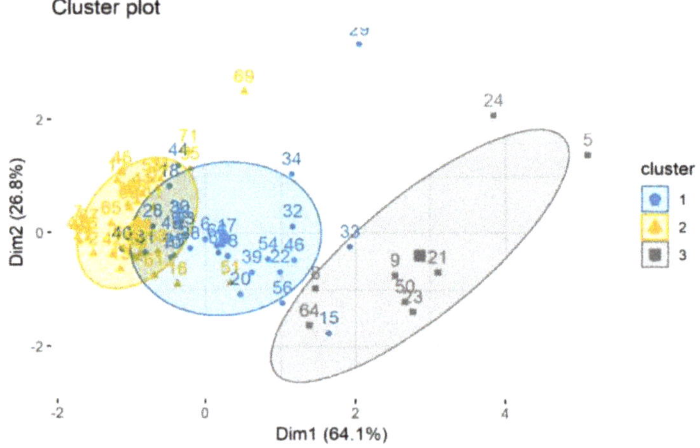

Figure 7. The three cluster layout.

Table 6. Statistical summary of the cluster results.

Variable	Total	Mean	Standard Error Mean	Standard Deviation	Variance	Sum	Minimum	Q1	Median	Q3	Maximum	IQR Interquartile Range
WUI_C1	30	1.01	0.15	0.8	0.63	30.3	0.2	0.6	0.87	1.02	4.4	0.38
EUI_C1	30	308.2	13.1	71.9	5170.9	9246	221.1	261.9	283.3	355.2	493.5	93.4
CUI_C1	30	0.02	0.002	0.01	0.0001	0.5	0.002	0.01	0.017	0.02	0.043	0.014
WUI_C2	33	0.8	0.12	0.7	0.5	26.3	0.06	0.292	0.5	1.3	3.22	0.97
EUI_C2	33	127.4	6.58	37.8	1429.6	4204	59.5	89.20	131.3	160.1	189.1	70.92
CUI_C2	33	0.008	0.001	0.007	0.00005	0.3	0.0006	0.00198	0.008	0.01	0.035	0.008
WUI_C3	8	1.7	0.5	1.5	2.3	13.7	0.2	0.754	1.1	3.4	4.16	2.62
EUI_C3	8	690.2	51	144.3	20833	5521	515.8	554.7	652.4	840.0	889.5	285.3
CUI_C3	8	0.03	0.003	0.007	0.00005	0.28	0.03	0.03	0.03	0.04	0.04	0.01

4. Discussion

The fuzzy clustering approach classifies buildings with similar WEC performances into the same group. It is found that buildings with a large portion of their areas allocated for laboratories are among the worst performing buildings in the university. Table 7 summarizes the area allocation per function and the averages per cluster with respect to the flows studied. The results show that the buildings in cluster 3 are predominantly laboratory-intensive with a higher percentage of the area dedicated to the laboratories than cluster 1. While cluster 2 was found to be the cluster with the highest environmental performance and the lowest area of laboratories per buildings on average.

Table 7. Benchmarking results.

Cluster	No. of Buildings	Average Area	Avg WUI	Avg EUI	Avg CUI	Office %	Library %	Laboratory %	Classroom %
C1	30	11,528	1.01	308.21	0.02	33.33	6.67	43.33	0
C2	33	10,051	0.8	127.4	0.01	30.3	12.12	12.12	9.08
C3	8	5,843	1.72	690.16	0.04	12.5	0	75	0

Laboratory-intensive buildings are reported to be the assets with the highest environmental impact in a university, due to the high energy consumption [20,21]. Some research has attributed the high energy consumption to inefficient HVAC systems which are either outdated or providing flows more than needed as a precautionary measure [53]. The results of the present study also show that laboratory-intensive buildings are also responsible for high water consumption and carbon emissions.

The national grid highly influences universities grid in the jurisdiction. Figure 8 illustrates the GHG growth per province and the number of universities in that province. By collecting the data (provided in Appendix B) for 36 Canadian HEIs in the STARS database, the baseline GHG values set by Canadian HEIs. The growth rate was calculated, aggregated by province, and plotted in the graph with the number of HEIs per province. Only one HEI obtained the platinum ranking (i.e., the most sustainable rank in STARS), while 16 HEIs obtained the gold, 17 HEIs obtained the silver, and two HEIs obtained the bronze rating from the STARS reporting system. Provinces that use renewables as their major source of energy (i.e., BC hydro projects and QB) have shown better performance in terms of GHG reductions, while AB and SK highly depended on fossil fuels as their main source of energy, thus, have shown little improvements from their baseline values. This could be attributed to the offset programs available in provinces like BC as well. For example, the University of Calgary and the University of Alberta is located in the same province (i.e., Alberta), and use similar electric grid systems. The former managed to reduce GHG emissions by 24%, while the latter had a 34% increase in emissions.

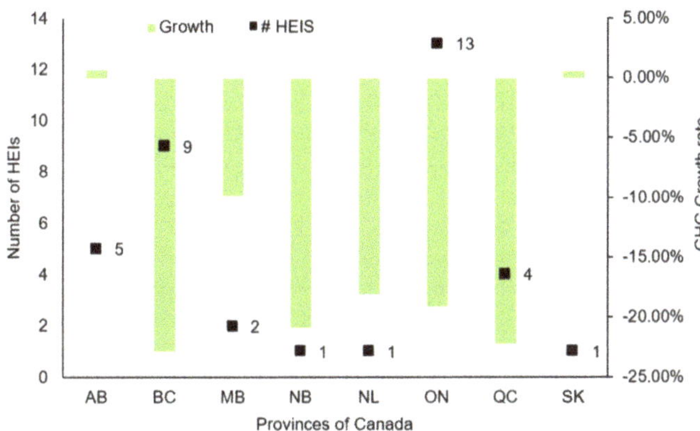

Figure 8. GHG Growth rate per province and the number of HEIs per province.

Finally, using holistic reporting systems, such as STARS, to communicate the overall sustainability may not yield the desirable momentum needed for universities to reach the desired goal of a sustainable campus. For instance, both universities (the University of Calgary and the University of Alberta) are ranked gold in the STARS reporting system. This is not to neglect the importance of holistic reporting, but to provide a mechanism within the holistic reporting systems for infrastructure management improvements. Furthermore, most reporting systems fail to capture this or reflect this into their weighing structure.

5. Conclusions

Holistic reporting systems, communicating the overall sustainability performance, may result in the same (high) performance for different universities based on meeting their overall socio-economic sustainability goals. While for technical level decision-making to practically optimize the WEC flows in HEIs, the environmental performance of individual academic buildings needs to be benchmarked.

Benchmarking academic buildings in HEIs is facing two main challenges. The first challenge is the lack of available national academic building database that is required to compare and determine a set of best practices in academic buildings; the second challenge is relevant to the conventional benchmarking methods that may yield misleading benchmarking results. By determining the environmental performance of academic buildings, the proposed fuzzy clustering-based framework allows efficient resource allocation for buildings that are identified with low environmental performance.

The proposed framework was applied to benchmark 71 academic buildings in two different campuses of UBC. The academic buildings were grouped into three clusters based on the reported MF in terms of energy and water consumption, as well as carbon emissions. Cluster 2 (33 buildings) is the group of buildings with the best environmental performance, followed by cluster 1 (30 buildings), and eight buildings associated with the lowest environmental performance are grouped into cluster 3. The average area of buildings per cluster is 11,528 m^2 for cluster 1, 10,051 m^2 for cluster 2, and 5,843 m^2 for cluster 3. The average WUI and EUI per cluster are 1.01 m^3/m^2 and 308.21 kWh/m^2 for cluster 1, 0.8 m^3/m^2 and 127.4 kWh/m^2 for cluster 2, and 1.72 m^3/m^2 and 690.16 kWh/m^2 for cluster 3, respectively. By comparing the results, the average EUI of buildings in cluster 3 is roughly four times higher than that of buildings in cluster 2 and nearly 120% more than that of cluster 1.

By grouping academic buildings into three clusters, and identifying a set of best performers and least performers (laboratory buildings), this study also identified the inner characteristics of academic buildings. The clustering analysis results showed that the environmental performances of predominant laboratory buildings are generally low, and this is in line with the results discovered in other studies.

There are several limitations to the proposed benchmarking methodology. The carbon emission factors for converting electricity consumption in two campuses are derived from the BC Best Practices Methodology for Quantifying GHG. The factor for the city of Kelowna, where UBCO is located, is reported as 0.719 kgCO2e/GJ, while the factor for the city of Vancouver is reported to be 2.964 kgCO2e/GJ. However, the values of carbon emission factors vary significantly from year to year. This could result in variations in the benchmarking results for the same buildings in different years. Moreover, the benchmarking results cannot provide detailed solutions to help HEIs improve the aspects that buildings are associated with low performance. Future research can apply a more aggressive data collection program to report detailed energy and water use behavior in the buildings which are identified poorly performed in the benchmarking. Based on the collected big data, system dynamic modeling and optimization can be used to help improve the performance of the buildings.

The developed methodology represents a new approach to track, assess, and aid retrofitting and/or decision making that best allocates the resources available in order to achieve low-impact infrastructure management in HEIs. By identifying a set of building performance, decision-makers can manage their resources more efficiently for further investigations and planning of interventions. Moreover, by identifying critical buildings, further information may be collected per floor or functional systems within a building. The flexible nature of the proposed framework allows the decision-makers to include further information for developing a more detailed decision support tool. The clustering results may also be used to help set attainable goals and plan future environmental commitments accordingly.

Author Contributions: A.A. collected the data, developed the methodology, performed detailed analysis and prepared the initial draft of the paper. G.H. contributed to the development of methodology and paper writing. H.H. contributed to conceptualization of the framework, development of methodology, and writing the final draft of the paper. K.H. contributed to the literature review and definition of the scope of the work. R.S. contributed to the conceptualization of the paper, methodology refinement, and the statistical analysis. All authors have read and agreed to the published version of the manuscript.

Funding: This research received no external funding.

Acknowledgments: The author gratefully acknowledges the supports of the STARS database and the sustainability office at the campus of University of British Columbia are highly appreciated.

Conflicts of Interest: The authors declare no conflict of interest.

Appendix A

Table A1. A detailed fuzzy clustering analysis results.

S/N	Name	Code	Cluster	WUI	EUI	CUI	DOM1	DOM2	DOM3	Area m2	Class	Lab	Library	Office	Office	Library	Lab	Class
1	Alumni Centre	V6	1	0.97	235.05	0.01	0.66	0.33	0.02	4106	0	0	0	0.18	Office			
2	AMS Nest	V2	1	0.93	259.53	0.02	0.86	0.13	0.01	22,933	0	0	0	0.07	Office			
3	ASC	O3	1	0.88	316.82	0.01	1.00	0.00	0.00	7801	0.11	0.38	0	0.12			Lab	
4	Brimacombe-QMI	V19	1	0.40	375.39	0.00	0.89	0.07	0.04	13,781	0	0.28	0	0.05			Lab	
5	CCM	V25	1	0.24	427.85	0.04	0.75	0.12	0.13	10,367	0	0.52	0	0.06			Lab	
6	CEME	V26	1	1.00	270.76	0.02	0.92	0.07	0.01	9361	0.06	0.29	0	0.32				
7	Centre for Brain Health	V31	1	1.43	245.79	0.00	0.76	0.23	0.01	15,441	0	0.14	0	0.2	Office			
8	Chem Bio	V34	1	0.30	379.72	0.02	0.89	0.07	0.04	14,030	0.06	0.33	0	0.15			Lab	
9	Chem East	V36	1	0.85	348.95	0.03	0.96	0.03	0.01	3573	0.07	0.49	0	0.03			Lab	
10	CICSR	V27	1	0.49	262.79	0.01	0.88	0.11	0.01	10,097	0	0.47	0	0.16			Lab	
11	EME	O6	1	0.76	288.11	0.00	0.98	0.02	0.00	16,520	0.11	0.28	0	0.22				
12	EOS	V45	1	4.44	374.11	0.02	0.90	0.06	0.03	10,799	0.01	0.55	0	0.15			Lab	
13	ESB	V46	1	0.98	272.83	0.01	0.93	0.06	0.01	17,755	0.06	0.22	0	0.25				
14	Fipke	O7	1	0.35	308.63	0.00	1.00	0.00	0.00	6725	0.17	0.34	0	0.12			Lab	
15	FNH	V48	1	1.53	397.79	0.02	0.84	0.09	0.07	5962	0.08	0.26	0	0.17			Lab	
16	Forest Sci	V50	1	1.54	476.18	0.03	0.58	0.14	0.28	22,459	0.07	0.31	0	0.17			Lab	
17	Frank Forward	V51	1	2.30	287.50	0.02	0.98	0.02	0.00	7880	0.06	0.29	0	0.22			Lab	
18	Henry Angus	V62	1	0.93	262.63	0.01	0.88	0.11	0.01	16,922	0.11	0.01	0	0.42	Office			
19	ICICS	V65	1	0.68	294.79	0.01	0.99	0.01	0.00	10,583	0	0.28	0	0.3	Office			
20	J.B. MacDonald	V69	1	0.69	332.43	0.03	0.98	0.01	0.00	7328	0.02	0.2	0	0.28	Office			
21	Jack Bell	V70	1	0.30	221.14	0.00	0.51	0.48	0.02	2712	0.14	0.05	0.38	0		Lib		
22	Klinck	V72	1	0.73	236.09	0.01	0.67	0.32	0.02	10,720	0.11	0.02	0	0.33	Office			
23	Koerner Library	V73	1	1.79	235.58	0.00	0.66	0.32	0.02	7303	0	0	0.29	0.12		Lib		
24	Life Sci	V76	1	1.02	493.52	0.02	0.51	0.13	0.36	52,177	0.04	0.39	0	0.13			Lab	
25	Longhouse	V78	1	0.79	278.62	0.02	0.96	0.04	0.00	2352	0	0	0.09	0.34	Office			

Table A1. *Cont.*

S/N	Name	Code	Cluster	WUI	EUI	CUI	DOM1	DOM2	DOM3	Area m2	Class	Lab	Library	Office	Office	Library	Lab	Class
26	NSDC-UBC	V86	1	1.01	224.54	0.03	0.55	0.44	0.02	3714	0	0	0	0				
27	Pond East	V94	1	0.45	313.90	0.04	1.00	0.00	0.00	11,080	0	0.18	0	0.05			Lab	
28	RHS	O11	1	0.79	278.58	0.02	0.96	0.04	0.00	5021	0.17	0.19	0	0.23				
29	Strangway	V105	1	0.86	267.52	0.01	0.91	0.08	0.01	12,403	0	0	0	0.25	Office			
30	University Centre	V117	1	0.88	279.16	0.02	0.96	0.04	0.00	3944	0.11	0	0	0.19	Office			
31	AERL	V1	2	0.33	142.44	0.01	0.01	0.99	0.00	5368	0.06	0.05	0	0.46	Office			
32	Allard Hall	V5	2	0.11	138.36	0.00	0.00	1.00	0.00	14,909	0.08	0	0.2	0.2				
33	Asian Centre	V9	2	0.42	59.48	0.00	0.07	0.92	0.01	4926	0	0.03	0.42	0.14		Lib		
34	Buchanan A,B,C	V21	2	1.09	147.11	0.01	0.01	0.99	0.00	10,936	0.31	0.04	0	0.18				Class
35	Buchanan D,E	V22	2	1.28	89.76	0.00	0.03	0.96	0.00	7134	0.28	0.01	0	0.25				Class
36	C.K. Choi	V24	2	0.06	124.68	0.01	0.00	1.00	0.00	2912	0.01	0.03	0	0.42	Office			
37	Cassiar	O18	2	1.56	80.19	0.01	0.05	0.95	0.01	3951	0	0	0	0				
38	CCS	O5	2	0.20	111.31	0.03	0.01	0.99	0.00	4797	0	0.44	0	0.23			Lab	
39	Chan Centre	V32	2	0.30	162.56	0.01	0.05	0.95	0.00	11,440	0	0	0	0.02	Office			
40	CIRS	V28	2	0.30	104.95	0.00	0.02	0.98	0.00	5454	0.08	0.04	0	0.27	Office			
41	Cunningham	V42	2	0.51	153.17	0.01	0.02	0.98	0.00	4901	0	0	0	0.07	Office			
42	Geography	V57	2	0.98	160.57	0.00	0.04	0.96	0.00	5525	0.17	0.22	0	0.35				
43	I.K. Barber	V64	2	0.36	177.77	0.01	0.11	0.88	0.01	27,316	0.05	0.01	0.43	0.12		Lib		
44	Kalamalka	O19	2	1.24	79.77	0.01	0.05	0.95	0.01	4835	0	0	0	0				
45	Kenny	V71	2	0.21	131.32	0.01	0.00	1.00	0.00	9613	0.02	0.31	0	0.16			Lab	
46	Lasserre	V75	2	1.42	88.64	0.00	0.03	0.96	0.00	4710	0.16	0.23	0	0.25				
47	Liu	V77	2	0.26	73.22	0.00	0.05	0.94	0.01	1729	0	0	0.53	0		Lib		
48	Mathematics	V83	2	0.22	87.09	0.00	0.04	0.96	0.00	6140	0.15	0	0	0.28	Office			
49	MWO	O10	2	0.52	123.90	0.04	0.00	1.00	0.00	1681	0	0	0	0				
50	Neville Scarfe	V88	2	0.56	146.85	0.01	0.01	0.99	0.00	19,382	0.12	0.02	0.15	0.19				

Table A1. *Cont.*

S/N	Name	Code	Cluster	WUI	EUI	CUI	DOM1	DOM2	DOM3	Area m2	Class	Lab	Library	Office	Office	Library	Lab	Class
51	Nicola	O21	2	1.48	91.53	0.01	0.03	0.97	0.00	5667	0	0	0	0				
52	Orchard Commons	V90	2	1.83	179.87	0.01	0.13	0.86	0.01	43,194	0.04	0	0	0.03				Class
53	Pond North	V95	2	1.38	153.86	0.01	0.02	0.98	0.00	27,922	0.02	0.01	0	0.05	Office			
54	Pond West	V96	2	0.50	177.47	0.01	0.11	0.88	0.01	18,779	0	0.03	0	0			Lab	
55	Purcell	O22	2	0.98	81.13	0.01	0.04	0.95	0.01	6208	0	0	0	0				
56	Sing Tao	V103	2	0.14	189.10	0.01	0.19	0.80	0.01	1571	0.09	0.15	0	0.24				
57	SPPH	V100	2	0.35	173.27	0.01	0.09	0.90	0.01	8442	0.05	0.06	0	0.48	Office			
58	Totem Infill	V113	2	0.64	130.24	0.00	0.00	1.00	0.00	15,756	0	0	0	0				
59	USB	V116	2	0.41	96.07	0.00	0.03	0.97	0.00	11,598	0	0	0	0.2	Office			
60	Valhalla	O26	2	1.11	78.84	0.01	0.05	0.95	0.01	4797	0	0	0	0				
61	Wesbrook Building	V130	2	3.22	170.72	0.01	0.08	0.92	0.01	10,272	0.06	0.29	0	0.08			Lab	
62	Woodward IRC	V133	2	0.29	159.68	0.01	0.04	0.96	0.00	12,049	0.08	0.01	0	0.15	Office			
63	Woodward Library	V134	2	2.09	139.34	0.01	0.00	1.00	0.00	7777	0	0	0.43	0		Lib		
64	Aquatic Centre	V8	3	3.99	889.49	0.04	0.08	0.05	0.87	8041	0	0	0	0				
65	Bio Sci West	V16	3	0.70	609.39	0.02	0.11	0.04	0.85	8021	0.04	0.36	0	0.1			Lab	
66	Biomed	V17	3	1.33	624.05	0.04	0.07	0.03	0.90	4407	0	0.47	0	0.1			Lab	
67	Chem Centre	V35	3	1.52	847.20	0.03	0.06	0.03	0.91	7274	0.06	0.37	0	0.16			Lab	
68	Chem North	V37	3	0.91	680.79	0.04	0.01	0.00	0.99	2739	0	0.53	0	0.09			Lab	
69	Chem South	V39	3	4.16	536.41	0.04	0.34	0.11	0.55	5373	0.1	0.42	0	0.09			Lab	
70	Michael Smith	V84	3	0.94	818.22	0.03	0.04	0.02	0.94	8477	0.03	0.41	0	0.15			Lab	
71	Tennis Centre—Old	V109	3	0.20	515.76	0.03	0.42	0.12	0.46	2409	0	0	0	0.01	Office			

Appendix B

Table A2. Data collected for Canadian HEIs in the STARS database.

HEI	Code	Province	Rank	Baseline YR	Baseline Value	Performance Year	Performance Value	Growth
MacEwan College	AB1	Alberta	Silver	13/14	30,754.00	16/17	28,068.00	−8.7%
MRU	AB2	Alberta	Silver	14/15	29,294.60	16/17	34,393.31	17.4%
NAIT	AB3	Alberta	Silver	2007/2008	68,190.37	2013/2014	54,128.58	−20.6%
University of Alberta	AB4	Alberta	Gold	2005/2006	212,190.20	2015/2016	285,815.00	34.7%
University of Calgary	AB5	Alberta	Gold	2008/2009	239,954.60	2017/2018	182,112.28	−24.1%
Camosun College	BC1	BC	Silver	2010	1,826.26	2014	1371.32	−24.9%
Okanagan College	BC2	BC	Silver	2015	1686.00	2018	1236.00	−26.7%
Royal Roads University	BC3	BC	Gold	2010	1460.00	2017	1016.00	−30.4%
Selkirk College	BC4	BC	Silver	2008	1698.80	2018	949.12	−44.1%
Simon Fraser University	BC5	BC	Gold	2007	18,934.00	2017	15,235.43	−19.5%
TRU	BC6	BC	Platinum	2016	3359.00	2018	3715.00	10.6%
UBC	BC7	BC	Gold	2007	60,100.00	2017	42,786.00	−28.8%
UNBC	BC8	BC	Silver	2010	5688.73	2018	7199.00	26.5%
University of Victoria	BC9	BC	Gold	2010	15,545.90	2018	11,603.00	−25.4%
University of Manitoba	MB1	MB	Gold	1990/1991	38,442.00	2016/2017	35,304.00	−8.2%
University of Winnipeg	MB2	MB	Silver	2009/2010	3883.00	2017/2018	2860.40	−26.3%
UNBF	NB1	NB	Silver	2007/2008	39,070.00	2015/2016	30,947.00	−20.8%
Dalhousie University	NS1	NS	Gold	2009/2010	106,178.00	2016/2017	87,056.00	−18.0%
Carleton University	ON1	ON	Silver	2012	26,729.00	2015	26,203.00	−2.0%
Durham College	ON2	ON	Silver	2012	5600.00	2013	5369.00	−4.1%
Fanshawe College	ON3	ON	Gold	2005	5366.76	2017	5200.04	−3.1%
Fleming College	ON4	ON	Gold	2012/2013	4614.00	2017/2018	4506.00	−2.3%
George Brown College	ON5	ON	Silver	2006/2007	3703.00	2012/2013	4192.00	13.2%
Loyalist College	ON6	ON	Bronze	2013	2488.10	2014	2484.60	−0.1%
Mohawk College	ON7	ON	Gold	2007	8521.00	2017	3235.00	−62.0%

Table A2. *Cont.*

HEI	Code	Province	Rank	Baseline YR	Baseline Value	Performance Year	Performance Value	Growth
Sheridan College	ON8	ON	Silver	2010/2011	8136.00	2016/2017	7306.90	–10.2%
St Lawrence College	ON9	ON	Bronze	2010	4078.50	2017	3373.30	–17.3%
University of Ottawa	ON10	ON	Silver	2005/2006	37,317.00	2016	27,918.94	–25.2%
University of waterloo	ON11	ON	Silver	2010	38,710.92	2017	39,495.34	2.0%
Western University	ON12	ON	Gold	2009	84,417.50	2016	55,524.00	–34.2%
Wilfrid Laurier University	ON13	ON	Gold	2009	10,875.61	2018	9901.96	–9.0%
Concordia University	QC1	QC	Gold	2010/2011	10,362.09	2014/2015	9665.33	–6.7%
HEC Montreal	QC2	QC	Silver	2005/2007	1909.66	2016	1105.57	–42.1%
McGill University	QC3	QC	Gold	2002/2003	57,590.00	2014	43,249.00	–24.9%
Polytechnique Montreal	QC4	QC	Gold	2004/2005	3596.90	2016/2017	3152.60	–12.4%
University of Saskatchewan	SK1	SK	Silver	2005/2006	151,541.50	2015/2016	152,453.80	0.6%

References

1. IPCC. *Climate Change 2014: Mitigation of Climate Change*; Cambridge University Press: Cambridge, UK, 2014.
2. Canada, E. Pan-Canadian Framework on Clean Growth and Climate Change: Canada's Plan to Address Climate Change and Grow the Economy. Environment and Climate Change Canada 2016: Gatineau, Quebec. 2016. Available online: http://publications.gc.ca/site/eng/9.828774/publication.html (accessed on 15 January 2020).
3. Pérez-Lombard, L.; Ortiz, J.; Pout, C. A review on buildings energy consumption information. *Energy Build.* **2008**, *40*, 394–398. [CrossRef]
4. Papadopoulos, S.; Kontokosta, C.E. Grading buildings on energy performance using city benchmarking data. *Appl. Energy* **2019**, *233–234*, 244–253. [CrossRef]
5. Collinge, W.; Landis, A.; Jones, A.; Schaefer, L.; Bilec, M. Dynamic life cycle assessment: Framework and application to an institutional building. *Int. J. Life Cycle Assess.* **2013**, *18*, 538–552. [CrossRef]
6. Junnila, S.; Horvath, A.; Guggemos, A. Life-Cycle Assessment of Office Buildings in Europe and the United States. *Journal of Infrastructure Systems* **2006**, *12*, 10–17. [CrossRef]
7. Scheuer, C.; Keoleian, G.A.; Reppe, P. Life cycle energy and environmental performance of a new university building: modeling challenges and design implications. *Energy and Buildings* **2003**, *35*, 1049–1064. [CrossRef]
8. Li, X.; Tan, H.; Rackes, A. Carbon footprint analysis of student behavior for a sustainable university campus in China. *J. Cleaner Prod.* **2015**, *106*, 97–108. [CrossRef]
9. Robinson, J.; Berkhout, T. *The University as an Agent of Change for Sustainability*; Policy Horizons Canada: Ottawa, ON, Canada, 2011.
10. Verhoef, L.; Graamans, L.; Gioutsos, D.; van Wijk, A.; Geraedts, J.; Hellinga, C. ShowHow: A flexible, structured approach to commit university stakeholders to sustainable development. In *Handbook of Theory and Practice of Sustainable Development in Higher Education*; Springer: Berlin/Heidelberg, Germany, 2017; pp. 491–508.
11. Lo-Iacono-Ferreira, V.G.; Torregrosa-López, J.I.; Capuz-Rizo, S.F. Use of Life Cycle Assessment methodology in the analysis of Ecological Footprint Assessment results to evaluate the environmental performance of universities. *J. Clean. Prod.* **2016**, *133*, 43–53. [CrossRef]
12. Abidin, N.I.; Zakaria, R.; Aminuddin, E.; Abdul Hamid, A.R.; Munikanan, V.; Sahamir, S.R.; Shamsuddin, S.M. Factor Analysis on Criteria Affecting Lean Retrofit for Energy Efficient Initiatives in Higher Learning Institution Buildings. *MATEC Web Conf.* **2017**, *138*, 02025. [CrossRef]
13. Bouscayrol, A.; Castex, E.; Delarue, P.; Desreveaux, A.; Ferla, O.; Frotey, J.; German, R.; Klein, J.; Lhomme, W.; Sergent, J.F. Campus of University with Mobility Based on Innovation and Carbon Neutral. In Proceedings of the 2017 IEEE Vehicle Power and Propulsion Conference (VPPC), Belfort, France, 11–14 December 2017; pp. 1–5.
14. NRC. Consumption of Energy Survey for Universities, Colleges and Hospitals 2003. 2004. Available online: http://oee.nrcan.gc.ca/corporate/statistics/neud/dpa/data_e/consumption03/universities.cfm?attr=0 (accessed on 17 September 2017).
15. BC. 2018 Carbon Neutral Government Year in Review 2018: Summary, Environment, B.M.o., Ed. 2018. Available online: https://www2.gov.bc.ca/assets/gov/environment/climate-change/cnar/2018/347953_attachment_cng_annual_report_summary_2018.pdf (accessed on 12 February 2019).
16. EPA. EnviroAtlas. In *Educational Facilities*; 2012. Available online: https://www.epa.gov/watersense/types-facilities (accessed on 10 January 2020).
17. Wang, J.C. A study on the energy performance of school buildings in Taiwan. *Energy Build.* **2016**, *133*, 810–822. [CrossRef]
18. Tan, H.; Chen, S.; Shi, Q.; Wang, L. Development of green campus in China. *J. Clean. Prod.* **2014**, *64*, 646–653. [CrossRef]
19. Larsen, H.N.; Pettersen, J.; Solli, C.; Hertwich, E.G. Investigating the Carbon Footprint of a University—The case of NTNU. *J. Clean. Prod.* **2013**, *48*, 39–47. [CrossRef]
20. Khoshbakht, M.; Goumartinbur, Z.; Dupre, K. Energy use characteristics and benchmarking for higher education buildings. *Energy Build.* **2018**, *164*, 61–76. [CrossRef]
21. Chihib, M.; Salmerón-Manzano, E.; Manzano-Agugliaro, F. Benchmarking Energy Use at University of Almeria (Spain). *Sustainability* **2020**, *12*, 1336. [CrossRef]

22. Tan, H.; Chen, S.; Zhang, N. *Annual Report of Green Campus Development in China*. 2012. China Green University Network. Available online: http://www.cgun.org/down.aspx?info_lb=120%26flag=120 (accessed on 11 April 2016).

23. Ward, I.; Ogbonna, A.; Altan, H. Sector review of UK higher education energy consumption. *Energy Policy* **2008**, *36*, 2939–2949. [CrossRef]

24. Alghamdi, N. Calm Before the Storm: Assessing Climate Change and Sustainability in Saudi Arabian Universities. In *Handbook of Climate Change Communication: Volume 2: Practice of Climate Change Communication*; Leal Filho, W., Manolas, E., Azul, A.M., Azeiteiro, U.M., McGhie, H., Eds.; Springer International Publishing: Cham, Switzerland, 2018; pp. 317–340. [CrossRef]

25. Alghamdi, A.; Haider, H.; Hewage, K.; Sadiq, R. Inter-University Sustainability Benchmarking for Canadian Higher Education Institutions: Water, Energy, and Carbon Flows for Technical-Level Decision-Making. *Sustainability* **2019**, *11*, 2599. [CrossRef]

26. AASHE. The Sustainability Tracking, Assessment & Rating System. Available online: https://reports.aashe.org/institutions/participants-and-reports/?sort=country (accessed on 11 March 2019).

27. NSF. What's an Engineered System? Available online: http://erc-assoc.org/content/what%E2%80%99s-engineered-system (accessed on 28 November 2018).

28. Alshuwaikhat, H.M.; Abubakar, I. An integrated approach to achieving campus sustainability: Assessment of the current campus environmental management practices. *J. Clean. Prod.* **2008**, *16*, 1777–1785. [CrossRef]

29. Calder, W.; Clugston, R.M. *Progress Towards Sustainability in Higher Education*; Enviornmental Law Institute: Washington, DC, USA, 2003; Available online: http://www.ulsf.org/pdf/dernbach_chapter_short.pdf (accessed on 18 November 2017).

30. Fischer, D.; Jenssen, S.; Tappeser, V. Getting an empirical hold of the sustainable university: A comparative analysis of evaluation frameworks across 12 contemporary sustainability assessment tools. *Assess. Eval. High. Educ.* **2015**, *40*, 785–800. [CrossRef]

31. Grindsted, T.S. Sustainable universities–from declarations on sustainability in higher education to national law. *Environ. Econ.* **2011**, *2*, 29–36. Available online: https://forskning.ruc.dk/en/publications/sustainable-universities-from-declarations-on-sustainability-in-h (accessed on 21 March 2015). [CrossRef]

32. Clarkson, R.E.; Samson, K.; Bekessy, S.A. The failure of non-binding declarations to achieve university sustainability: A need for accountability. *Int. J. Sustain. High. Educ.* **2007**, *8*, 301–316. [CrossRef]

33. Lozano, R. A tool for a Graphical Assessment of Sustainability in Universities (GASU). *J. Clean. Prod.* **2006**, *14*, 963–972. [CrossRef]

34. Shi, H.; Lai, E. An alternative university sustainability rating framework with a structured criteria tree. *J. Clean. Prod.* **2013**, *61*, 59–69. [CrossRef]

35. Adams, C.A. Sustainability reporting and performance management in universities: Challenges and benefits. *Sustain. Account. Manag. Policy J.* **2013**, *4*, 384–392. [CrossRef]

36. Renard, Y.J.S.; Maria, C.A.B.; David, G.C. A review of building/infrastructure sustainability reporting tools (SRTs). *Smart Sustain. Built Environ.* **2013**, *2*, 106–139. [CrossRef]

37. AASHE. STARS Participants & Reports. Available online: https://reports.aashe.org/institutions/participants-and-reports/?sort=country (accessed on 12 January 2020).

38. Urbanski, M.; Filho, W.L. Measuring sustainability at universities by means of the Sustainability Tracking, Assessment and Rating System (STARS): Early findings from STARS data. *Environ. Dev. Sustain.* **2015**, *17*, 209–220. [CrossRef]

39. Shriberg, M. Institutional assessment tools for sustainability in higher education: Strengths, weaknesses, and implications for practice and theory. *Int. J. Sustain. High. Educ.* **2002**, *3*, 254–270. [CrossRef]

40. McIntosh, M.; Cacciola, K.; Clermont, S.; Keniry, J. State of the Campus Environment: A National Report Card on Environmental Performance and Sustainability in Higher Education. Available online: https://www.nwf.org/EcoLeaders/Campus-Ecology-Resource-Center/Reports/State-of-the-Campus-Environment (accessed on 2 February 2020).

41. Martin, J.; Samels, J.E. *The Sustainable University: Green Goals and New Challenges for Higher Education Leaders*; Johns Hopkins University Press: Baltimore, MD, USA, 2012.

42. Gao, X.; Malkawi, A. A new methodology for building energy performance benchmarking: An approach based on intelligent clustering algorithm. *Energy Build.* **2014**, *84*, 607–616. [CrossRef]

43. Hong, S.-M.; Paterson, G.; Burman, E.; Steadman, P.; Mumovic, D. A comparative study of benchmarking approaches for non-domestic buildings: Part 1—Top-down approach. *Int. J. Sustain. Built Environ.* **2014**, *2*, 119–130. [CrossRef]

44. Burman, E.; Hong, S.-M.; Paterson, G.; Kimpian, J.; Mumovic, D. A comparative study of benchmarking approaches for non-domestic buildings: Part 2—Bottom-up approach. *Int. J. Sustain. Built Environ.* **2014**, *3*, 247–261. [CrossRef]

45. Camp, R.C. Benchmarking—The search for industry best practices that lead to superior performance. *Qual. Prog.* **1989**, *22*, 66–68.

46. Sartor, D.; Piette, M.A.; Tschudi, W. *Strategies for Energy Benchmarking in Cleanrooms and Laboratory-Type Facilities*; Lawrence Berkeley National Laboratory: Berkeley, CA, USA, 2000.

47. Turner, D. Benchmarking in universities: League tables revisited. *Oxf. Rev. Educ.* **2005**, *31*, 353–371. [CrossRef]

48. Yang, J.; Santamouris, M.; Lee, S.E.; Deb, C. Energy performance model development and occupancy number identification of institutional buildings. *Energy Build.* **2016**, *123*, 192–204. [CrossRef]

49. Nikolaou, T.; Kolokotsa, D.; Stavrakakis, G. Review on methodologies for energy benchmarking, rating and classification of buildings. *Adv. Build. Energy Res.* **2011**, *5*, 53–70. [CrossRef]

50. Pérez-Lombard, L.; Ortiz, J.; González, R.; Maestre, I.R. A review of benchmarking, rating and labelling concepts within the framework of building energy certification schemes. *Energy Build.* **2009**, *41*, 272–278. [CrossRef]

51. Mills, E.; Bell, G.D.S.; Chen, A.; Avery, D.; Siminovitch, M.; Greenberg, S.; Marton, G.; de Almeida, A.; Lock, L.E. *Energy Efficiency in California Laboratory-Type Facilities*; Berkeley Lab: Berkeley, CA, USA, 1997; p. 359.

52. Federspiel, C.; Zhang, Q.; Arens, E. Model-based benchmarking with application to laboratory buildings. *Energy Build.* **2002**, *34*, 203–214. [CrossRef]

53. Huizenga, C.; Liere, W.; Bauman, F.; Arens, E. *Development of Low-Cost Monitoring Protocols for Evaluating Energy Use in Laboratory Buildings*; Center for Environmental Design Research, University of California: Berkeley, CA, USA, 1996.

54. Baboulet, O.; Lenzen, M. Evaluating the environmental performance of a university. *J. Clean. Prod.* **2010**, *18*, 1134–1141. [CrossRef]

55. He, Y.; Kvan, T.; Liu, M.; Li, B. How green building rating systems affect designing green. *Build. Environ.* **2018**, *133*, 19–31. [CrossRef]

56. Lauder, A.; Sari, R.F.; Suwartha, N.; Tjahjono, G. Critical review of a global campus sustainability ranking: GreenMetric. *J. Clean. Prod.* **2015**, *108*, 852–863. [CrossRef]

57. Boer, P. Assessing Sustainability and Social Responsibility in Higher Education Assessment Frameworks Explained. In *Sustainability Assessment Tools in Higher Education Institutions: Mapping Trends and Good Practices Around the World*; Caeiro, S., Filho, W.L., Jabbour, C., Azeiteiro, U.M., Eds.; Springer International Publishing: Cham, Switzerland, 2013; pp. 121–137. [CrossRef]

58. Davis, J.A.; Nutter, D.W. Occupancy diversity factors for common university building types. *Energy Build.* **2010**, *42*, 1543–1551. [CrossRef]

59. Santamouris, M.; Mihalakakou, G.; Patargias, P.; Gaitani, N.; Sfakianaki, K.; Papaglastra, M.; Pavlou, C.; Doukas, P.; Primikiri, E.; Geros, V.; et al. Using intelligent clustering techniques to classify the energy performance of school buildings. *Energy Build.* **2007**, *39*, 45–51. [CrossRef]

60. Chung, W.; Hu, Y.V.; Lam, Y.M. Benchmarking the energy efficiency of commercial buildings. *Appl. Energy* **2004**, *83*, 1–14. [CrossRef]

61. Iliadis, L.S.; Vangeloudh, M.; Spartalis, S. An intelligfent system employing an enhanced fuzzy c-means clustering model: Application in the case of forest fires. *Comput. Electron. Agric.* **2010**, *70*, 276–284. [CrossRef]

62. Krajnic, D.; Mele, M.; Glavic, P. Fuzzy Logic Model for the performance benchmarking of sugar plants by considering best available techniques. *Resour. Conserv. Recycl.* **2007**, *52*, 314–330. [CrossRef]

63. Kouloumpis, V.; Azapagic, A. Integrated life cycle sustainability assessment using fuzzy inference: a novel FELICITA model. *Sustain. Product. Consumption* **2018**, *15*, 25–34. [CrossRef]

64. Haider, H.; Hewage, K.; Umer, A.; Ruparathna, R.; Chhipi-Shrestha, G.; Culver, K.; Holland, M.; Kay, J.; Sadiq, R. Sustainability Assessment Framework for Small-sized Urban Neighbourhoods: An Application of Fuzzy Synthetic Evaluation. *Sustain. Cities Soc.* **2018**, *36*, 21–32. [CrossRef]

65. Hu, G.; Kaur, M.; Hewage, K.; Sadiq, R. An integrated chemical management methodology for hydraulic fracturing: A fuzzy-based indexing approach. *J. Clean. Product.* **2018**, *187*, 63–75. [CrossRef]

66. UBC. SkySpark. Available online: https://skyspark.energy.ubc.ca/user/login (accessed on 1 January2019).

67. Climatemps. Vancouver British Columbia and Kelowna BC Climate & Distance in Between. Available online: http://www.vancouver.climatemps.com/vs/kelowna.php (accessed on 18 January 2020).

68. CER. Canada's Renewable Power Landscape 2016-Energy Market Analysis- British Columbia. Available online: https://www.cer-rec.gc.ca/nrg/sttstc/lctrct/rprt/2016cndrnwblpwr/prvnc/bc-eng.html (accessed on 19 January 2020).

69. Environment, B.M. *2016/2017 B.C. Best Practices Methodology for Quanitifying Greenhouse Gas Emissions*; Ministry of Environment: Victoria, BC, Canada, 2016.

70. FortisBC. Public Sector Organizations: Reducing GHF emissions with RNG. Available online: https://www.fortisbc.com/services/sustainable-energy-options/renewable-natural-gas/public-sector-organizations-reducing-ghg-emissions-with-rng (accessed on 9 January 2020).

71. Hu, G.; Kaur, M.; Hewage, K.; Sadiq, R. Fuzzy clustering analysis of hydraulic fracturing additives for environmental and human health risk mitigation. *Clean Technol. Environ. Policy* **2019**, *21*, 39–53. [CrossRef]

72. Cox, E. Chapter 7—Fuzzy Clustering. In *Fuzzy Modeling and Genetic Algorithms for Data Mining and Exploration*; Cox, E., Ed.; Morgan Kaufmann: San Francisco, CA, USA, 2005; pp. 207–263. [CrossRef]

73. Sadiq, R.; Rodriguez, M.J.; Imran, S.A.; Najjaran, H. Communicating human health risks associated with disinfection by-products in drinking water supplies: A fuzzy-based approach. *Stoch. Environ. Res. Risk Assess.* **2007**, *21*, 341–353. [CrossRef]

74. Lu, R.-S.; Lo, S.-L.; Hu, J.-Y. Analysis of reservoir water quality using fuzzy synthetic evaluation. *Stoch. Environ. Res. Risk Assess.* **1999**, *13*, 327–336. [CrossRef]

sustainability

MDPI

Article

Flood Resilience of Housing Infrastructure Modeling and Quantification Using a Bayesian Belief Network

Mrinal Kanti Sen [1], Subhrajit Dutta [1] and Golam Kabir [2,*]

[1] Department of Civil Engineering, National Institute of Technology Silchar, Assam 788010, India; mksen88@gmail.com (M.K.S.); subhrajit.nits@gmail.com (S.D.)
[2] Industrial Systems Engineering, University of Regina, Regina, SK S4S 0A2, Canada
* Correspondence: golam.kabir@uregina.ca

Abstract: Resilience is the capability of a system to resist any hazard and revive to a desirable performance. The consequences of such hazards require the development of resilient infrastructure to ensure community safety and sustainability. However, resilience-based housing infrastructure design is a challenging task due to a lack of appropriate post-disaster datasets and the non-availability of resilience models for housing infrastructure. Hence, it is necessary to build a resilience model for housing infrastructure based on a realistic dataset. In this work, a Bayesian belief network (BBN) model was developed for housing infrastructure resilience. The proposed model was tested in a real community in Northeast India and the reliability, recovery, and resilience of housing infrastructure against flood hazards for that community were quantified. The required data for resilience quantification were collected by conducting a field survey and from public reports and documents. Lastly, a sensitivity analysis was performed to observe the critical parameters of the proposed BBN model, which can be used to inform designers, policymakers, and stakeholders in making resilience-based decisions.

Keywords: resilience; housing infrastructure; Bayesian belief network; flood hazard and sensitivity analysis

Citation: Sen, M.K.; Dutta, S.; Kabir, G. Flood Resilience of Housing Infrastructure Modeling and Quantification Using a Bayesian Belief Network. *Sustainability* **2021**, *13*, 1026. https://doi.org/10.3390/su13031026

Academic Editor: Abdollah Shafieezadeh
Received: 14 December 2020
Accepted: 18 January 2021
Published: 20 January 2021

Publisher's Note: MDPI stays neutral with regard to jurisdictional claims in published maps and institutional affiliations.

1. Introduction

Resilience is defined as the capability of a system to sustain against any hazard and return to its desired performance level after the occurrence of the hazard [1]. Hosseini et al. and Meerow et al. reviewed the definition of resilience in different disciplines [2,3], and its meaning has been discussed and evaluated in the existing literature [3,4]. The reliability and recovery of infrastructure are the two key dependent parameters of infrastructure resilience; furthermore, these two key parameters depend on four additional parameters: robustness, redundancy, rapidity, and resourcefulness, as shown in Figure 1 [1,5–7]. Robustness refers to the sustainability of a system against the effects of the disaster, redundancy refers to the duplication of any critical components or functions of a system that are intended to increase the reliability of the system, rapidity refers to the length of time required to return to its desired position after the occurrence of the hazard, and resourcefulness refers to the availability of resources for recovery. Reliability depends on the robustness and redundancy of the infrastructure, whereas the recovery process depends on rapidity and resourcefulness. Therefore, determining the reliability of infrastructure involves considering parameters based on robustness and redundancy, and similarly, determining the recoverability of infrastructure involves considering parameters based on rapidity and resourcefulness.

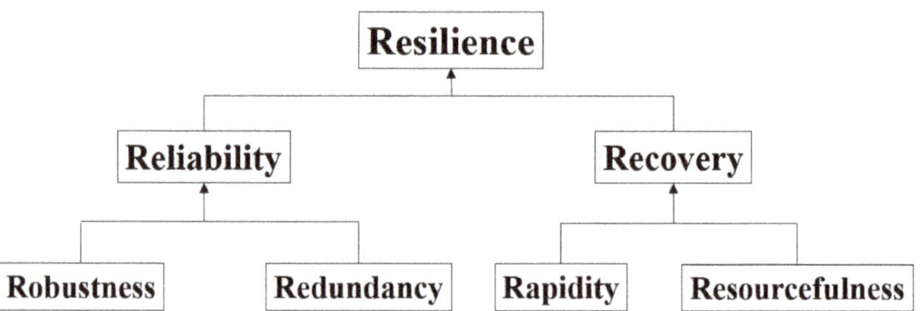

Figure 1. Resilience flowchart of the dependency parameters.

Figure 2 represents the generalized performance of a system/infrastructure over its service life [7,8], where A represents the initial condition of a system (which is generally considered to be 100% performance); AB and DE represent the gradual degradation in system performance due to operational conditions; BC represents a sudden drop in system performance due to a disaster, which is also known as loss; CT_1 represents the robustness of a system; T_1T_2 represents the time required for the recovery of the system; CD represents the recovery profile of the system. Figure 2 shows that, initially, the system/infrastructure performance degrades with time due to natural causes. Then, due to the occurrence of a disaster, the performance level sharply declines. The loss that is shown in the figure mainly depends on the impact of the disaster and the robustness of the system/infrastructure, which means that if the resistance ability of the system is very high, then the losses due to the disaster will be very low. The losses can be estimated using the Hazus technical manual created by the Federal Emergency Management Agency [9]. This manual provides several methodologies for multihazard loss estimation. After the loss, the system tries to recover to its baseline performance by following a recovery profile, which is uncertain and dependent on the type of infrastructure system. There are three types of recovery profiles: linear, non-linear, and stepped. The restoration of roads and bridges, for example, typically follows a stepped recovery pattern.

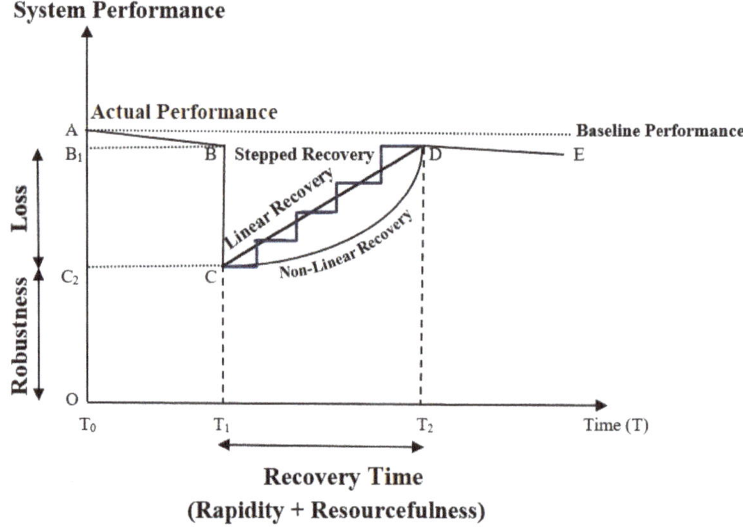

Figure 2. System Performance over its service Life.

1.1. Socio-Physical Infrastructure

Over the past decade, resilience quantification for communities has been an active area of research for both scientists and engineers. Engineering resilience is relatively new and currently developing, and valuable resources are available for the development of new engineering practices, codes, and regulations [10]. A community is defined as a group of people living in a given geographical area and mainly comprising two key infrastructure systems, namely, social and physical, as shown in Figure 3a [7]. The cross-dependency between different infrastructure systems is shown in Figure 3b [11,12].

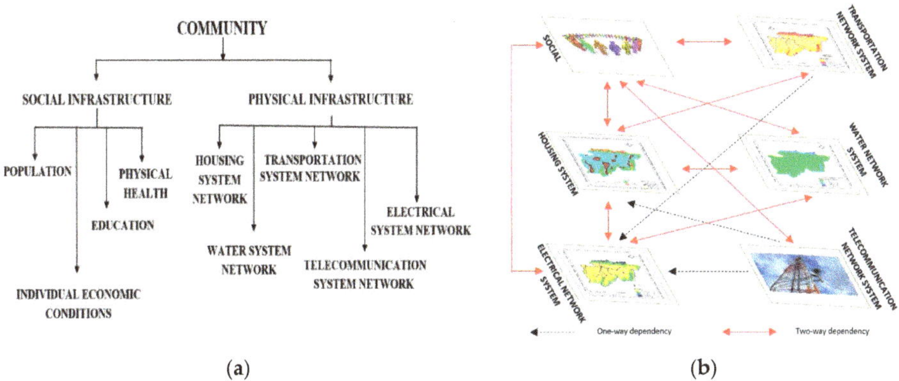

Figure 3. Representation of community infrastructure and multilayer interdependency between different infrastructure systems: (**a**) Community infrastructure; (**b**) Interdependency.

Social infrastructure resilience is the ability of societies to resist the effects of a disaster and mainly depends on the population of the community, physical health conditions, literacy or education level, and economic conditions. The physical infrastructure consists of networks for transportation, electricity, water, and telecommunications [7]. The dependencies between infrastructures enhance the complexity of infrastructure resilience quantification, which can be modeled using different approaches [12–14].

Quantification of resilience is very challenging because of many factors, including non-linear relationships between the dependent parameters, a lack of mathematically proven equations or studies to represent these relationships, a requirement for both qualitative and quantitative data, a need for the involvement of experts, data scarcity, data from different sources, and missing data. To overcome these challenges, a resilience measurement scale has been developed for performing quality assessments [15]. Mahmoud and Chulahwat developed a mathematical model to reduce the effect of hazards and proposed a new resilience model for resilience quantification [16]. An infrastructure resilience model plays a crucial role in the proper operation of an infrastructure system during and post disaster in terms of satisfying societal needs [17]. Resilience quantification for water and telecommunication networks was performed using the Resilience-compositional demand/supply (Re-CoDeS) framework [18].

Various frameworks and models have been proposed for quantification and studying resilience in different fields [19–21]. Several methodologies have been proposed for the proper quantification of resilience, such as probabilistic methods [14,22,23], graph theory methods [24,25], fuzzy logic methods [26], and analytical methods [27,28]. A "PEOPLES" (Population and demographics, Environmental and ecosystem, Organized governmental services, Physical infrastructure, Lifestyle and community competence, Economic development, Social cultural capital) factor-based framework for resilience quantification at different scales was also proposed [7,8]. To keep the sustainability of a structure against future hazards, the structure should be resilient enough. Resilience quantification needs a well-formatted past event dataset; however, the biggest hindrance in quantifying resilience is the lack of availability of properly formatted data for the damage and recovery for infrastructure systems. Inappropriate datasets for previous damage and restoration can lead to inaccurate probability estimations for future disasters and hamper sustainable development. Additionally, dependencies between infrastructure systems play a crucial role in resilience quantification [29]. A virtual system was formed based on an interdisciplinary system to improve resilience and identify the impacts of post-disaster recovery efforts [30]. The codes and standards for designing resilient systems were updated to consider both physical and non-physical infrastructure systems, and a new model for system resilience quantification was developed that considers dependencies and cross-dependencies between the networks, which makes the system more resilient [11]. Resilience has also been discussed and quantified in various networks, such as housing [31], the transportation network [32–34], the electrical network [35,36], the water network [37,38], and the telecommunication network [39,40], but there is a lack of literature that is directly focused on the flood resilience of housing infrastructure systems. Sen et al. studied the resilience of housing infrastructure by using the variable elimination method, but interdependencies between the resilience parameters were not considered in that study, which is a major drawback, as dependency plays a vital role in resilience [31]. This present work is novel in that it directly addressed the housing infrastructure system by considering the dependencies between the resilience parameters against flood hazards.

The main objectives of this work were as follows: (i) to perform a comprehensive study/survey of a community and its socio-physical infrastructure to identify the most influential factors affecting the flood resilience of its housing infrastructure system, (ii) to develop a probabilistic graphical model (Bayesian network model) for the flood resilience quantification of a housing infrastructure system, (iii) to quantify the flood resilience for housing infrastructure against flood hazards, and (iv) to check the sensitivity of each dependent parameter of reliability and recovery.

1.2. Socio-Physical Infrastructure of the Barak Valley Community

In this research, the case study region selected was the Barak Valley region of Northeast India. This valley is one of the most important regions of Northeast India as it connects many neighboring states of India. The longitude of this region ranges from $92°15'$ E to $93°15'$ E, and the latitude ranges from $24°8'$ N to $25°8'$ N. The total surface area of this valley is approximately 262,230 km^2, with a population of more than 3.6 million [41]. The climate of Barak Valley is sub-tropical, warm, and humid, the average rainfall of this valley is 3180 mm, and due to the high intensity of rainfall, floods and landslides are common in the valley. Per the Assam State Disaster Management Authority (ASDMA), in 2017, due to flooding, many water sources were severely damaged, with an estimated restoration cost of more than 277 million USD and an additional 150,000 USD sanctioned for housing system recovery. In 2018, approximately 200,000 people were affected, more than 1300 hectares of agricultural land were damaged, and a main national highway (NH-53, 44) and several other highways remained non-functional for several days. With each year, the damage and costs due to flooding increase [42]. The occurrences of such disasters are frequent in this region, and hence, the associated risk is high [42]. In this valley, nearly 11% of the population do not live in a house, and only 1% of the population live in a house with three or more rooms. The average annual per capita income of the valley is generally low and is in the range of 205 to 342 USD [41,43]. In this study, the housing infrastructure system of Barak Valley was used as the basis for the case study.

Barak Valley is a developing community with mixed demographics and economic conditions. Per the census report, only 30.75% of households use electricity and 0.84% of households use internet services [41]. The elevation of the valley varies from -58 m to 1694 m from mean sea level (MSL), as shown in Figure 4 [44]. In Figure 4, the outlines signify the administrative divisions of an Indian state, which is known as a district; this valley consists of three districts, namely, Cachar, Karimganj, and Hailakandi. Most of the population-dense areas of this valley are located in low-elevation zones, as shown in

Figure 5 [41,44]. Hence, from a flood risk perspective, the valley can be expected to incur significant socio-economic losses.

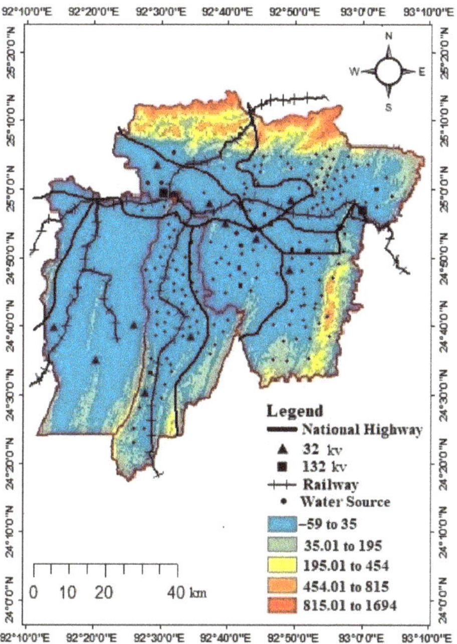

Figure 4. Details of Barak Valley overlapping with a digital elevation model (DEM).

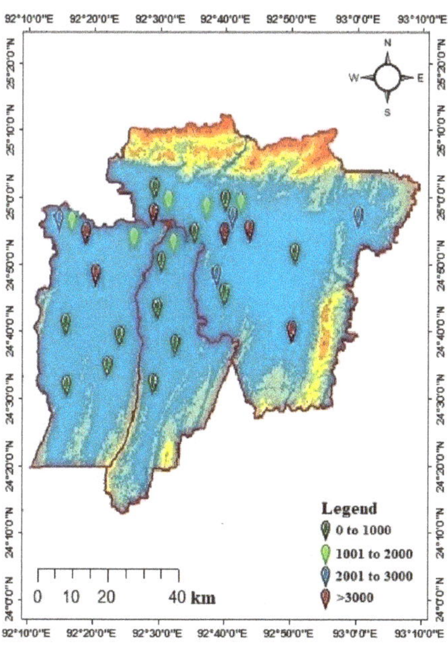

Figure 5. Population densities of different areas.

The housing system of this valley comprises various building typologies with a wide range of construction materials. Most of the residential and industrial buildings of this valley are located in low elevations, in the range of −59 to 35 m from MSL, as shown in Figure 6 [44]. These buildings are expected to have a high level of exposure with significant losses during a flood-related disaster. In this valley, traditional single-family houses, also known as Assam-type houses, were found to be the most common type of construction for both urban and rural areas. This type of house is constructed in flat and sloped terrains. The roof is mainly erected using high gables and the walls are made up of timber frames that are plastered with cement and the flooring is made up of either wood or concrete. This type of construction is less reliable and robust compared to RCC construction. More recently, RC construction has increased significantly in urban localities.

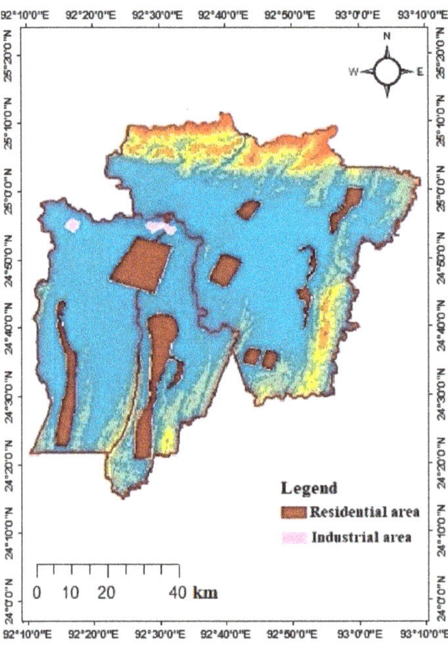

Figure 6. Categorization of buildings based on occupancy.

The water system plays a vital role in a community, as its primary function is to provide potable water to residential and commercial buildings. Proper functioning of the overall system depends on the working conditions of individual components, such as the supply source, water pipeline, treatment plant, water tanks, and reservoirs, along with their dependencies. Most of the water supply sources of this valley are located at lower elevations, as shown in Figure 4.

There are three aspects of the electrical power supply network: generation, transmission, and distribution [45]. This network plays a role that is as critical as other infrastructure systems, such as housing, and the water network depends on the electrical network to function. The electrical network comprises five components, which include transmission towers, substations, transformers, electric towers, and electric poles. It can be seen in Figure 7 that the majority of the substations in this valley are located at low elevations, leading to a higher risk of being damaged by floods [46]. On the other hand, the telecommunications network is another important infrastructure system in a community, as the number of phone calls increases during and after any disaster. Most of the population in rural areas do not use internet services, which increases the lack of awareness and communication. For instance, many small communication towers are installed on building roofs, which may lead to low supply connectivity in an area with a high connectivity demand. Hence, it is expected that the resilience of the communication network and other interdependent systems in Barak Valley will be relatively low during and after disasters.

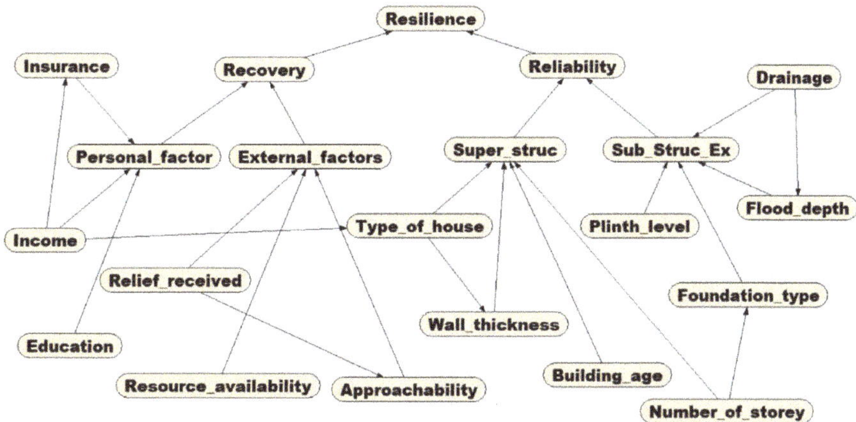

Figure 7. Bayesian belief network (BBN) model for flood resilience quantification of a housing infrastructure system.

The transportation network of this valley connects several major states within India. The transportation network comprises four modes: the railway, the roadway, the airway, and the waterway. In the roadway network, bridges are lifelines, as they are considered the most sensitive points of failure during a disaster. Flexible pavement is found to be the most common type of pavement in this valley. Recently, rigid pavement with RC and paver block has become preferred in construction for mitigating repetitive damage due to floods. The airport serving this valley is located at a relatively high elevation, 107 m from the MSL, with a total area of 36.70 acres. Due to the small-scale operation of the airport with limited aircraft, helicopters, and cargo vehicles, it is expected to have a low impact on post-disaster recovery, rescue, and relief operations.

The remaining sections of this paper are organized as follows. In Section 2, the proposed BBN model based on the housing infrastructure system is described in detail. In Section 3, the data collection process for the flood resilience study is described. In Section 4, the proposed model is verified, the sensitivities of the parameters of the proposed BBN model are evaluated, and the proposed BBN model is applied to assess the reliability, recovery, and resilience values of Barak Valley for the housing infrastructure system. Finally, in Section 5, conclusions, limitations of the study, and recommendations for further research are discussed.

2. Probabilistic Graphical Model

To develop an effective resilience framework for the housing infrastructure system, it is necessary to utilize different types of data, such as damage and recovery data from multiple sources. Expert judgment should also be obtained for the data interpretation, as the data can often be incomplete. Therefore, it is necessary to consider uncertainties in resilience assessments for the housing infrastructure. To address these uncertainties, different network-based models, such as artificial neural networks (ANNs), an analytic network process (ANP), a Bayesian belief network (BBN), and fuzzy cognitive maps (FCMs) are used. ANN provides insights into uncertainties if a comprehensive post-disaster dataset is available. In the case of insufficient data, techniques such as an ANP, BBN, or FCM can mitigate such uncertainties. It becomes very difficult for experts to generate a supermatrix, as found in the ANP method, where the representation of the relationship between parameters is performed using pairwise comparisons [47]. FCMs allow for the expression of dependence between the nodes with an influence degree ranging from "+1" to "−1" [48]. BBN assigns effective relationships between the nodes by considering a conditional probability table (CPT). A comparison of these different techniques is shown in Table 1 [4]. Note that, in the table, VH—very high, H—high, M—medium, and L—low. Based on this comparison, the BBN tool was selected for this study.

Table 1. Comparison between different techniques.

Attributes	ANN	ANP	BBN	FCM
Capability to express causality	N	L	VH	H
Capability to control qualitative inputs	N	VH	H	VH
Capability to control quantitative inputs	VH	L	M	L
Capability to control dynamic data	H	M	H	M
Capability to model complex systems	VH	M	VH	H
Learning/training capability	VH	H	H	H

ANN: artificial neural network, ANP: analytic network process, BBN: Bayesian belief network, FCM: fuzzy cognitive map; L: low, M: medium, H: high, VH: very high, N: Negligible.

2.1. Bayesian Belief Network

BBN is an extensive probabilistic model that is used to characterize the uncertainty that is associated with variables that constitute the model [49]. BBN is a graph-based model comprising nodes and edges, where nodes represent model variables and edges that represent the relationship between the nodes and also the conditional dependencies [50,51]. BBN can also be defined as a class of graphical models that presents a brief representation of the probabilistic dependencies between a given set of random variables [52]. The BBN model generally deals with discrete probabilities; therefore, each node is categorized into a finite set of variables with their probability values [53]. The CPT quantifies the dependencies between the child node and the parent node, as shown in Figure 8, where node A represents the child node and nodes B and C represent the parent nodes. The CPT for the parent nodes transfers to the unconditional probability (UP) and can be attained via an expert decision [53] and/or training from the dataset [54].

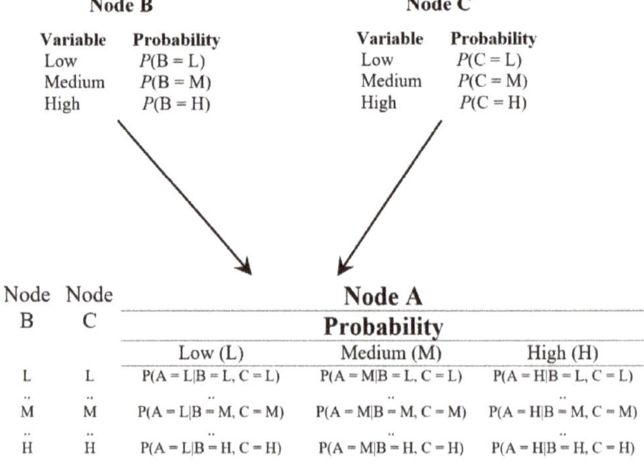

Figure 8. Representation of a BBN.

Probabilistic methods, such as a BBN, have been used to model various types of networks, such as transportation networks [32,55–57], water networks [14,38], telecommunications networks [58,59], and electrical networks [60]. An infrastructure system can be depicted as a BBN by using the dependent parameters and their interdependencies. BBN modes have also been used in the assessment of risks in infrastructure systems [61,62]. In this model, variables of any parameter can be updated when the values of evidence variables are available [63]. BBN comprises two types of inference, such as forward inference and backward inference, where the forward inference method is used to determine the system's probability value based on its component probability values and the backward inference method is used to determine the value of the updated probability of a system or its components based on the system state. It can also simulate the dependence of a network of uncertain variables and compute the inferences of any event based on evidence. For example, consider a BBN in which the probability occurrence of an event A depends on the occurrences of any other event, such as E_i ($i = 1, \ldots, n$). The posterior probability of the ith event, E_i, given the occurrence of A, is given by Bayes' theorem, as shown in Equation (1) [64]:

$$P(E_i|A) = \frac{P(A|E_i)P(E_i)}{P(A)},$$

(1)

where $P(A \mid E_i)$ is the conditional probability of the event A given the occurrence of E_i and $P(E_i)$ is the prior probability of E_i.

2.2. BBN Model for Flood Resilience Quantification of Housing Infrastructure

In this section, the development of the BBN model for flood resilience is discussed. Initially, several experts from various domains were selected. Then, based on the experts' knowledge and the literature, the resilience parameters were selected. The experts were selected based on the assumption that they have experience in the field of utility infrastructure management, risk management, Bayesian analysis, or catastrophes. A total of ten experts were selected for this work; their experience was as follows: (i) two field officers from the District Disaster Management Authority (DDMA) with more than three years of experience and who also helped during the field survey for data collection; (ii) one district project officer (DPO) from DDMA with more than ten years of experience and expertise in disaster management monitoring and policy implementation; (iii) two academic experts from different institutions having more than two and ten years of experience, respectively, with both obtaining a Ph.D. in Civil Engineering and are experts in reliability, risk, and resilience assessments; (iv) five flood catastrophe modelers from different industries with experiences of 16, 7, 5, 5, and, 4 years, respectively, where all of whom obtained a Ph.D. in Civil Engineering. Several meetings were conducted with the experts for the selection of resilience parameters and the identification of interdependencies. As resilience depends on the two key factors of reliability and recovery, 14 resilience parameters for housing infrastructure against flood hazards were selected based on the experts' knowledge and a literature review. The selected resilience parameters were as follows: (i) Type_of_house (robust types of houses are expected to perform better during floods) [31,65]; (ii) Wall_thickness (increasing wall thickness increases the resistance capability and reliability) [31,66]; (iii) Building_age (newly constructed buildings were found to be in better condition than older building) [31,67]; (iv) Number_of_stories (as stories increase, casualties decrease during floods) [31,67]; (v) Drainage (adequate drainage systems prevent damage to infrastructure) [31,68]; (vi) Flood_depth (this produces water pressure, which reduces the reliability of the infrastructure) [31,68]; (vii) Foundation_type [65,69]; (viii) Plinth_level (increasing the plinth height of the house to the top level of the road reduces vulnerability) [31]; (ix) Insurance (insurance enhances the recovery rate as the insurer can pay off the insurance claims) [31,70]; (x) Income (for households with higher income levels, the recovery process will be faster after a disaster) [31,69,71]; (xi) Education (education enhances the preparedness against disaster) [67,69,72]; (xii) Relief_received (whether during/after the disaster relief is received or not) [31,73,74]; (xiii) Approachability (whether connectivity from resource location to vulnerable site is disturbed or not) [31,69]; (xiv) Resource_availability (whether during/after the disaster raw materials for construction are available locally or not) [31].

To reduce the complexity during the construction of the CPT, reliability was subdivided into two parameters: (i) superstructure condition (Super_Struc) and (ii) substructure and external conditions (Sub_Struc_Ex). Next, these two parameters were linked with different resilience parameters, such that Super_Struc was linked with Type_of_house, Wall_thickness, Building_age, and Number_of_stories, while Sub_Struc_Ex was linked with Drainage, Flood_depth, Foundation_type, and Plinth_level. Similarly, recovery was divided into two parameters: (i) Personal_factor and External_factor. Personal_factor was then linked with Insurance, Income, and Education, while External_factor was linked with Relief_received, Approachability, and Resource_availability. Finally, the two key parameters, namely, reliability and recovery, were linked with resilience, as dependency plays a crucial role in resilience quantification. Therefore, based on the experts' judgements, the dependency was constructed and the dependencies between the parameters were as follows: (i) Type_of_house depended on Income, (ii) Wall_thickness depended on Type_of_house, (iii) Flood_depth depended on Drainage, (iv) Relief_received depended on Approachability, (v) Insurance depended on Income, and (vi) Foundation_type depended on Number_of_stories. After the selection of resilience parameters and the assignment of interdependencies, a BBN model was developed.

The developed BBN model for the flood resilience quantification of housing infrastructure with dependencies is shown in Figure 7. In the model, resilience was categorized into three different probability states—low (L), medium (M), and high (H)—where low means that immediate attention should be given by the stakeholders to the housing infrastructure of the community for strengthening or reconstruction, medium means that the housing infrastructure of the community can act as functional for the long-term, and high means safety exists in the housing infrastructure of the community for future hazards [75]. Reliability was categorized into four different probability states: DS1, DS2, DS3, and DS4, as discussed in Table 2 [76].

Table 2. Damage state (DS) descriptions.

Damage State	Category	Description
DS1	Low	No-damage condition, where floodwater touches the foundation but has no contact with electrical systems with a water height that is about 2.5 cm from ground level and damage occurs to carpets and flooring.
DS2	Medium	Drywall damage up to a 30 cm water level from the ground level and damage occurs to household furniture and other major equipment on the floor; doors need to be replaced.
DS3	High	Electrical panels, bathroom/kitchen cabinets and electrical appliances, lighting fixtures on walls, ceiling lighting, and studs got damaged.
DS4	Very High	The structure is fully damaged.

DS4 means that damage due to the occurrence of the flood is very high, which indicates that the reliability of the housing infrastructure of the community is very low; on the other end of the range, DS1 means damage due to the occurrence of the flood is very low, which indicates that the reliability of the housing infrastructure of the community is very high. Recovery is categorized into three different probability states—low (L), medium (M), and high (H)—where H is less than 10 days, which corresponds to the 25th percentile for recovery time; M is between 11 and 35 days, which corresponds to the 26th to 75th percentile for recovery time; L is more than 35 days, which corresponds to the 76th to 100th percentile for recovery time. The recovery of the housing infrastructure is discussed based on the amount of time required for the infrastructure to fully recover, where high means that the infrastructure took little time to fully recover and low means that the infrastructure took a very long time to fully recover. The recovery time is mainly based on the recovery of the infrastructure; it does not include the recovery of housing essentials, such as kitchen essentials, bathroom essentials, furniture, and utilities. The different probability states of each parent and intermediate parameter, as shown in Tables 3 and 4, were assigned based on the field survey data and experts' knowledge.

Table 3. Variable details for the parent parameters.

Parameter	Scale	Parameter	Scale
Annual Income (Indian Rupees)	Less than 10,000, 10,000–20,000, or more than 20,000	Education	Below 10th standard, 10th or 12th standard passed, or graduate
Insurance	Yes or no	Relief_received	Yes or no
Resource_availability	No, yes with a 0 to 10% increase, or yes with more than a 10% increase	Approachability	Yes or no
Building_age	Less than 10 years, 10 to 20 years, or more than 20 years	Plinth_level of house w.r.t the road top level	Up to 1 m, 1 to 2 m, and above 2 m
Drainage availability	Yes or no	Number_of_stories	1 or more than 1

Table 4. Variable details for the intermediate parameters.

Parameter	Scale	Parameter	Scale
Personal_factor	Good or bad	Super_Struc	Good or bad
External_factor	Good or bad	Sub_Struc_Ex	Good or bad
Foundation_type	Shallow or deep	Flood_depth	Less than 30 cm, 30 to 90 cm, or more than 90 cm
Type_of_house	Bamboo, masonry (Assam type), or Reinforced Cement Concrete (RCC)	Wall_thickness	Less than 5 cm, 5 to 10 cm, or more than 10 cm

3. Data Collection for Flood Resilience Study

In Barak Valley, flood hazards occur at regular intervals, which mainly affect the infrastructure systems and a considerable rise in annual rainfall in the past few years has contributed to flooding. It has already been stated that for accurate resilience quantification, a properly formatted dataset of past disasters should be utilized; however, the collection of such data is often a difficult task.

As India is a developing country, the proper collection and management of pre- and post-disaster data for housing should be readily available for most governmental agencies. Post-disaster data has been provided by some governmental agencies, but the volume of data provided is inadequate as their variables differ from those in this work. For example, the datasets divided the housing infrastructure into two types, namely, pucca and kutcha, and the number of damaged houses in terms of pucca and the kutcha type was provided. In our work, however, we required the data in terms of bamboo, masonry, and RC type. Moreover, information on the maximum considered nodes in this BBN model is unavailable.

To overcome this challenge and to acquire the necessary data, an extensive field survey was performed in various flood-affected areas in this valley regarding housing infrastructure systems. A flood resilience assessment form was prepared for the survey, as shown in Figure 9. In this process, we visited 23 vulnerable places and 1 non-vulnerable place. Post-disaster data were collected for more than 500 houses, with each survey taking around 20–25 min.

FLOOD-RESILIENCE ASSESSMENT FORM **DATE:** *09/02/19*

GPS location:	*24.7421°N, 92.6115°E*	**Elevation:**	*25 m*

Place Name: *Algapur (Hailakandi District)*

RELIABILITY INFORMATION

Type of House:	Bamboo ✓	Assam Type/Masonry	R.C.C.

Year of Construction:	*2006 (12 yrs 8 months)*		
Numbers of stories:	*01*	**Floor Height:**	*2.5 m*

Foundation type:	Pile or deep	Shallow	Crawlspace ✓	other: Explain

Height of plinth w.r.t. to road:	*1 m*	**Flood Depth:**	*3 ft.*
Wall Thickness:	*4 cm*	**Drainage Available:**	*NO*
Overall damage state:	DS1	DS2	DS3 ✓	DS4

FLOOD INFORMATION

Reason of flood:

- ✓ High intensity rainfall in saturated soil
- ✓ Poor drainage system
- Failure of flood embankment
- Overflow of water from water bodies, such as river, lake or ocean
- Others

RECOVERY INFORMATION

Annual Income (Indian Rupees):	Less than 1 Lacs ✓	1 lacs – 2 Lacs	More than 2 Lacs
Resource Availability:	No	Yes	Yes at increased rate ✓ *(10%)*

Approachability/Accessibility:	*Yes (10 days)*		
Insurance:	*NO*	**Relief Received:**	*Yes*
Expected Recovery time:	*3 weeks (21 days)*		

Education	Below 10ᵗʰ Std	Matriculation ✓	12ᵗʰ Std passed	Graduation

Figure 9. Flood resilience assessment form.

During the survey, respondents provided recommendations for future preparedness, which we discuss in the last section of this document. The form was designed in a generalized manner, such that it can be used for resilience quantification of housing infrastructure systems in different communities. This collected data and associated data-driven resilience analysis are beneficial for improving the preparedness of a housing infrastructure system for future flood disasters, increasing its structural reliability, and enabling a thorough risk assessment against flood hazards.

4. Results and Discussion

4.1. Model Validation

The validation of a model is critical, as an inaccurate model will always provide erroneous results. Therefore, it is very important to validate the proposed BBN model such that it can provide accurate information. In this study, two qualitative validation approaches, namely, an extreme condition test and a scenario analysis, were performed to validate the proposed BBN model [77]. The experts' judgment played a critical role in developing and validating the proposed BBN model due to data scarcity. Initially, the prior probability of each parent node was assigned from the collected field survey dataset. As an example, consider $P(Y)$, the prior probability of "Y" and, say, there are two variables for Y, namely, "yes" and "no." If X is the total number of data collected, out of which, "A" is the number of data for the "no" variable and "B" is the number of data for the "yes" variable, then the prior probability of "Y" is assigned based on Equation (2) [78]:

$$P(Y = \text{no}) = A/X \text{ and } P(Y = \text{yes}) = B/X, \tag{2}$$

and the CPT between the parameters is obtained based on expert judgment and the collected data. During the development of the CPT, the DPO of DDMA and other industry experts generated the CPT values for each intermediate and child node based on their knowledge and the available literature. Next, the academic experts modified the CPT values according to their experience, and finally, CPT values were assigned between the parent and child node.

In this study, 332 CPT values were generated for the proposed BBN model. The recovery, reliability, and resilience values for different vulnerable places were computed using Netica software [79].

4.1.1. Extreme Condition Analysis

In the extreme condition analysis, two extreme conditions, namely, extreme 1 (E-1) and extreme 2 (E-2), were considered for the analysis in this process. E-1 represented one of the most vulnerable places (Burunga) in this valley, where all the parent nodes of resilience were in the worst condition states, while E-2 represented a non-vulnerable place (Tarapur), where all the parent nodes were in favorable condition states. The proposed BBN model for resilience based on the housing infrastructure system was applied to these two extreme conditions. Here, for the E-1 and E-2 conditions, the recovery for Burunga and Tarapur was estimated as being (low, medium, high) = (75.9, 15.7, 8.38) and (24.5, 26.6, 48.9), respectively. Similarly, the reliability for Burunga and Tarapur was estimated as (DS1, DS2, DS3, DS4) = (12.2, 15.2, 21.6, 51.0) and (33.2, 24.6, 22.1, 20.1), respectively. The resilience for Burunga and Tarapur was estimated as (low, medium, high) = (66.9, 18.4, 14.7) and (29.3, 25.7, 45.0), respectively. The evaluated values indicated that E-1 and E-2 had the highest probabilities of 66.9% and 45.0% at the low and high resiliencies, respectively. These indicated that for the E-1 condition, the probability of resilience in the low state was higher than other states, but for the E-2 condition, the probability of resilience was in the high state was higher. The E-1 and E-2 tests showed that the proposed BBN model based on the housing infrastructure system performed according to the assumed model behavior, which also indicated that the proposed BBN model was valid.

4.1.2. Scenario Analysis

In the scenario analysis, five different types of scenarios were considered for the quantification of reliability and recovery. The probability states of all resilience parameters for all five scenarios are presented in Tables 5 and 6. Scenario 1 represents all the parent nodes in the severe condition; with subsequent progress of the parent nodes, the conditions improved gradually to ultimately reach the best condition, as represented by scenario 5. The probability states for the resilience parameters for each scenario were assigned based on the condition of the scenario. Based on the probability states, the reliability and recovery for all the scenarios were calculated, as shown in Tables 5 and 6.

Table 5. Scenario analysis for recovery.

Parameter	Scenario 1	Scenario 2	Scenario 3	Scenario 4	Scenario 5
Insurance	No	No	No	No	Yes
Income	Less than 10,000	10,000 to 20,000	10,000 to 20,000	More than 20,000	More than 20,000
Education	Below 10th standard	10th or 12th standard passed	10th or 12th standard passed	Graduate	Graduate
Relief_received	No	No	Yes	Yes	Yes
Resource_availability	No	Yes with more than a 10% increase	Yes with more than a 10% increase	Yes with more than a 10% increase	Yes with an increase by 0 to 10%
Approachability	Yes	Yes	No	No	No
Recovery					
Low	74.3	50.9	32.9	26.5	1.7
Medium	18.3	29	35.1	33	11.9
High	7.4	20.1	31.9	40.5	86.4

Table 6. Scenario analysis for reliability.

Parameter	Scenario 1	Scenario 2	Scenario 3	Scenario 4	Scenario 5
Type_of_house	Bamboo	Assam Type	RCC	RCC	RCC
Wall_thickness	Less than 5 cm	5 to 10 cm	More than 10 cm	More than 10 cm	More than 10 cm
Building_age	More than 20	More than 20	Up to 20	Up to 10	Up to 10
Number_of_floor	1	1	More than 1	More than 1	More than 1
Plinth level	More than 2 m	More than 2 m	More than 2 m	Up to 1 m	Up to 1 m
Foundation_type	Shallow	Shallow	Shallow	Deep	Deep
Flood_depth	More than 90 cm	More than 90 cm	More than 90 cm	30 to 90 cm	Less than 30 cm
Drainage	No	No	No	No	Yes
Reliability					
DS1	4.5	9.9	17.5	28.3	87.2
DS2	7.8	16.1	27.8	28.7	7.8
DS3	30.9	29.4	27.3	23	3.3
DS4	56.8	44.6	27.4	20	1.7

In scenario 1, the probabilities of recovery and reliability were (low, medium, high) = (74.3, 18.3, 7.4) and (DS1, DS2, DS3, DS4) = (4.5, 7.8, 30.9, 56.8); in scenario 2, the probabilities of a low state recovery and DS4 state reliability decreased from 74.3 to 50.9 and from 56.8 to 44.6, respectively, while the probabilities of a high state recovery and DS1 state reliability increased from 7.4 to 20.1 and from 4.5 to 9.9, respectively; in scenario 3, the probabilities of a low state recovery and DS4 state reliability decreased to 32.9 and 27.4, respectively, while the probabilities of a high state recovery and DS1 state reliability increased to 31.9 and 17.5, respectively; in scenario 4, the probabilities of a low state recovery and DS4 state reliability decreased to 26.4 and 20.0, respectively, while the probabilities of a high state recovery and DS1 state reliability increased to 40.5 and 28.3, respectively; in scenario 5, the probabilities of a low state recovery and DS4 state reliability decreased to 1.7 and 1.7, respectively, while the probabilities of a high state recovery and DS1 state reliability increased to 86.4 and 87.2, respectively. All five scenarios represented the desired model behavior. Similarly, different combinations of parameters were considered to generate different scenarios and their recovery and reliability probability distributions were tested to perform the model validation. Based on the results of the analysis and the discussion presented above, it is believed that the proposed BBN model was validated.

The validation of the proposed BBN model using different approaches also indicated that the constructed CPTs based on the experts' knowledge were correct. Hence, this model can be used for resilience quantification for housing infrastructure against flood hazards.

4.2. Sensitivity Analysis

Sensitivity analysis in the BBN is broadly concerned with understanding the relationship between local network parameters and global conclusions drawn from the network [80–84]. A sensitivity analysis was performed to achieve the following objectives: (i) to identify the critical parameters for reliability and recovery of a housing infrastructure system and (ii) to identify the possible changes of the dependent parameters in the BBN model that can ensure the satisfaction of a query constraint for the target reliability, recovery, or resilience. Sensitivity analysis provides essential information about the results and their variance according to a very small change in the input value with uncertainty [53,80]. This analysis included an investigation of the effect of changes in uncertain input parameters on the uncertainty of the response of interest. The sensitivity analysis also reduced the predicted uncertainty as it identified the high-impact parameters. Here, the variance reduction (VR) method was utilized to identify the sensitivity of the parameters of the proposed BBN model based on the housing infrastructure system [85,86]. This method computes the VR of the expected real value of a query node R, for example, reliability and recovery, due to a result that was caused by changing variable node P, such as Drainage, Type_of_house, Income, or Resource_availability. Therefore, the variance of the real value of R given evidence on P, namely, $V(R\,|\,q)$, can be computed using Equation (3) [84]:

$$V(R|q) = \sum_z p(r|q)[Y_r - E(R|q)]^2,\tag{3}$$

where r is the state of the query node R, q is the state of the varying variable node P, $p(r\,|\,q)$ is the conditional probability of r given q, Y_r is the value corresponding to state r, and $E(R\,|\,q)$ is the expected real value of R, after the new finding q for node P.

The VR and percentage of VR of the parent nodes for the child node recovery and reliability are shown in Figure 10. For the recovery, Insurance showed the highest contribution (2.87%) to the percentage of VR, followed by Relief_received (2.13%), Income (1.05%), Approachability (0.91%), Resource_availability (0.79%), and Education (0.08%). It can be observed that the parameters Education and Resource_availability were far less sensitive for recovery. Similarly, regarding reliability, Type_of_house showed the highest contribution (5.06%) to the percentage of VR, followed by Wall_thickness (4.69%), Drainage (1.53%), Flood_depth (1.04%), Building_age (0.63%), Number_of_stories (0.29%), Foundation_type (0.26%), and Plinth_level (0.19%). It can be observed that the parameters Plinth_level, Foundation_type, and Number_of_stories were far less sensitive regarding reliability.

The sensitivity analysis aligned with the expert statements, as recovery was highly dependent on External_factor (24.1%) and Personal_factor (15%), followed by Insurance and Relief_received, as it is known that the recovery for insured houses is relatively fast; similarly, after a disaster, if a stakeholder provides relief to vulnerable places, then the recovery process can be fast. Reliability was highly dependent on Super_Struc (37.3%) and Sub_Struc_Ex (28.4%), followed by Type_of_house and Wall_thickness, as the resisting ability of RC houses (Type_of_house) against flood hazards is greater compared to bamboo houses, and with an increase of wall thickness, the withstanding capability against hazards increases. Lower wall thickness impacts the Super_Struc, which directly impacts the reliability of the housing infrastructure system, and finally, affects the resilience of the system. It can be observed from the outcome of this analysis that the sensitivity of the child node highly depended on the variability of the parent nodes. This technique also provided information for optimal changes to parameters that were required to obtain a targeted recovery, reliability, and resilience of an infrastructure system. The crucial parameters to recovery and reliability were identified, thereby providing information for tuning those sensitive parameters for increasing the reliability of systems and to speed up the recovery process.

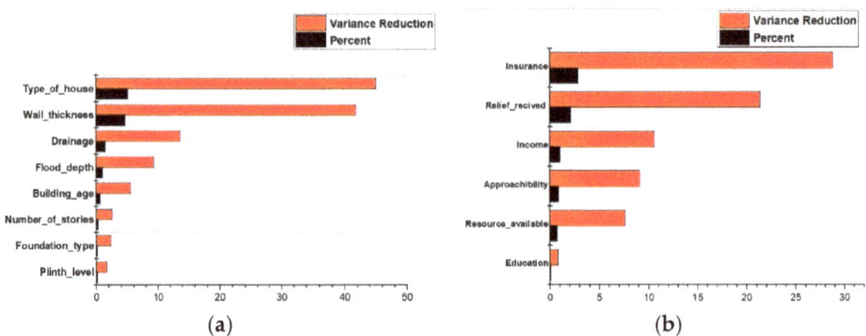

Figure 10. Sensitivity analysis for (**a**) reliability and (**b**) recovery.

4.3. Reliability, Recovery, and Resilience Values

In this section, the evaluated reliability, recovery, and resilience of the housing infrastructure for Barak Valley is discussed. In this work, the evaluated resilience values considering different parameters represent the resistance to flood hazards and recoverability after the occurrence of the hazard. Tables 7 and 8 show the reliability, recovery, and resilience of Barak Valley. It can be observed from Table 7 that the reliability of more than 50% of the housing infrastructure of Burunga (one of the locations visited) fell under the DS4 state, which indicates that the reliability of that location was extremely low. Similarly, locations such as Poschim Kumrapara, Algapur, and Rajnagar fell under the same probability state. The housing infrastructure recovery of Burunga was also not positive, as can be observed from Table 8, as more than 75% of the housing infrastructure recovery fell under the low probability state. It has been stated that resilience is a combination of recovery and reliability.

It can be observed from Table 8 that the housing infrastructure resilience of Burunga had a maximum probability in the low state. It can be noted, by including the dependencies between the resilience parameters in the model, the results changed compared to the evaluated results by Sen et al. [31]. According to Sen et al., the probabilities of housing infrastructure (Algapur) were 0, 0.1, 0.6, and 0.3, but in this study, the evaluated values were 12.7, 15.3, 21.8, and 50.1 [31]. During the field visit, it was observed that the maximum of the housing infrastructure was in an extremely hazardous situation, which means that the evaluated values of this study were more reliable than the earlier study. Similarly, for Tarapur, it was observed that there were some houses that needed immediate attention in terms of strengthening, but in the earlier study, the probability of housing infrastructure in the DS4 state was zero. In this study, 20.1% of the housing infrastructure of this place was in the DS4 state, which indicated that this study provided more accurate results. Overall, this indicates that the BBN approach was better than the VE method.

Table 7. Reliability values of all flood vulnerable areas.

Place Name	DS1	DS2	DS3	DS4
Algapur	12.7	15.3	21.8	50.1
Amjurghat	13.6	16.1	21.8	48.5
Anipur Grant	16.4	18	21.9	43.7
Baleswar	16.9	18.2	21.9	43
Bhatirkupa	19.2	18.9	22	39.8
Borbond	17.5	18.3	22	42.2
Burunga	12.2	15.2	21.6	51
Dullabcherra	21.1	20.3	21.7	36.9
Dwarbond	19.8	19	25	36.1
Fanai Cherra Grant	16.9	19	21.9	42.9
Hailakandi Town	20.7	21	21.7	36.6
Jamira	17.2	18	22	42.8
Kanakpur	18.8	18.6	22.1	40.5
Katlicherra	16.6	18.5	21.8	43.1
Lalaghat	11.3	14.5	21.5	52.8
Rajnagar	12.7	15.6	21.7	50.1
Panchgram	19.8	19.7	21.8	38.7
Poschim Kumarpara	11.1	15	21.2	52.8
Rakhal Khalerpaar	13.5	17	21.6	47.6
Rangirghat	17.5	18.7	21.8	41.9
Ratnapur	16.3	18	21.9	43.8
Silchar Municipality	15.7	17.4	21.9	45
Tarapur	33.2	24.6	22.1	20.1
Uttar Krishnapur	18.2	18.9	21.9	41

Table 8. Recovery and resilience values of all flood-vulnerable areas.

Place Name	Recovery			Resilience		
	Low	Medium	High	Low	Medium	High
Algapur	54.2	25.5	20.2	54.9	23.2	21.9
Amjurghat	57	24.3	18.7	55.9	22.7	21.4
Anipur Grant	52.1	22.4	21.9	51.5	23.8	24.7
Baleswar	53.8	25.1	21.1	52.2	23.6	24.2
Bhatirkupa	60.8	23	16.2	54.7	23	22.2
Borbond	59.8	23.8	16.4	55.1	23.1	21.8
Burunga	75.9	15.7	8.38	66.9	18.4	14.7
Dullabcherra	54.8	24.3	20.9	50.4	24	25.6
Dwarbond	56.4	24.1	19.5	51.6	23.8	24.6
Fanai Cherra Grant	56.2	23.9	19.9	53.3	23.2	23.4
Hailakandi Town	46.8	23.6	26.9	46.1	25	28.9
Jamira	46.6	26.3	27.1	48.1	24.5	27.4
Kanakpur	60	23.4	16.6	54.6	23.1	22.4
Katlicherra	53.3	24.7	22	51.9	23.6	24.5
Lalaghat	56.2	24.9	18.9	56.9	22.5	20.5
Rajnagar	60.1	22.7	17.2	58.1	21.8	20.1
Panchgram	66.3	20.3	13.4	57.1	22	20.9
Poschim Kumarpara	68.1	19.5	12.4	63.5	19.8	16.7
Rakhal Khalerpaar	54.8	24.3	20.9	54.4	22.9	22.7
Rangirghat	48.8	25.8	25.3	59	24.3	26.7
Ratnapur	49.1	25.8	25.1	49.8	24.1	26.1
Silchar Municipality	53.6	24.9	21.4	52.8	23.5	23.8
Tarapur	24.5	26.6	48.9	29.3	25.7	45
Uttar Krishnapur	59.2	23	17.9	54.2	23	22.8

As most of the surveyed areas are vulnerable to flood hazards, it is clear that the recovery, reliability, and resilience values for the housing infrastructure system were categorized as low, DS4, and low, respectively. To obtain more detailed information for the reliability, recovery, and resilience of the housing infrastructure of Barak Valley, the "low" state of recovery and resilience, and the DS4 state of reliability were further sub-divided into four additional categories according to the percentile of the total low state values (higher to lower): extremely low ($\geq$75th percentiles), very low (50th to 74th percentile), moderate–low (26th to 49th percentile), and low ($\leq$25th percentile). Finally, three types of flood models—a flood recovery model, flood reliability model, and flood resilience model—of the valley were prepared based on the categorization of the low state, as shown in Figures 11–13.

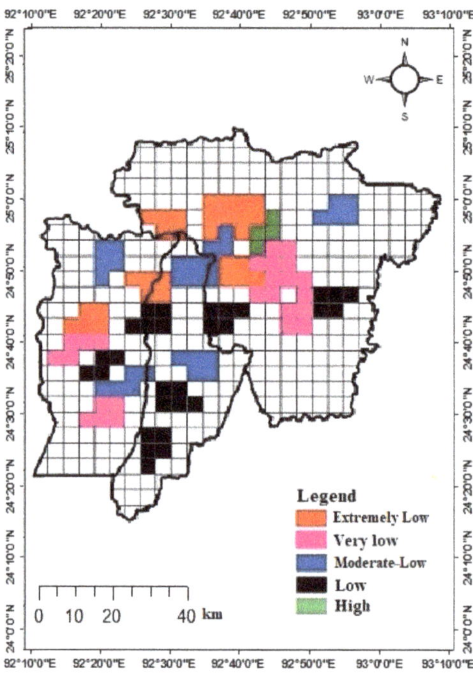

Figure 11. The flood recovery model of the housing infrastructure in Barak Valley.

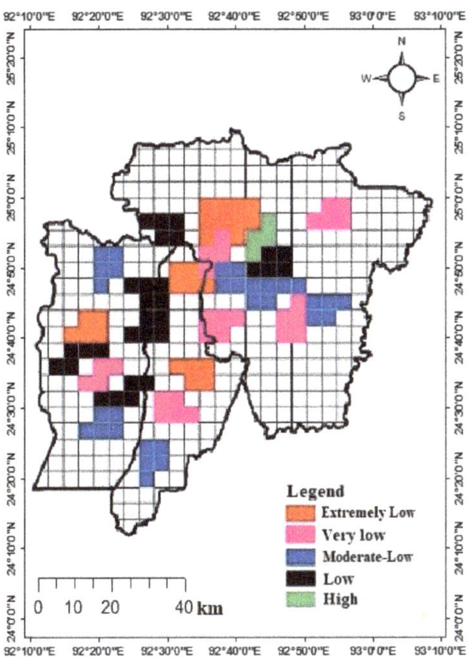

Figure 12. The flood reliability map of the housing infrastructure in Barak Valley.

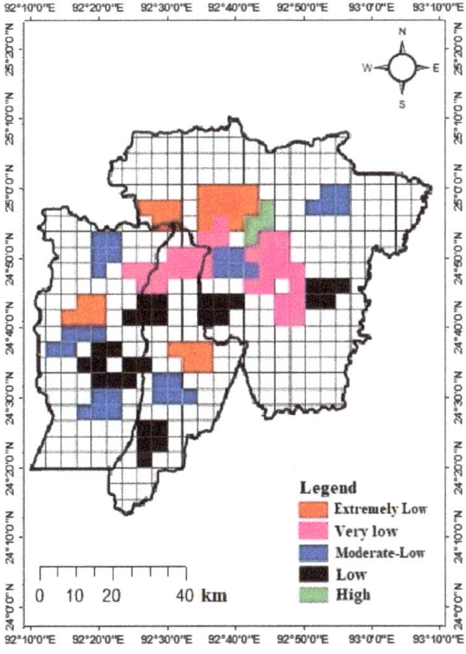

Figure 13. The flood resilience map of the housing infrastructure in Barak Valley.

5. Conclusions, Limitations, and Future Research Direction

This work illustrates a BBN model for flood resilience quantification of the housing infrastructure system in Barak Valley of Northeast India. The main challenge faced during the quantification of resilience values was a lack of proper post-disaster data. To overcome the challenge, a flood resilience assessment form was developed and a detailed survey was performed in various vulnerable places according to the DDMA, and the resilience values were evaluated for those places. Those quantified values provide a realistic scenario of the housing infrastructure system of this valley, which supports the planners, designers, policymakers, and stakeholders of this valley to become involved in the detailed identification of resilience-influencing parameters for housing systems and the preparedness of these vulnerable places against flood hazards. In this study, influential parameters for resilience quantification were considered and localized and could be mapped with global resilience quantification, which may include a new study area. Lastly, a sensitivity analysis was performed to find the most crucial parent nodes of recovery and reliability, which can also help decision-makers of this valley to focus on the most sensitive parent parameters; to improve a child node, it was not necessary to improve all the associated parent nodes because small changes in the most sensitive parent parameters may lead to a targeted probability of the child node. This analysis will also help in the decision-making process for preparedness against future hazards [86]. As this method is generalized, it can be integrated with any kind of infrastructure system and can be used by a public authority for the resilience quantification of any infrastructure system against flood hazards. The main contributions of the proposed resilience model were: (i) the BBN model provided a more realistic scenario of housing infrastructure for this valley based on collected real data and can be updated by including more uncertain parameters and associated data, and (ii) the sensitivity analysis helped to identify the crucial parent parameters of recovery and reliability against flood hazards.

The following are the recommendations for improving the resilience of the housing/building infrastructure system of this valley against flood hazards based on the discussions with affected householders during the field survey: (i) construction of building infrastructure should follow engineering principles; (ii) people should have a solid understanding of reliability and recovery processes related to housing infrastructure in preparedness for future disasters in the valley; (iii) stakeholders should immediately give attention to the housing infrastructure of Burunga, Poschim Kumrapara, Rajnagar, Panchgram, and Lalaghat, as the resilience of these places was extremely low. Moreover, it was observed that the parent parameters, such as Type_of_house and Wall_thickness were most sensitive regarding reliability, and Insurance and Relief_received were the most vulnerable parent parameters for the recovery of the housing infrastructure against flood hazards. Therefore, decision-makers should strengthen these sensitive parameters to make the infrastructure more resilient against future floods. There are some noted limitations in this work, such as (i) the consideration of more factors for a comprehensive framework is required, (ii) more detailed information about the factors or more data collection is required, and (iii) the involvement of multiple experts from various disciplines is required. In the future, resilience scenarios for other infrastructure systems, such as water, electrical, transportation, and telecommunication systems, as well as critical housing infrastructure systems, such as hospitals, markets, and schools, can be evaluated. Similarly, resilience values against other natural disasters, such as earthquakes and landslides, can be computed [87–89]. Outcome comparisons with other hierarchical-based methods, such as fuzzy AHP (Analytic hierarchy process) and Dempster–Shafer theory, can also be performed. As infrastructure resilience changes with time, these variabilities can be captured with the help of a dynamic Bayesian network.

Author Contributions: Conceptualization, M.K.S., S.D. and G.K.; methodology, M.K.S., S.D. and G.K.; software, M.K.S., S.D. and G.K.; validation M.K.S., S.D. and G.K.; formal analysis, M.K.S.; investigation, M.K.S., S.D. and G.K.; data curation, M.K.S.; writing—original draft preparation,

M.K.S.; writing—review and editing, S.D. and G.K.; visualization, M.K.S.; supervision, S.D. and G.K. All authors have read and agreed to the published version of the manuscript.

Funding: The first author acknowledges the student's scholarship received from the Ministry of Human Resource and Development, Government of India. The third author acknowledges the financial support through Natural Science Engineering Research Council, Canada Discovery Grant Program (RGPIN-2019-04704) for the professional editing, proofreading, and article processing fees.

Institutional Review Board Statement: Not applicable.

Informed Consent Statement: Informed consent was obtained from all subjects involved in the study.

Data Availability Statement: The sample data collection sheet and the data used for analysis, along with the associated computer programs, can be available on request from the corresponding author.

Acknowledgments: The authors acknowledge the support of the District Disaster Management Authority (DDMA), Assam, India, for providing the information of flood-vulnerable places of Barak Valley in Northeast India, as well as their valuable role in providing experts judgment.

Conflicts of Interest: The authors declare no conflict of interest.

References

1. Bruneau, M.; Chang, S.E.; Eguchi, R.T.; Lee, G.C.; O'Rourke, T.D.; Reinhorn, A.M.; Von Winterfeldt, D. A framework to quantitatively assess and enhance the seismic resilience of communities. *Earthq. Spectra* **2003**, *19*, 733–752. [CrossRef]
2. Hosseini, S.; Barker, K.; Ramirez-Marquez, J.E. A review of definitions and measures of system resilience. *Reliab. Eng. System Saf.* **2016**, *145*, 47–61. [CrossRef]
3. Meerow, S.; Newell, J.P.; Stults, M. Defining urban resilience: A review. *Landsc. Urban Plan.* **2016**, *147*, 38–49. [CrossRef]
4. Liu, W.; Song, Z. Review of studies on the resilience of urban critical infrastructure networks. *Reliab. Eng. System Saf.* **2020**, *193*, 106617. [CrossRef]
5. Bruneau, M.; Reinhorn, A. Exploring the concept of seismic resilience for acute care facilities. *Earthq. Spectra* **2007**, *23*, 41–62. [CrossRef]
6. Cimellaro, G.P.; Reinhorn, A.M.; Bruneau, M. Seismic resilience of a hospital system. *Struct. Infrastruct. Eng.* **2010**, *6*, 127–144. [CrossRef]
7. Cimellaro, G.P.; Renschler, C.; Reinhorn, A.M.; Arendt, L. PEOPLES: A framework for evaluating resilience. *J. Struct. Eng.* **2016**, *142*, 04016063. [CrossRef]
8. Kammouh, O.; Zamani Noori, A.; Cimellaro, G.P.; Mahin, S.A. Resilience assessment of urban communities. *ASCE-ASME J. Risk Uncertain. Eng. Syst. Part A Civ. Eng.* **2019**, *5*, 04019002. [CrossRef]
9. FEMA. Multi-Hazard Loss Estimation Methodology: Flood Model. HAZUS-MH MR4 Technical Manual. 2009. Available online: www.fema.gov/plan/prevent/hazus (accessed on 16 July 2019).
10. Farsangi, E.N.; Takewaki, I.; Yang, T.Y.; Astaneh-Asl, A.; Gardoni, P. *Resilient Structures and Infrastructure*; Springer: Berlin, Germany, 2019.
11. Masoomi, H.; van de Lindt, J.W. Community-resilience-based design of the built environment. *ASCE-ASME J. Risk Uncertain. Eng. Syst. Part A Civ. Eng.* **2019**, *5*, 04018044. [CrossRef]
12. Sen, M.K.; Dutta, S.; Gandomi, A.H.; Putcha, C. Case Study for Quantifying Flood Resilience of Interdependent Building–Roadway Infrastructure Systems. *ASCE-ASME J. Risk Uncertain. Eng. Syst. Part A Civ. Eng.* **2021**. (Accepted, forthcoming). [CrossRef]
13. Ghorbani-Renani, N.; González, A.D.; Barker, K.; Morshedlou, N. Protection-interdiction-restoration: Tri-level optimization for enhancing interdependent network resilience. *Reliab. Eng. Syst. Saf.* **2020**, 106907. [CrossRef]
14. Hossain, N.U.I.; El Amrani, S.; Jaradat, R.; Marufuzzaman, M.; Buchanan, R.; Rinaudo, C.; Hamilton, M. Modeling and assessing interdependencies between critical infrastructures using Bayesian network: A case study of inland waterway port and surrounding supply chain network. *Reliab. Eng. Syst. Saf.* **2020**, *198*, 106898. [CrossRef]
15. Windle, G.; Bennett, K.M.; Noyes, J. A methodological review of resilience measurement scales. *Health Qual. Life Outcomes* **2011**, *9*, 8. [CrossRef] [PubMed]
16. Mahmoud, H.; Chulahwat, A. Spatial and temporal quantification of community resilience: Gotham City under attack. *Comp.-Aided Civ. Infrastruct. Eng.* **2018**, *33*, 353–372. [CrossRef]
17. Gardoni, P.; Murphy, C. Society-based design: Promoting societal well-being by designing sustainable and resilient infrastructure. *Sustain. Resili. Infrastruct.* **2020**, *5*, 4–19. [CrossRef]
18. Didier, M.; Baumberger, S.; Tobler, R.; Esposito, S.; Ghosh, S.; Stojadinovic, B. Seismic resilience of water distribution and cellular communication systems after the 2015 Gorkha earthquake. *J. Struct. Eng.* **2018**, *144*, 04018043. [CrossRef]
19. Zhang, M.; Yang, Y.; Li, H.; van Dijk, M.P. Measuring urban resilience to climate change in three chinese cities. *Sustainability* **2020**, *12*, 9735. [CrossRef]

20. Cajete, G.A. Indigenous Science, Climate Change, and Indigenous Community Building: A Framework of Foundational Perspectives for Indigenous Community Resilience and Revitalization. *Sustainability* **2020**, *12*, 9569. [CrossRef]
21. Skondras, N.A.; Tsesmelis, D.E.; Vasilakou, C.G.; Karavitis, C.A. Resilience–Vulnerability Analysis: A Decision-Making Framework for Systems Assessment. *Sustainability* **2020**, *12*, 9306. [CrossRef]
22. Miller-Hooks, E.; Zhang, X.; Faturechi, R. Measuring and maximizing resilience of freight transportation networks. *Comp. Operat. Res.* **2012**, *39*, 1633–1643. [CrossRef]
23. Queiroz, C.; Garg, S.K.; Tari, Z. A probabilistic model for quantifying the resilience of networked systems. *IBM J. Res. Dev.* **2013**, *57*, 1–3. [CrossRef]
24. Berche, B.; Von Ferber, C.; Holovatch, T.; Holovatch, Y. Resilience of public transport networks against attacks. *Eur. Phys. J. B* **2009**, *71*, 125–137. [CrossRef]
25. Dorbritz, R. Assessing the resilience of transportation systems in case of large-scale disastrous events. In Proceedings of the 8th International Conference on Environmental Engineering, Vilnius Gediminas Technical University, Vilnius, Lithuania, 19–20 May 2011; pp. 1070–1076.
26. Heaslip, K.; Louisell, W.; Collura, J.; Urena Serulle, N. A sketch level method for assessing transportation network resiliency to natural disasters and man-made events. In Proceedings of the Conference of Transportation Research Board 89th Annual Meeting, Washington, DC, USA, 10–14 January 2010; pp. 10–3185. Available online: https://trid.trb.org/view/910940 (accessed on 15 January 2019).
27. Cimellaro, G.P.; Reinhorn, A.M.; Bruneau, M. Framework for analytical quantification of disaster resilience. *Eng. Struct.* **2010**, *32*, 3639–3649. [CrossRef]
28. Tamvakis, P.; Xenidis, Y. Comparative evaluation of resilience quantification methods for infrastructure systems. *Proc. Soc. Behav. Sci.* **2013**, *74*, 339–348. [CrossRef]
29. Koliou, M.; van de Lindt, J.W.; McAllister, T.P.; Ellingwood, B.R.; Dillard, M.; Cutler, H. State of the research in community resilience: Progress and challenges. *Sustain. Resil. Infrastruct.* **2020**, *5*, 131–151. [CrossRef] [PubMed]
30. Ellingwood, B.R.; Cutler, H.; Gardoni, P.; Peacock, W.G.; van de Lindt, J.W.; Wang, N. The centerville virtual community: A fully integrated decision model of interacting physical and social infrastructure systems. *Sustain. Resil. Infrastruct.* **2016**, *1*, 95–107. [CrossRef]
31. Sen, M.K.; Dutta, S.; Laskar, J.I. A Hierarchical Bayesian Network Model for Flood Resilience Quantification of Housing Infrastructure Systems. *ASCE-ASME J. Risk Uncertain. Eng. Syst. Part A Civ. Eng.* **2021**, *7*, 04020060. [CrossRef]
32. Sen, M.K.; Dutta, S. An Integrated GIS-BBN Approach to Quantify Resilience of Roadways Network Infrastructure System against Flood Hazard. *ASCE-ASME J. Risk Uncertain. Eng. Syst. Part A Civ. Eng.* **2020**, *6*, 04020045.
33. Liao, T.Y.; Hu, T.Y.; Ko, Y.N. A resilience optimization model for transportation networks under disasters. *Nat. Hazards* **2018**, *93*, 469–489. [CrossRef]
34. Markolf, S.A.; Hoehne, C.; Fraser, A.; Chester, M.V.; Underwood, B.S. Transportation resilience to climate change and extreme weather events–Beyond risk and robustness. *Transp. Policy* **2019**, *74*, 174–186. [CrossRef]
35. Raoufi, H.; Vahidinasab, V.; Mehran, K. Power Systems Resilience Metrics: A Comprehensive Review of Challenges and Outlook. *Sustainability* **2020**, *12*, 9698. [CrossRef]
36. Panteli, M.; Mancarella, P.; Trakas, D.N.; Kyriakides, E.; Hatziargyriou, N.D. Metrics and quantification of operational and infrastructure resilience in power systems. *IEEE Trans. Power Syst.* **2017**, *32*, 4732–4742. [CrossRef]
37. Baroud, H.; Ramirez-Marquez, J.E.; Barker, K.; Rocco, C.M. Stochastic measures of network resilience: Applications to waterway commodity flows. *Risk Analysis* **2014**, *34*, 1317–1335. [CrossRef] [PubMed]
38. Hosseini, S.; Barker, K. Modeling infrastructure resilience using Bayesian networks: A case study of inland waterway ports. *Comp. Ind. Eng.* **2016**, *93*, 252–266. [CrossRef]
39. Omer, M.; Nilchiani, R.; Mostashari, A. Measuring the resilience of the trans-oceanic telecommunication cable system. *IEEE Syst. J.* **2009**, *3*, 295–303. [CrossRef]
40. Tipper, D. Resilient network design: Challenges and future directions. *Telecommunicat. Syst.* **2014**, *56*, 5–16. [CrossRef]
41. Census. District Census 2011 Report. 2011. Available online: https://www.census2011.co.in/district.php (accessed on 16 February 2019).
42. ASDMA Report. Available online: http://sdmassam.nic.in/ (accessed on 12 May 2019).
43. District Report. Available online: https://www.icssr.org. (accessed on 12 April 2019).
44. Bhuvan. Satellite Image. Available online: https://bhuvan-app1.nrsc.gov.in/bhuvan2d/bhuvan/bhuvan2d.php (accessed on 2 July 2019).
45. Grigsby, L.L. *Electric Power Generation, Transmission, and Distribution*; CRC Press: Boca Raton, FL, USA, 2016.
46. De Risi, R.; Jalayer, F.; De Paola, F.; Lindley, S. Delineation of flooding risk hotspots based on digital elevation model, calculated and historical flooding extents: The case of Ouagadougou. *Stoch. Environ. Res. Risk Assess.* **2018**, *32*, 1545–1559. [CrossRef]
47. Saaty, T.L. *Decision Making with Dependence and Feedback: The Analytic Network Process*; RWS Publications: Pittsburgh, PA, USA, 1996.
48. Kosko, B. Fuzzy cognitive maps. *Int. J. Man Mach. Stud.* **1986**, *24*, 65–75. [CrossRef]
49. Tien, I.; Der Kiureghian, A. Reliability assessment of critical infrastructure using Bayesian networks. *J. Infrastruct. Syst.* **2017**, *23*, 04017025. [CrossRef]

50. Gehl, P.; Cavalieri, F.; Franchin, P. Approximate Bayesian network formulation for the rapid loss assessment of real-world infrastructure systems. *Reliab. Eng. Syst. Saf.* **2018**, *177*, 80–93. [CrossRef]

51. Cavalieri, F.; Franchin, P.; Gehl, P.; D'Ayala, D. Bayesian networks and infrastructure systems: Computational and methodological challenges. In *Risk and Reliability Analysis: Theory and Applications*; Gardoni, P., Ed.; Springer: Cham, Switzerland, 2017; pp. 385–415.

52. Koller, D.; Friedman, N. *Probabilistic Graphical Models: Principles and Techniques*; MIT Press: Cambridge, MA, USA, 2009.

53. Tien, I.; Der Kiureghian, A. Algorithms for Bayesian network modeling and reliability assessment of infrastructure systems. *Reliab. Eng. Syst. Saf.* **2016**, *156*, 134–147. [CrossRef]

54. Kabir, G.; Demissie, G.; Sadiq, R.; Tesfamariam, S. Integrating failure prediction models for water mains: Bayesian belief network based data fusion. *Knowl. Based Syst.* **2015**, *85*, 159–169. [CrossRef]

55. Sun, S.; Zhang, C.; Yu, G. A Bayesian network approach to traffic flow forecasting. *IEEE Trans. Intell. Transp. Syst.* **2006**, *7*, 124–132. [CrossRef]

56. Zhao, L.; Wang, X.; Qian, Y. Analysis of factors that influence hazardous material ransportation accidents based on Bayesian networks: A case study in China. *Saf. Sci.* **2012**, *50*, 1049–1055. [CrossRef]

57. Castillo, E.; Menéndez, J.M.; Sánchez-Cambronero, S. Predicting traffic flow using Bayesian networks. *Transp. Res. Part B Methodol.* **2008**, *42*, 482–509. [CrossRef]

58. Khanafer, R.M.; Solana, B.; Triola, J.; Barco, R.; Moltsen, L.; Altman, Z.; Lazaro, P. Automated diagnosis for UMTS networks using Bayesian network approach. *IEEE Tran. Veh. Technol.* **2008**, *57*, 2451–2461. [CrossRef]

59. Barco, R.; Nielsen, L.; Guerrero, R.; Hylander, G.; Patel, S. Automated Troubleshooting of a Mobile Communication Network Using Bayesian Networks. In Proceedings of the 4th International Workshop on Mobile and Wireless Communications Network, Stockholm, Sweden, 9–11 September 2002; IEEE: Piscataway, NJ, USA, 2002; pp. 606–610.

60. Yongli, Z.; Limin, H.; Jinling, L. Bayesian networks-based approach for power systems fault diagnosis. *IEEE Trans. Power Deliv.* **2006**, *21*, 634–639. [CrossRef]

61. Marrone, S.; Nardone, R.; Tedesco, A.; D'Amore, P.; Vittorini, V.; Setola, R.; Mazzocca, N. Vulnerability modeling and analysis for critical infrastructure protection applications. *Int. J. Crit. Infrastruct. Protect.* **2013**, *6*, 217–227. [CrossRef]

62. John, A.; Yang, Z.; Riahi, R.; Wang, J. A risk assessment approach to improve the resilience of a seaport system using Bayesian networks. *Ocean Eng.* **2016**, *111*, 136–147. [CrossRef]

63. Saliminejad, S.; Gharaibeh, N.G. A spatial-Bayesian technique for imputing pavement network repair data. *Comp. Aided Civ. Infrastruc. Eng.* **2012**, *27*, 594–607. [CrossRef]

64. Ang, A.H.S.; Tang, W.H. *Probability Concepts in Engineering: Emphasis on Application to Civil and Environmental Engineering*; Wiley: Hoboken, NJ, USA, 2007.

65. Tagg, A.; Laverty, K.; Escarameia, M.; Garvin, S.; Cripps, A.; Craig, R.; Clutterbuck, A. A new standard for flood resistance and resilience of buildings: New build and retrofit. *E3S Web Conf. EDP Sci.* **2016**, *7*, 13004. [CrossRef]

66. Jones, P. Housing resilience and the informal city. *J. Reg. City Plan.* **2017**, *28*, 129–139. [CrossRef]

67. van de Lindt, J.W.; Peacock, W.G.; Mitrani-Reiser, J.; Rosenheim, N.; Deniz, D.; Dillard, M.; Harrison, K. Community Resilience-Focused Technical Investigation of the 2016 Lumberton, North Carolina, Flood: An Interdisciplinary Approach. *Nat. Hazards Rev.* **2020**, *21*, 04020029. [CrossRef]

68. Golz, S.; Schinke, R.; Naumann, T. Assessing the effects of flood resilience technologies on building scale. *Urban Water J.* **2015**, *12*, 30–43. [CrossRef]

69. Cutter, S.L.; Burton, C.G.; Emrich, C.T. Disaster resilience indicators for benchmarking baseline conditions. *J. Homel. Secur. Emerg. Manag.* **2010**, *7*. [CrossRef]

70. De Iuliis, M.; Kammouh, O.; Cimellaro, G.P.; Tesfamariam, S. Downtime estimation of building structures using fuzzy logic. *Int. J. Dis. Risk Reduct.* **2019**, *34*, 196–208. [CrossRef]

71. Cavallaro, M.; Asprone, D.; Latora, V.; Manfredi, G.; Nicosia, V. Assessment of urban ecosystem resilience through hybrid social–physical complex networks. *Comp. Aided Civ. Infrastruct. Eng.* **2014**, *29*, 608–625. [CrossRef]

72. De Luna, F.; Fuentes-Mariles, O.; Ramos, J.; Sanchez, J.G. *Flood Risk Assessment in Housing under an Urban Development Scheme Simulating Water Flow in Plains. Chapters*; IntechOpen: London, UK, 2019.

73. Proverbs, D.; Lamond, J. Flood resilient construction and adaptation of buildings. In *Oxford Research Encyclopedia of Natural Hazard Science*; University of the West of England: Bristol, UK, 2017.

74. Pham, T.M.; Hagman, B.; Codita, A.; Van Loo, P.L.P.; Strömmer, L.; Baumans, V. Housing environment influences the need for pain relief during post-operative recovery in mice. *Physiol. Behav.* **2010**, *99*, 663–668. [CrossRef]

75. GOAL. Analysis of the Resilience of Communities to Disasters, Arc-D Toolkit User Guidance Manual. 2019. Available online: https://www.goalglobal.org/wp-content/uploads/2019/11/ARC-D-Toolkit-User-Manual-2016.pdf (accessed on 21 February 2019).

76. NIST. Available online: http://nist.gov (accessed on 2 July 2019).

77. Kleemann, J.; Celio, E.; Fürst, C. Validation approaches of an expert-based Bayesian Belief Network in Northern Ghana, West Africa. *Ecol. Model.* **2017**, *365*, 10–29. [CrossRef]

78. Kotu, V.; Bala, D. *Data Science: Concepts and Practice*; Morgan Kaufmann: Burlington, MA, USA, 2018.

79. Norsys Software Corp. Netica Version 4.16. Available online: www.norsys.com (accessed on 6 March 2019).

80. Bensi, M.; Der Kiureghian, A.; Straub, D. Bayesian network modeling of correlated random variables drawn from a Gaussian random field. *Struct. Saf.* **2011**, *33*, 317–332. [CrossRef]
81. Castillo, E.; Gutiérrez, J.M.; Hadi, A.S. Sensitivity analysis in discrete Bayesian networks. *IEEE Trans. Syst. Man Cybernet. Part A Syst. Hum.* **1997**, *27*, 412–423. [CrossRef]
82. Bensi, M.; Der Kiureghian, A.; Straub, D. Efficient Bayesian network modeling of systems. *Reliab. Eng. Syst. Saf.* **2013**, *112*, 200–213. [CrossRef]
83. Kjærulff, U.; van der Gaag, L.C. Making sensitivity analysis computationally efficient. *arXiv* **2013**, arXiv:1301.3868, 2013.
84. Pearl, J. *Probabilistic Reasoning in Intelligent Systems: Networks of Plausible Inference*; Elsevier: Amsterdam, The Netherlands, 2014.
85. Cheng, R.C. Variance reduction methods. In Proceedings of the 18th Conference on Winter Simulation, Washington, DC, USA, 8–10 December 1986; pp. 60–68.
86. Kabir, G.; Tesfamariam, S.; Francisque, A.; Sadiq, R. Evaluating risk of water mains failure using a Bayesian belief network model. *Eur. J. Operat. Res.* **2015**, *240*, 220–234. [CrossRef]
87. Bensi, M.; Der Kiureghian, A.; Straub, D. Framework for post-earthquake risk assessment and decision making for infrastructure systems. *ASCE-ASME J. Risk Uncertain. Eng. Syst. Part A Civ. Eng.* **2015**, *1*, 04014003. [CrossRef]
88. Marasco, S.; Cardoni, A.; Noori, A.Z.; Kammouh, O.; Domaneschi, M.; Cimellaro, G.P. Integrated platform to assess seismic resilience at the community level. *Sustain. Cities Soc.* **2020**, *64*, 102506. [CrossRef]
89. Franchin, P.; Cavalieri, F. Probabilistic assessment of civil infrastructure resilience to earthquakes. *Comp. Aided Civ. Infrastruct. Eng.* **2015**, *30*, 583–600. [CrossRef]

 *sustainability*

Article

Sustainability Assessment of Construction Technologies for Large Pipelines on Urban Highways: Scenario Analysis using Fuzzy QFD

Majed Alinizzi, Husnain Haider *, Meshal Almoshaogeh, Fawaz Alharbi, Saleh M. Alogla and Gamal A. Al-Saadi

Department of Civil Engineering, College of Engineering, Qassim University, 51452 Buraydah, Qassim, Saudi Arabia; alinizzi@qec.edu.sa (M.A.); m.moshaogeh@qec.edu.sa (M.A.); f.a@qec.edu.sa (F.A.); s.alogla@qec.edu.sa (S.M.A.); gamal_alsaadi@qec.edu.sa (G.A.A.-S.)
* Correspondence: husnain@qec.edu.sa

Received: 29 February 2020; Accepted: 22 March 2020; Published: 26 March 2020

Abstract: Urban highways users frequently face disruptions due to construction and maintenance of buried infrastructure. In conventional open cut construction, social costs (vehicle operating and traffic delay costs) are generally high at work zone construction areas (WZCA). Municipalities also bear additional costs due to early maintenance of alternate routes, i.e., non-work zone construction area (NWZCA). Besides, work zone and non-work zone areas together experience significant potential socio-economic and environment impacts. In addition to minimal disturbance to existing socioenvironmental setting and user cost savings, trenchless construction result in agency cost savings by avoiding early maintenance at NWZCA. Past studies primarily focused on social costs associated to WZCA. In present research, a sustainability assessment framework has been developed that includes agency and user costs at both the work zone and non-work zone area. The framework evaluates various traffic detoured scenarios (for open cut construction) and trenchless technology scenario based on all three dimensions of sustainability. Fuzzy Quality Function Deployment (Fuzzy QFD) method has been used to incorporate the interaction between the agency's sustainability objectives and public expectations for large-sized pipeline construction projects in urban areas. The framework effectively handles the uncertainties associated to data limitations and vagueness in expert opinion for subjective assessment criteria. To evaluate the pragmatism of proposed framework, it was applied on the case of a storm sewer construction project in Qassim Region, Saudi Arabia. Trenchless technology was found to be the most sustainable construction scenario followed by the open cut scenario with 50% traffic detoured to NWCA. The proposed methodology is also sought to enhance decision making process pertaining to the viability of trenchless technologies in KSA and elsewhere.

Keywords: trenchless technologies; pipelines; construction; sustainability; social cost; fuzzy QFD

---

1. Introduction

Most countries around the world, including Kingdom of Saudi Arabia (KSA), are facing major challenging in managing their buried infrastructure [1]. The majority of these buried infrastructure are constructed under the paved roads in urban areas [2]. Construction phase of the large-sized pipeline projects spans over long period and affect both the natural environment and socio-economic setting of the surrounding area. The projects are owned either by the municipalities (in case of public infrastructure) or the developers (in case of privately owned neighbourhoods). In their individual capacity, both of them intend to achieve all the three dimensions of sustainability, i.e., economic, environment, and social, throughout the construction period of these projects.

Typically, conventional open cut (OC) methods are the most widely used methods for constructing and repairing the underground utilities because of their low cost compared to trenchless technology (TT). When the OC methods are used on paved roads, they typically cut the pavements that might be in a good condition. Open cutting might adversely impact pavement condition and ultimately result in high pavement restoration cost. The OC method also cause significant disruption to the traffic due to its need for a large work zone construction areas (WZCA). In addition to environmental impacts (e.g., air and noise pollution), OC methods affect the businesses and other socioeconomic activities at and around the project location [3].

Most underground utilities are constructed in urban settings and can raise significant environmental and socioeconomic concerns by the users. In the recent past, the municipalities have been shifting to the wide adoption of TT for large-sized pipelines construction projects where social costs and environmental concerns are of high importance. Trenchless construction methods have proven to minimize traffic disruptions significantly and to cause less impact on pavement condition resulting in major agency and user cost savings [3,4]. These cost savings typically are directly related to the minimization of WZCA. However, TT methods may result in cost savings (benefits) that possibly result from the non-work zone construction area (NWZCA) such as the saving in maintenance expenditures and vehicle operating cost on alternate routes. When alternate routes (urban highways) are faced with unexpected high traffic volumes due to the traffic avoiding WZCA, these roads are prone to premature pavement deterioration, which possibly increases the needs for maintenance and repair activities. Furthermore, the trenchless technology minimizes environmental impacts directly in WZCA as well as in the NWZCA by avoiding the diverted traffic in case of OC construction.

Hundreds of studies have been conducted in the past on sustainability assessment of buildings while the literature on the sustainability of core public infrastructure is relatively scanter [4,5]. In particular, very limited studies on sustainability of pipelines during the construction phase are available to date. In some of the studies, TT methods were compared with the conventional OC methods on the basis of engineering, capital, and social costs for WZCA. Detailed review of all these studies is out of scope of present research; here some relevant studies, conducted in the last two decades or so, are briefly outlined to identify the research gaps.

We found three types of studies on this topic. The first type of the studies focused on estimating the agency and user costs at the WZCA. The main limitation of such studies is that various qualitative social and environmental impacts cannot be included in the assessment process. Najafi and Kim [6] reported that TT methods are more cost-effective than traditional OC methods based on their analysis, including social costs, for construction of pipelines in urban centers. Qualitative social impacts, such as loss of business and nuisance in the NWZCA due to diverted traffic, were not considered. Matthews and Allouche [7] developed a social cost calculator, with a user friendly interface, to evaluate the construction projects involving TT methods. This work gathered useful information at a single platform and facilitates decision-making for evaluating the feasibility of TT method. However, the calculator only estimates the agency and user costs for WZCA which primarily addresses the economic dimension of sustainability.

The second type of studies conduct an overall sustainability performance assessment of infrastructure projects over their lifecycle. In these studies, typically the selected indicators, covering the three dimensions of sustainability, are aggregated into indices (or key performance indicators) to assess the overall (including operations) sustainability of infrastructure projects [8,9]. Yu et al. [10] developed a construction project sustainability assessing system for Taiwan which essentially is a four level hierarchical framework consisting of main pillars of sustainability at Level 1 and their categories, sub-categories, and indicators in the three subsequent levels. Recently, Alnoaimi and Rahman [11] developed a methodology for assessing the sustainability of sewer pipeline projects over their lifecycle. Such studies are not focused on various agency and user costs at WZCA and NWZCA during the construction phase and thus are not directly applicable to evaluate different construction methods.

Third types of studies focused on sustainability of pipeline projects during the construction phase and are most relevant to present research. Matar et al. [5] developed a sustainability assessment model for infrastructure construction projects. Their methodology was based on the system engineering concepts using flow balance of materials and energy during the project's timeline. They applied the proposed model on a TT method named horizontal directional drilling for a pipeline construction project. De la Fuente et al. [12] developed a decision tree consisting of 10 criteria (indices) and 14 indicators for sustainability assessment of sewer construction projects. They developed value functions for all the indicators and aggregated them to develop a global sustainability index. They also performed scenario analysis for eight alternatives. However, the uncertainties associated to data limitations and vagueness in expert judgment on qualitative impacts were not addressed in these studies. Hojjati et al. [4] carried out a review of sustainability assessment tools available for infrastructure projects, including TT and OC, and highlighted the need for a more comprehensive sustainability costing model. Koo et al. [13] covered all three dimensions of sustainability to evaluate different design alternatives of underground utilities' construction projects. They used Analytical Hierarchical Process (AHP) for estimating the indicators' weights and used weighted sum method for finding the overall performance index for an alternative.

Based on the review of past efforts, we have identified the following three main research gaps. Firstly, past studies were primarily focused on estimating social cost savings associated with the WZCA. NWZCA cost savings may be significant and affect the TT viability assessment process but these costs have not yet been considered in the overall cost-benefit analysis. Secondly, the uncertainties associated to qualitative environmental and social impacts have not been adequately considered in most of the methodologies. Finally, the interaction between the user (public) expectations and the agency's defined sustainability objectives has not been included in in the decision-making process. Hence, the main goal of this paper is to develop a sustainability assessment methodology which can address the above stated issues for large-sized pipeline construction projects on urban highways. We used Fuzzy Quality Function Deployment Method (Fuzzy-QFD) for evaluating different possible scenarios of 'detoured traffic with OC method' and the 'TT method'. Eventually, we applied the developed methodology on an actual storm sewer construction project in Unayzah city, Qassim, KSA.

2. Methodology

2.1. Sustainability Assessment Framework

The methodological framework developed for sustainability assessment of pipeline construction projects is presented in Figure 1. The framework initiates with the development of different traffic movement scenarios for both the OC and TT options for construction of a large-sized storm sewer. After defining the boundaries of the study area, traffic surveys were conducted to find out annual average daily traffic (AADT) which is an important input to scenario analysis. In the subsequent steps social costs (both agency and user) were estimated for both the WZCA and NWZCA. Data of construction costs was obtained from the relevant agencies. Finally the framework culminates are scenario analysis using Fuzzy-QFD. This methodology is expected to overcome the limitations of previous studies by accounting for the impact of work zone areas, during an open cut, on the performance of adjacent roads in NWZCA. The method also was found effective in terms of handling uncertainties associated to vagueness in expert judgment while assessing the qualitative parameters. All the steps of the proposed framework are discussed in detail in the ensuing sections.

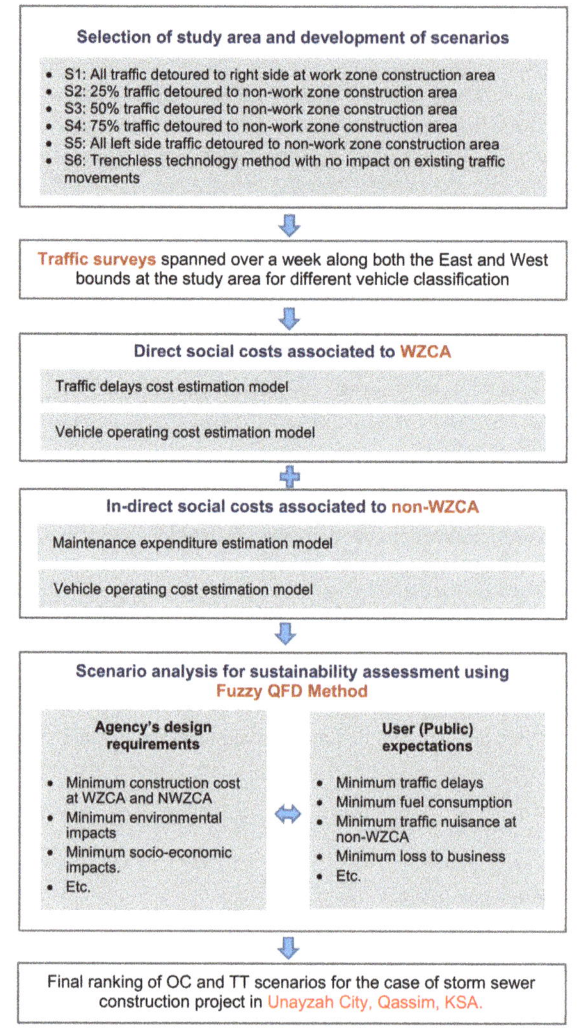

Figure 1. Methodology for sustainability evaluation of large-size storm sewer construction projects.

2.2. Social Costs at Work Zone Construction Area

Performing a utility cut requires the establishment of work zones, which are known to influence traffic mobility. In most cases, work zones delay traffic due to lane closures or block access to certain locations, forcing traffic to reroute and travel longer distances. These impacts often result in costs to the traffic users. Such costs depend mainly on the characteristics of the established work zone, for example, the lane closures and speed limits necessitated by the work zone. Existing models can be used to estimate the travel delay costs and vehicle operating costs, such as the model for travel delay costs in Equation (1) [14–16].

$$UC_{ttd} = WZ_d * \sum_{j}^{J} \left(V_j * TTD_j * DC_j \right) \tag{1}$$

where UC_{ttd} is work zone travel delay cost, WZ_d is work zone duration in days, V_j is the number of vehicles delayed by the speed change for vehicle class j, TTD_j is the travel time difference resulting from the speed changes due to the work zone for vehicle class j in hours, DC_j is the delay cost rate for vehicle class j in dollars/mile, and j is vehicle class (truck or auto).

The work zone vehicle operating costs incurred due to increased fuel consumption can be estimated using AASHTO methodology, as shown in Equation (2) [14].

$$UC_{voc} = WZ_d * \sum_{j}^{J}\left(V_j * TTD_j * Fg_j * Fp_j\right) \tag{2}$$

where UC_{voc} is work zone vehicle operating cost, Fg_j is the amount of fuel consumed due to delay in gallons/hour for vehicle class j, and Fp_j is the average fuel price in dollars/gallon consumed by vehicle class j.

2.3. Social Costs at Non Work Zone Construction Area

Pavement performance models are statistical models developed to represents the deterioration mechanism of pavements under different factors. These performance models are functions of the significant factors that are believed to have an influence on pavement condition. One of these factors that is proven to have an impact on pavement condition is the cumulative traffic loading. The amount of cumulative traffic loading, correlates positively with the accelerated pavement deterioration rates. Accordingly, unexpected high traffic volume on alternate routes can potentially cause these roads to have a higher deterioration rate. Therefore, to assess this reduction in pavement condition caused by increasing traffic volume, a pavement performance model needs to be developed. The general form of the performance model is termed in Equation (3).

$$P = f\ (Xi; \beta i) \tag{3}$$

where P is the response variable of the performance model that represent the level of pavement condition and likely denoted in term of performance indices, $Xi = (X1, X2, \ldots, Xi)$ is a vector of predictor variables (such as material type, accumulated traffic loading, weather severity), $\beta i = (\beta 1 + \beta 2, \ldots, \beta i)$ is a set of parameters to be estimated and $f\ (Xi; \beta i)$ is a mathematical form used to describe the variation of the response as function of the predictor variables.

The cost of maintaining pavements relies significantly on its condition or state. The worse the pavement conditions get, the more likely agencies are to spend money on maintaining it [17]. Therefore, in order to estimate maintenance expenditures, the conditions of the pavement need to be assessed using Equation (3). Then, associated maintenance expenditures corresponding to those pavement conditions are estimated. The general form of the maintenance expenditures model is termed in Equation (4).

$$ME(i) = f(Xi;\ Pi;\ \beta i) \tag{4}$$

where ME is the maintenance expenditures for pavement section (i); $Xi = (x1, x2, \ldots, xi)$ is a vector of predictor variables (such as traffic level and weather severity); $Pi = (p1, p2, \ldots, pi)$ is pavement performance level at treatment application time; $\beta i = (\beta 1, \beta 2, \ldots, \beta i)$ is a set of parameters to be estimated and $f(Xi;\ Pi;\ \beta i)$ is a mathematical form used to describe the variation of the response as a function of the predictor variables.

User costs, typically, are those incurred by users due to their normal use of the asset over its service life. An example of these user costs is vehicle operating costs (VOC). VOC is typically represented by the amount of money (typically is expressed in cents) spent for each mile traveled by each vehicle. These costs consist of fuel, tires, repairs and, car depreciation cost due to high mileages. One of the significant factors that increases these costs is the pavement roughness level [18]. An example of the effect of poor pavement condition (very rough pavement) on VOC is that road users might drive slowly which possibly results in high fuel consumption [19]. Typically, pavement roughness is measured in term of Present Serviceability Rating (PSR) and International Roughness Index (IRI), which are typically modeled as the response (dependent) variable of a pavement performance model using Equation (3). Having pavement roughness determined, the corresponding VOC can be then estimated using the VOC model presented in Equation (5).

$$VOC(i) = f(Xi; Pi; \beta i) \tag{5}$$

where *VOC* is vehicle operating cost estimated for vehicle (i); $Xi = (x1, x2, \ldots, xi)$ is a vector of predictor variables (such as vehicle type and age); $Pi = (p1, p2, \ldots, pi)$ is pavement performance level; $\beta i = (\beta 1, \beta 2, \ldots, \beta i)$ is a set of parameters to be estimated and $f(Xi; Pi; \beta i)$ is a mathematical form used to describe the variation of the response as a function of the predictor variables.

2.4. Social Cost Savings with Trenchless Technology

The reduction in maintenance expenditures could cause significant agency cost savings. An example of these can be the possible reduction in maintenance expenditures that result from employing TT methods. Adopting TT, which requires no or minimal work zone areas, results in almost no traffic increase on adjacent local streets. Conversely, open cut methods typically require large work zone areas which often cause/force traffic to detour to adjacent roads resulting in unexpected traffic volume on those roads. Since the pavement deterioration rate is impacted significantly by the amount of traffic volume, the needs for maintenance and rehabilitation activities on those adjacent roads might increase. Therefore, by adopting TT, transportation agencies might minimize their maintenance expenditures on these adjacent roads.

To estimate reduction/savings in an agency's maintenance expenditures on alternative roads, the comparison of alternative construction techniques (i.e., TT vs. open cut methods) needs to be conducted. In the case of employing TT, traffic volume on alternative roads are expected to remain unchanged. On the other hand, open cut method is expected to increase traffic volume on adjacent roads. Therefore, the extra maintenance expenditures incurred due to unexpected high traffic volume are quantified by taking the difference between the estimated maintenance expenditures on alternative roads for the open cut and TT scenarios. This difference in maintenance expenditures represents agency cost savings and can be calculated using Equation (6).

$$ACSs = \sum_{i=1}^{n} [G_{NOC}]_i - \sum_{i=1}^{n} [G_{HTV}]_i \tag{6}$$

where *ACSs* is agency cost savings; G_{NOC} is maintenance expenditures under normal operating condition corresponding to route i; G_{HTV} is maintenance expenditures under high traffic volume corresponding to route i; $i = 1, 2, \ldots, n$ represents number of adjacent routes.

Savings in VOC can be seen as a type of user cost savings. User cost savings can be estimated as the reduction in VOC when TT is implemented compared to the case where open cut method is considered. The general form of estimating user cost savings is presented in the following equation:

$$UCSs = \sum_{i=1}^{n} [VOC_{NOC}]_i - \sum_{i=1}^{n} [G_{HTV}]_i \tag{7}$$

where *UCSs* is user cost savings; VOC_{NOC} is vehicle operating cost under normal operating condition corresponding to vehicle i; VOC_{HTV} is vehicle operating cost under high traffic volume corresponding to vehicle i; $i = 1, 2, \ldots, n$ represents number of vehicles impacted by pavement condition.

2.5. Fuzzy Quality Function Deployment Method

During the recent past, QFD has been widely recognized as a decision-making tool for site selection process [20]. According to the fundamental idea of QFD, consumer needs are translated into technical requirements of the product by developing a House of Quality (HoQ) (see Figure 2). The HoQ is essentially a combinations of various matrices that translate the data or information obtained by the experts on: (i) What should be done related to the customer requirements?, (ii) How to relate customer requirements with the product requirements?, and (iii) What is the relationship between product requirements and the desired benchmarks? [21].

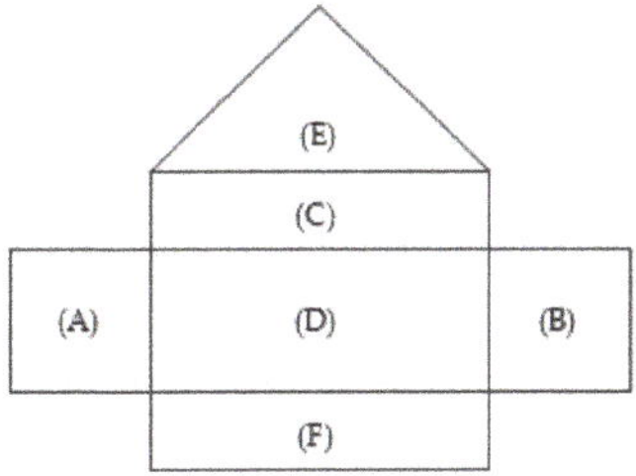

Figure 2. House of Quality for Quality Function Deployment Method.

In present research, the agency's objectives are essentially the design requirements (DR) while the customer requirements (CR) are the expectations of public (e.g., road users, visitors of the commercial area, and residents of both the WZCA and non-WZCA) during the period of construction. Subjective assessment of some of the agency requirements and public expectations introduce uncertainty in the decision-making process due to vagueness in expert judgment. In present research, this limitation has been addressed by integrating the fuzzy set theory with conventional QFD.

Following is the step-by-step procedure of Fuzzy-QFD for analyzing different scenarios for traffic movements due to open cut and trenchless technologies [20]:

Step 1: The customer requirements (CR) or the public expectations (PE) are defined as: (i) minimum travel time, (ii) minimum fuel consumption, (iii) minimum air and noise pollution, (iv) minimum traffic nuisance at NWZCA (e.g., increase risk of accidents and stress level in residents of a calm neighbourhood), and v) minimum loss in business. This step correspond to the component "A" of HoQ shown in Figure 2.

Step 2: The design requirements (DR), i.e., the component "C" of HoQ shown in Figure 2, are defined below as the following sustainability objectives (SO) which the agency or the private contractor require to obtain:

Low Agency Cost—this includes total cost of material, labour, and other miscellaneous costs. This also includes the pavement restoration cost in case of open cut construction along with the pavement maintenance cost at NWZCA.

Low User Cost at WZCA—This includes travel delay and vehicle operating costs at the work zone construction area.

Low User Cost at non-WZCA—It includes vehicle operating cost at NWZCA.

Low Environmental Impacts—This includes the possible increase in impacts of air and noise pollution at both the work zone and non-work zone during the construction period.

Low Socio-economic Impacts—This includes the impacts on socio-economics setting of the area, such as loss of business and increase in traffic congestion during the construction period.

Step 3: Determine the priorities of public expectations (PE) by using Triangular Fuzzy Numbers (TFNs) for defining the priority of each requirement.

Step 4: Determine the weight of each decision-maker. As all the decision-makers in present research were highly qualified and experienced with ample indigenous know-how of the study area, equal weights were allocated to all of the four decision-makers.

Step 5: Estimate the aggregated weights (ωp) assigned by the decision-makers for PE using the following Equation (8):

$$\omega p = (r1 \otimes \omega p1) \oplus (r2 \otimes \omega p2) \oplus \ldots \oplus (rN \otimes \omega pN) \tag{8}$$

where p is the number of CR (p = 1, 2,..., P).

Step 6: Ascertain the relationships among the PE and SO using linguistic scales for the TFNs illustrated in Figure 3. This step completes the component "D" of HoQ in Figure 2.

Step 7: Estimate the aggregated weights (a_{pm}) between PE and SO with the help of Equation (9).

$$a_{pm} = \left(r_1 \otimes a_{pm1}\right) \oplus \left(r_2 \otimes a_{pm2}\right) \oplus \ldots \oplus \left(r_N \otimes a_{pmN}\right) \tag{9}$$

where N denotes the number of decision makers *(n = 1, 2, ... , N)* and M is the number of *DR (m = 1, 2,..., M)*. This Equation (9) fills component (B) of Figure 3.

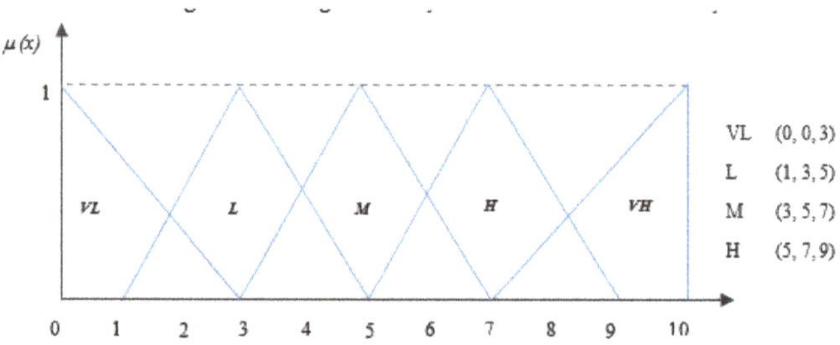

Figure 3. Triangular fuzzy numbers used in this study.

Step 8: Determine the prioritized technical descriptors to complete the matrix. Equation (10) estimates the weights of the SO (f_m) from the aggregated weight for PE (w_p) and aggregated weights between the SO and PE (a_{pm}).

$$f_m = (\omega_1 \otimes a_{1m}) \oplus \ldots \oplus \left(\omega_p \otimes a_{pm}\right) \tag{10}$$

Step 9: Evaluate each scenario (refer to Figure 1) against all the SO and the combine the results with the weight of each SO for final ranking of the scenarios. Subsequently, calculate the rating of each scenario (SR) using Equation (11).

$$SR_{km} = (r_1 \otimes lr_{km1}) \oplus (r_2 \otimes lr_{km2}) \oplus \ldots \oplus (r_N \otimes lr_{kmN}) \tag{11}$$

where K presents the number of scenarios *(k = 1, 2, ... , K)*.

Step 10: Calculate the fuzzy index (*FI*) using Equation (12) for expressing the degree to which a scenario complies with the desired SO.

$$FI_k = \frac{1}{M} \otimes (LR_{k1} \otimes f_1) \oplus \ldots \ldots \oplus (LR_{kM} \otimes f_M) \tag{12}$$

Step 11: Finally, use Equation (13) to de-fuzzify the TFNs obtained from Step 10 to rank all the pipeline construction scenarios.

$$Rk = \frac{a + 2b + c}{4} \tag{13}$$

where R_k is the priority rank of each SHP site and *a,b,&c* are the upper, middle, and lower bounds of the aggregated TFN.

3. Results and Discussion

3.1. Study Area

The selected study area presented in Figure 4 is Umar Ibn AlKhattab Road in Unayzah city, Qassim region, KSA. The area affected by the project is about 5 km^2 and the reach of the selected urban highway for application of proposed methodology is about 2.4 km. It can be seen in the figure that starting point is 670 m higher than the mean sea level while the ground elevation is 658 m at the end of sewer line. The horizontal drilling technique was used for constructing a storm sewer of 3900 mm diameter. Some salient features of the project are presented in Table 1. This data was obtained from the Ministry of Environment, Water and Agriculture in Saudi Arabia, i.e., the client of this project. As stated above that inclusion of the economic, social, and environmental impacts of OC construction on NWZA is an important contribution of this research, some possible routes of the diverted traffic to the NWZCA are also shown in light blue colour in Figure 4.

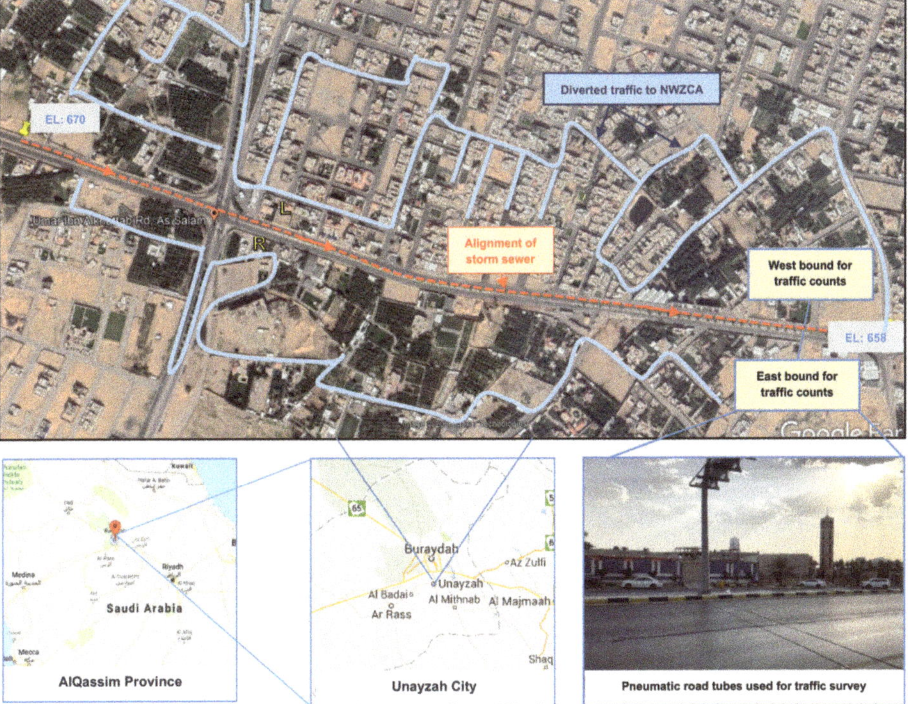

Figure 4. Study area in Unayzah city (Umar Ibn AlKhattab Road), EL stands for elevation at start and end of storm sewer.

Table 1. Project description in brief.

Item	Description
Client	Ministry of Environment, Water and Agriculture in Saudi Arabia
Type of Project	Storm Sewer
Total duration of the project	Approximately 2 years (720 days)
Total cost	77,000,000 SR
Type of construction	Trenchless technology (see Figure 5)
Location	Umar Ibn AlKhattab Road, Unayzah, AlQassim Region
Length of sewer	2.4 km
Depth of sewer	9–29 m
Sewer Size (diameter)	3900 mm
Pipe material	Reinforced Cement Concrete (RCC)
Weight of the machinery used	200 t
Area characteristics	Variation in depths up to 35m—high population density
Area affected	Approximately 5 km^2

(a)

(b)

Figure 5. Trenchless technology equipment used in construction of storm sewer at the urban highway in study area shown in Figure 4. (**a**) cross-sectional view, (**b**) side view.

Generally, TT methods are classified into two types. The first type is known as trenchless construction methods, such as utility tunneling, pipe jacking, horizontal auger boring, micro-tunneling, pipe ramming, horizontal directional drilling, and pilot tube micro-tunneling. The second type of TT methods are trenchless renewal methods which include cured-in-place pipe, pipe bursting, close-fit pipe, thermoformed pipe, slip-lining, in-line replacement, localized repair, lateral renewal, coating and linings, and manhole renewal [22]. Figure 5 shows cross-section and side views of the equipment of TT method used for the construction of storm sewer in the study area. In addition to opening all the road lanes for normal traffic operations and commercial activities, the use of trenchless technology reduced the load on the urban roads network in the vicinity of project area. Trenchless technology also improved the road safety by avoiding the possibility of accidents due to increased number of conflicts which could have occurred with open trench construction option. TT also reduced the impacts of air and noise pollution in the study area during the construction phase.

In this study, we developed six scenarios as illustrated in Figure 1. First four scenarios (S1 to S4) appraise various percentages of both sides traffic is detoured to NWZCA in case of the OC construction on left side of the urban highway under study (details given in the following sections). In the fifth scenario (S5), all the left side traffic is detoured to NWZCA while the right side traffic continue its routine operations with all the three lanes functional. Finally, S6 evaluates the trenchless technology with minimal disturbance to both sides traffic.

3.2. Traffic Count Survey

A screen line traffic count survey spanned over one week was conducted at the locations (west bound and east bound) highlighted in Figure 4. Pneumatic road tubes were used to count the number and types of the vehicles passed through the screen line. A burst of air pressure is sent through the sensors along a rubber tube when the tires of a specific vehicle class pass over tube. The pressure pulse produces an electrical signal which is transmitted to the analysis software. These portable pneumatic tubes were fixed on the road surface with the help of a specific gel. These tubes obtain energy from rechargeable batteries. Their main advantages include quick installation and easy maintenance [23]. A photograph of the tubes used at the study area is shown on the right bottom of Figure 4.

Traffic data was collected for the thirteen vehicles classes defined by American Association of State Highway and Transportation Officials (AASHTO) [23]. The results for both the west and east bounds are presented in Table 2.

Table 2. Summary of traffic counts.

Vehicle Class [1]	Vehicle Type	Traffic Counts at West Bound	Traffic Counts at East Bound
C1	Motorcycle	28	34
C2	Passenger car	10,749	8664
C3	4 tire single unit	283	624
C4	Bus	95	99
C5	Two-axel, six tire, single unit	47	40
C6	Three-axel single unit	15	25
C7	Four or more axel single unit	17	15
C8	Four or less axel single trailer	29	94
C9	5-axel tractor semi-trailer	122	139
C10	Six or more axel single trailer	5	8
C11	Five or less axel multi-trailer	1	2
C12	Six axel multi-trailer	1	1
C13	Seven or more axel multi-trailer	0	0

[1] Federal Highway Administration [23].

3.3. Agency and Social Costs Estimation

3.3.1. Work Zone Construction Area

In case of open cut construction, agency costs include the construction cost of pipeline and pavement restoration cost at the WZCA. A tentative cost for large-sized stormwater was obtained from the relevant agency working in the study area while the cost for TT case is given in Table 1. The first social cost at WZCA is travel delay cost (UC_{ttd}) which was estimated using Equation (1) for each scenario (see Figure 1). Work zone duration was 720 days (see Table 1). Instead of considering individual vehicle class listed in Table 2, C2 and C3 were classified under light transport vehicle (LTV) class and C4 to C13 under the heavy transport vehicle (HTV) class, i.e., V_j in Equation (1). At west bound, the ADDT for LTV was found to be 11,032 while the estimated HTV was 332. Similarly, the AADT for east bound were estimates as 9,300 for LTV and 423 for HTV. These values are used to estimate social costs at WZCA and NWZCA in the following sub-section. The travel time difference resulting from the speed changes due to the work zone (i.e., TTD_j) were estimated for LTV and HTV. The urban highway under study has a design speeds of 80 kph (for LTV) and 50 kph (for HTV) which were reduced to 50 kph and 30 kph during the constriction period. The delay cost rates for LTV and HTV were estimated based on average per hour salary in the country.

The second social cost at WZCA is the vehicle operating costs (UC_{VOC}) which was estimated using Equation 2. Based on reduced speeds at WZCA, the average amount of fuel consumed due to delay (i.e., Fg_j) was estimated to be 0.0264 gallon/km for LTV at 50 kph and 0.058284 gallons/km for HTV at 30 kph. The average fuel price (Fp_j) was obtained from market.

3.3.2. Non Work Zone Construction Area

The needed input data for estimating the social cost at NWZCA are presented in Table 3. This data mainly consists of data related to road physical and environment attributes. Road physical attributes include maintenance expenditures and pavement condition while the road environmental attributes composed of traffic loading and climate severity. A deterministic pavement performance model, capable of predicting pavement performance as a function of cumulative traffic loading and freezing index [16], is used to predict pavement condition in terms of IRI (in/mi). Therefore, pavement performance subject to normal operating conditions and high traffic volume can be estimated. Having determined pavement conditions, a cost model (Al-Mansour and Sinha 1994) capable of estimating agency maintenance expenditures as a function of pavement condition is employed. Therefore, maintenance expenditures for pavement under normal operating condition and under high traffic volume can be estimated.

Table 3. Summary of Input Variables.

Variables	Description	Values
Pavement current performance	Pavement current performance is estimated in terms of IRI (in/mi).	$PI_{current} = 150$ (in/mi)
α	Constant term of pavement performance model	(4.082)
β & γ	Estimated coefficients of pavement performance model	(0.017) & (0.054)
a	Constant term of pavement maintenance expenditures model.	4.0283
b	Estimated coefficient of pavement maintenance expenditures model.	-0.462
AADT	Average annual daily traffic volume (thousand)	The mean values of AADT = 11.3
ANDX	Average annual freezing index (thousands)	The mean values of ANDX = 0.40

The relation between pavement performance in terms of IRI (in/mi) and VOC (cents per vehicle-mile) developed by Opus [24] is implemented in this study. Thus, VOCs corresponding to pavement under normal operating condition and under high traffic volume can be estimated.

The savings in agency and user cost as a result of applying TT compared to the open cut method scenarios where estimated using Equation (4) and (5) respectively. Figure 6a,b shows the savings in terms of maintenance expenditures and vehicle operating cost, respectively. In order to determine the effect of varying traffic volume on maintenance expenditures and vehicle operating cost values, the above described methodology was tested by increasing the initial traffic volume each time by 1% until it reaches 10%, for different traffic scenarios (S1 to S6) while other factors are kept constant.

With increase in detoured traffic, both the maintenance and vehicle operating costs increase at the NWZA. With respect to VOC the traffic volume has no impact when using TT. The reason is that TT requires no or minimal work zone area and thus no traffic need to detour to adjacent roads. It can be seen that the highest values of maintenance expenditures and vehicle operating cost is when 100% of traffic is detoured and decrease as the percentage of detoured traffic is reduced. It is noteworthy that these values of user costs compared to agency costs are consistent with past studies in the literature [16,19,25].

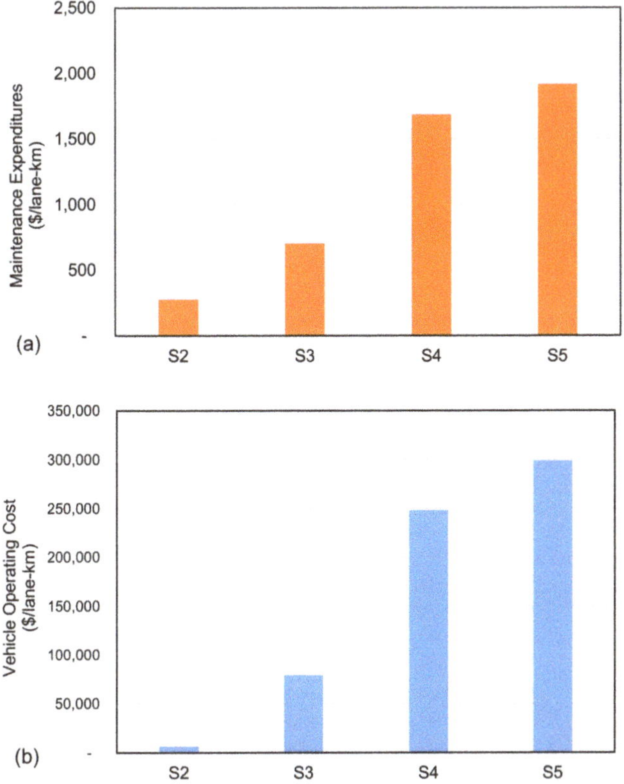

Figure 6. Non-work zone construction area (NWZCA) costs for different scenarios, (**a**) Maintenance expenditure, (**b**) Vehicle operating cost.

3.4. Scenario Analysis using Fuzzy QFD

The urban highway is a 3-lane dual carriageway (see photograph in Figure 4). Due to construction of a large size storm sewer (i.e., 3900 mm), the municipality had to close the entire left side of the highway. A hypothetical section of the urban highway showing WZCA area and left (westbound) traffic diverted to half of pavement width occupied from the right side is illustrated in Figure 7a. The photograph shown in the figure was taken for another open cut project in the city to describe the practical constraints of open cut scenarios. Figure 7b shows the total estimated cost including construction cost and social cost at both the WZCA and NWZCA for all the scenarios with different traffic detouring conditions. In first five scenarios of open cut construction, the agency cost includes construction cost of pipeline, pavement restoration cost at WZCA and early pavement maintenance requirements at NWZCA. User cost at WZCA includes the VOC and traffic delay costs (see Section 3.3.2 for details). User cost at NAZCA is primarily the VOC. Total social cost in the Figure 6a is essentially the sum of user costs at both the WZCA and NWZCA.

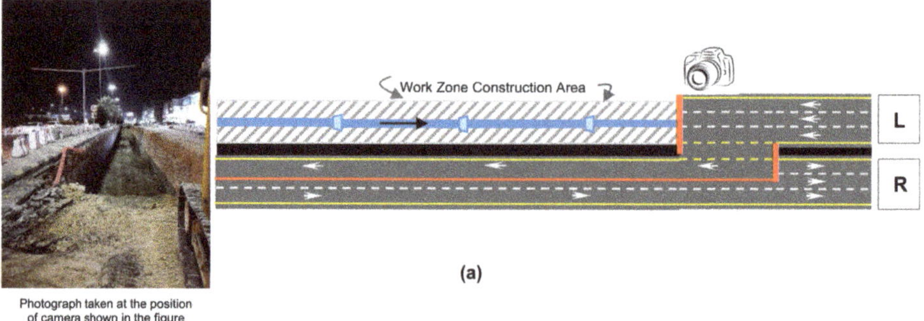

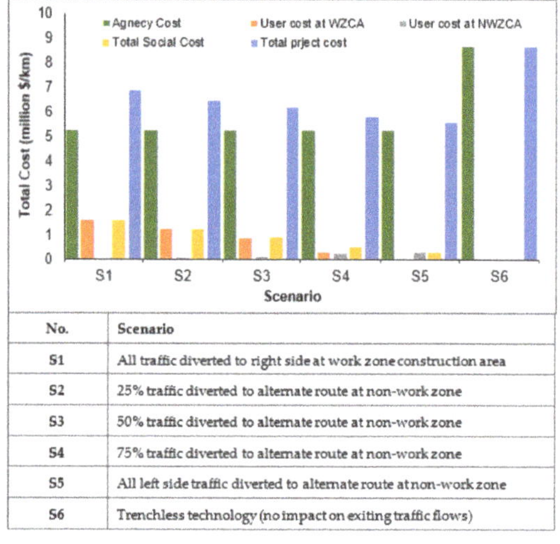

No.	Scenario
S1	All traffic diverted to right side at work zone construction area
S2	25% traffic diverted to alternate route at non-work zone
S3	50% traffic diverted to alternate route at non-work zone
S4	75% traffic diverted to alternate route at non-work zone
S5	All left side traffic diverted to alternate route at non-work zone
S6	Trenchless technology (no impact on exiting traffic flows)

(b)

Figure 7. Cost estimates for different scenarios, (**a**) Plan describing West bound traffic diverted to East bound due to work zone construction area resulted in speed reduction, congestion, and traffic delays (The photograph shown in the figure was taken for another open cut project in the city), (**b**) Summary of total cost considering both the work zone construction area (WZCA) and non-work zone construction area (NWZCA). The percentage of traffic (e.g., 50%) is a tentative value, actual percentage could vary between say 40–60%.

It can be seen in the scenarios defined in Figure 7b that a certain percentage of the traffic detoured in first five scenarios to NWZCA. There are two possible reasons of such diversion: i) drivers intentionally moved to other possible routes to avoid traffic delays and other possible hazards (e.g., dust and construction noise), and ii) drivers were compelled to use the alternate routes as WZCA had obstructed the direct access to their destinations.

As mentioned earlier, opinion of four decision-makers with indigenous knowhow of the project site was used for scenario analysis. Decision-makers had high educational levels and extensive experience in the fields of environmental management (DM1), asset management (DM2), construction management (DM3), and transportation management (DM4). Firstly, the customer requirements (CR) or the public expectations (PE) and the design requirements (DR) or the agency's sustainability objectives (SO) are defined (see Step 1 and Step 2 of fuzzy QFD methodology). In Step 3, the public expectations were prioritized by all the decision-makers. Equal importance weights were allocated to all the four decision-makers, i.e., $r_1 = r_2 = r_3 = r_4 = 0.25$, in Step 4 of the methodology. Next, the overall aggregated weights of the PE were estimated using Equation (8) and the results are presented in Table 4, i.e., Step 5.

Table 4. Aggregated weights of design objectives (DOs).

Public Expectations (PE)	DM1			DM2			DM3			DM4			Aggregated Weights (ω_p)		
Importance Weights (w_i)	0.25			0.25			0.25			0.25					
Minimum travel time	5	7	9	5	7	9	1	3	5	7	10	10	4.5	6.8	8.3
Minimum fuel consumption	5	7	9	3	5	7	0	0	3	5	7	9	3.3	4.8	7.0
Minimum air and noise pollution	7	10	10	0	0	3	3	5	7	5	7	9	3.8	5.5	7.3
Minimum traffic nuisance at NWZ	5	7	9	5	7	9	1	3	5	3	5	7	3.5	5.5	7.5
Minimum loss to business	3	5	7	1	3	5	5	7	9	3	5	7	3.0	5.0	7.0

Next in Step 6, the decision-makers established the relationships between all the PE and SO using the TFNs given in Figure 2. In the following Steps 7 and 8, we used Equation (9) to aggregate the weights (a_{pm}) between the PE and SO and Equation (10) to estimate the prioritized technical operators (f_m). The complete HoQ is illustrated in Table 5.

Finally, decision makers scored the SO for all the scenarios. The first three agency's sustainability objectives (low agency cost, low user cost at WZCA, and low user cost at NWZCA) were scored, using TFNs, based on the estimated costs given in Figure 7b. While the last two SO (low environmental impacts and low socio-economic impacts) were subjectively scored based on expert judgment. In the next Steps 9 and 10, Equation (11) was applied to determine the performance ratings of each scenario (SR_{km}) and Equation (12) estimated the fuzzy index (Fl_k). The fuzzified performance scores and the Fl_k are presented in Table 6. Finally, Equation (13) de-fuzzified the aggregated TFNs and all the scenarios were ranked. Final ranking of all the scenarios is presented in Table 7.

Table 5. Complete House of Quality (HoQ) for sustainability evaluation of pipeline construction scenarios.

	Low agency cost			Low user cost at WZCA			Low user cost at NWZCA			Low environmental impacts			Low socioeconomic impacts			Aggregated wt. (a_{mp})		
	a	b	c	a	b	c	a	b	c	a	b	c	a	b	c	a	b	c
Minimum travel time	0.5	1.5	4.0	6.5	9.3	9.8	6.0	8.5	9.5	4.5	6.8	8.3	5.5	8.0	9.0	4.5	6.8	8.3
Minimum fuel consumption	0.5	1.5	4.0	5.5	7.8	9.3	4.5	6.8	8.3	6.5	9.3	9.8	5.5	7.8	9.3	3.3	4.8	7.0
Minimum air and noise pollution	2.8	4.3	6.5	0.5	1.5	4.0	0.8	2.3	4.5	6.0	8.5	9.5	4.5	7.0	8.0	3.8	5.5	7.3
Minimum traffic nuisance at NWZCA	3.0	5.0	7.0	4.0	6.3	7.8	3.0	5.0	7.0	5.5	8.0	9.0	4.5	7.0	8.0	3.5	5.5	7.5
Minimum loss to business	0.8	2.3	4.5	0.5	1.5	4.0	2.0	4.0	6.0	1.0	2.0	4.5	6.5	9.3	9.8	3.0	5.0	7.0
	27	79	192	65	149	260	61	149	263	86	190	304	95	214	325			
							Prioritized technical operators (f_m)											

Table 6. Aggregated fuzzy performance matrix and fuzzy index for all the scenarios for pipeline construction project.

Scenarios [1]	Sustainability Objectives (SO)															Fuzzy Index (FI)		
	Low Agency Cost			Low User Cost at WZCA			Low User Cost at NWZCA			Low Environmental Impacts			Low Socioeconomic Impacts					
	a	b	c	a	b	c	a	b	c	a	b	c	a	b	c	a	b	c
S1	3.0	5.0	7.0	0.0	0.0	3.0	6.5	9.3	9.8	2.0	4.0	6.0	0.8	1.8	4.5	144	583	1596
S2	3.0	5.0	7.0	1.0	3.0	5.0	5.0	7.0	9.0	3.5	5.5	7.5	2.0	3.5	6.0	188	737	1850
S3	3.0	5.0	7.0	3.0	5.0	7.0	3.5	5.5	7.5	4.5	6.5	8.5	3.5	5.0	7.5	241	854	2033
S4	3.0	5.0	7.0	5.0	7.0	9.0	2.0	4.0	6.0	3.5	5.5	7.5	2.5	4.0	6.5	213	788	1933
S5	3.0	5.0	7.0	7.0	10.0	10.0	0.5	1.5	4.0	3.0	5.0	7.0	4.0	6.0	7.5	240	870	1914
S6	1.3	2.8	5.0	7.0	10.0	10.0	6.0	8.5	9.5	4.5	6.5	8.5	6.0	8.0	9.5	361	1186	2348

[1] Details of scenarios can be seen in Figure 7.

Table 7. Defuzzified Criteria Numbers of Sites and Ranking.

Scenarios	Defuzzified Performance Score	Final Ranking
S6: Trenchless technology	1270	1
S3: 50% traffic diverted to alternate route at NWZCA	996	2
S5: All left side traffic diverted to NWZCA	974	3
S4: 75% traffic diverted to alternate route NWZCA	931	4
S2: 25% traffic diverted to alternate route at NWZCA	878	5
S1: All traffic diverted to right side at WZCA	726	6

3.5. Discussion

It can be seen in Table 4 that almost equal aggregated weights (ωp) were assigned to all the five public expectations (or design objectives to meet such expectations). This shows that the decision-making team was well balanced with experts from the essential fields of expertise, i.e., environment, construction, traffic, and asset management, required for such multidisciplinary projects. The relationships between the agency's SO and PE are given in the complete HoD presented in Table 5. The scores shows that overall environmental and socioeconomic impacts are highly related to the public expectations. For example, 'minimum travel time' will not only result in low user cost and time savings, it will lead to low vehicular emissions as well. Likewise, traffic diversions to NWZCA can increase air pollution and noise levels in the clam neighbourhood environment during the construction phase (in case of OC method) and may also lead to other types of nuisances, such as increased risk of accidents and stress levels in the residents.

Actually, development of technical descriptors (component "F" of HoD shown in Figure 2) is the most important outcome of the complete matrix of HoD. These operators combine the information available on the aggregated weight for PE (w_p) and the aggregated weights between the SO and PE (a_{pm}). It can be seen in Table 5 that the highest numbers were obtained for environmental and socio-economic aspects of sustainability. In economic dimension, user costs at WZCA and NWZCA were found more important, with higher values of technical descriptors, than the agency cost. Subsequently, decision-makers scored all the scenarios against the agency's sustainability objectives and the 'Fuzzy Index' was developed by combining the outcomes of the HoD and the scoring of scenarios. Based on the defuzzified index, final ranks for all the options are given in Table 7.

It can be seen in Table 7 that trenchless technology was found to be the most sustainable option, i.e., top ranked scenario. The primary reasons of such a high rank are low user cost as well as low environmental and socioeconomic impacts at both the WZCA and NWZA. In case of open cut method, the scenario with 50% traffic (i.e., S3) diverted to NWZCA was found to be the most sustainable scenario; however, overall it is ranked at second place in Table 7. In the first two scenarios (S1 and S2), both the user cost and environmental and socioeconomic impacts were found to be very high at WZCA. While in case of S4 and S5, these costs and impacts were supposed to be shifted to the NWZA which essentially is a calm neighbourhood. The reason for S5 being at third place is that at least the right side traffic will perform its routine operations without delays which certainly can only be obtained at the cost of diverting the entire left side traffic to the NWZCA. These findings justify the need of evaluating the sustainability of traffic diversion scenarios for different construction technologies instead of taking decisions merely on economic basis. The results presented in Table 7 also manifest the reliability of proposed fuzzy-QFD based methodology for evaluation of construction technologies for buried infrastructure. The outcomes are also consistent with the past studies that emphasized on the importance of socio-economic aspects for sustainability assessment of construction projects [26].

Municipalities, water directorates, and the Ministry of Water, Agriculture, and Environment are responsible for construction of water supply, sewerage, and stormwater pipelines in the urban areas of Saudi Arabia. Decision-makers in these organizations can implement the proposed framework for sustainability assessment of construction technologies for all types of buried infrastructure. The proposed methodology can also take in additional site specific agency and user requirements. Fuzzy QFD can be integrated with Fuzzy AHP method in case of a large number of customer and technical requirements [27].

The responsible transportation agency can develop effective traffic management plan during the construction period of the pipeline projects in collaboration with other concerned agencies. Traffic counts data (including directional counts) can be used to identify the peak hours and the portion of traffic generated from nearby NWZCA. Then, the traffic from the nearby NWZCA along with some general traffic can be diverted to NWZCA through directional and informative sign (i.e., a legibly printed and very perceptible placard to inform the users that how they can minimize the traffic delays and user costs by using NWZCA). It is expected that the proposed framework will facilitate the decision makers in selection of more sustainable pipeline construction method and traffic management scenarios with minimal user costs and socioeconomic and environmental impacts.

4. Conclusions and Recommendations

Users of urban highways, particularly passing through central business districts, face disruptions during the long construction phases of large-sized buried infrastructure projects. Construction of these projects, using conventional open cut method, results in high social costs at work zone construction areas. In addition, municipalities are supposed to endure additional costs due to early maintenance of alternate routes in non-work zone construction area. Users of both the work zone and non-work zone construction areas also suffer through socioeconomic losses and environmental disturbances due to open cut construction. Overlooking these issues, municipalities still prefer to adopt these methods predominantly because of their low direct agency costs.

Sustainability assessment with the help of a more rational approach that value both the users' expectations and supposed agency's sustainability objectives produce more balanced and useful findings. Fuzzy Quality Function Deployment method allows the decision-makers to give appropriate attention to all the three dominations of sustainability instead of relying on pure economic analysis for selecting the construction technology for large-sized pipeline construction projects on urban highways. The method effectively incorporates the likely expectations of the users and residents, of both the WZCA and NWZCA, in the decision-making process.

In open cut construction, a significant portion of the highway (generally one full side in case of a large pipeline) is blocked for traffic movement at the work zone area. There are various possible scenarios of the percentages of total traffic detoured to non-work zone area, e.g., 0%, 25%, 50%, and 75%. Each of these scenarios results in varying degrees of social and environmental impacts on the WZCA and NWZCA. For example, 0% detoured traffic will impact only the work zone area while shifting 75% of the total traffic lead to significant disturbances in the NWZCA.

Scenario analysis using Fuzzy QFD ranked trenchless technology on the top due to its insignificant impacts on vehicle operating and traffic delay costs, and socioeconomic and environmental settings of both the WZCA and NWZCA, i.e., essentially negligible social costs. Among the scenarios of various percentages of detoured traffic to NWZCA due to open cut construction, 50% detoured traffic was found to be the second most sustainable scenario that somehow establishes a balance between socioeconomic and environmental impacts of work zone and non-work zone areas.

Comparison of the proposed method with other decision-making methods in future studies can further enhance the reliability of fuzzy-QFD applications to infrastructure projects. It is expected that the proposed methodology will facilitate the engineers, traffic planners, and construction manager in the municipalities and relevant agencies for more effective decision-making for sustainability assessment (particularly viability of trenchless technologies) of large sized pipeline construction projects in cities of Saudi Arabia, rest of gulf region, and elsewhere.

Author Contributions: M.A. developed the methodology, performed social cost estimation, and was involved in analysis, and prepared the manuscript. H.H. performed detailed cost estimates, developed methodology for multicriteria analysis, and prepared the manuscript. M.A and F.A conducted traffic surveys and performed traffic analysis. S.A reviewed the manuscript and was helped in data collection. All the authors shared their expert opinion and indigenous knowledge for performing scenario analysis.

Funding: The authors gratefully acknowledge Qassim University presented by Deanship of Scientific Research, on the material support for this research under the number (5094-qec-2018-1-14-S) during the academic year 1441 AH/2020 AD.

Acknowledgments: Authors acknowledge the municipalities in Qassim Region of Saudi Arabia for sharing the information of the trenchless pipeline construction project.

Conflicts of Interest: The authors declare no conflict of interest.

References

1. Too, E.G. A capability model to improve infrastructure asset performance. *J. Constr. Eng. Manag.* **2012**, *138*, 885–896. [CrossRef]
2. Jung, Y.J.; Sinha, S.K. Evaluation of trenchless technology methods for municipal infrastructure system. *J. Infrastruct. Syst.* **2007**, *13*, 144–156. [CrossRef]
3. Hunt, D.V.L.; Nash, D.; Rogers, C.D.F. Sustainable utility placement via multi-utility tunnels. *Tunn. Undergr. Sp. Technol.* **2014**, *39*, 15–26. [CrossRef]
4. Hojjati, A.; Jefferson, I.; Metje, N.; Rogers, C.D. Sustainability assessment for urban underground utility infrastructure projects. In *Proceedings of the Institution of Civil Engineers-Engineering Sustainability, March 2017*; Thomas Telford Ltd.: London, UK, 2017; Volume 171, pp. 68–80.
5. Matar, M.; Osman, H.; Georgy, M.; Abou-Zeid, A.; El-Said, M. 20 A systems engineering approach for realizing sustainability in infrastructure projects. *HBRC J.* **2017**, *13*, 190–201. [CrossRef]
6. Najafi, M.; Kim, K.O. Life-cycle-cost comparison of trenchless and conventional open-cut pipeline construction projects. *Pipe Eng. Cons: What's on the Horizon?* **2004**, 1–6.
7. Matthews, J.C.; Allouche, E.N. A social cost calculator for utility construction projects. In Proceedings of the North American Society for Trenchless Technology Paper F-4-03, Chicago, IL, USA, 2–7 May 2010.
8. Umer, A.; Hewage, K.; Haider, H.; Sadiq, R. Sustainability assessment of roadway projects under uncertainty using Green Proforma: An index-based approach. *Int. J. Sustain. Built Environ.* **2016**, *5*, 604–619. [CrossRef]
9. Shen, L.; Wu, Y.; Zhang, X. Key assessment indicators for the sustainability of infrastructure projects. *J. Constr. Eng. Manag.* **2011**, *137*, 441–451. [CrossRef]
10. Yu, W.D.; Cheng, S.T.; Ho, W.C.; Chang, Y.H. Measuring the Sustainability of construction projects throughout their lifecycle: A Taiwan lesson. *Sustainability* **2018**, *10*, 1523. [CrossRef]
11. Alnoaimi, A.; Rahman, A. Sustainability Assessment of Sewerage Infrastructure Projects: A Conceptual Framework. *Int. J. Environ. Sci. Dev.* **2019**, *10*. [CrossRef]
12. De la Fuente, A.; Pons, O.; Josa, A.; Aguado, A. Multi-Criteria Decision Making in the sustainability assessment of sewerage pipe systems. *J. Clean. Prod.* **2016**, *112*, 4762–4770. [CrossRef]
13. Koo, D.H.; Ariaratnam, S.T.; Kavazanjian, E. Development of a sustainability assessment model for underground infrastructure projects. *Can. J. Civ. Eng.* **2009**, *36*, 765–776. [CrossRef]
14. AASHTO (American Association of State Highway and Transportation Officials). *User Benefit Analysis for Highways*; AASHTO: Washington, DC, USA, 2003.
15. Labi, S.; Sinha, K.C. Life-cycle evaluation of flexible pavement preventive maintenance. *J. Transp. Eng.* **2005**, *131*, 744–751. [CrossRef]
16. Irfan, M.; Khurshid, M.B.; Labi, S.; Flora, W. Evaluating the cost-effectiveness of flexible rehabilitation treatments using different performance criteria. *J. Transp. Eng.* **2009**, *135*, 753–763. [CrossRef]
17. Al-Mansour, A.I.; Sinha, K.C. Economic analysis of effectiveness of pavement preventive maintenance. *Transp. Res. Rec.* **1994**, *1442*, 31–37.
18. Sinha, K.C.; Labi, S. *Transportation Decision Making Principles of Project Evaluation and Programming*; Wiley: New York, NY, USA, 2007.
19. Khurshid, M.B.; Irfan, M.; Labi, S. Optimal performance threshold determination for highway asset interventions: Analytical framework and application. *J. Transp. Eng.* **2011**, *137*, 128–139. [CrossRef]

20. Tavakkoli-Moghaddam, R.; Hassanzadeh Amin, S.H.; Zhang, G. A proposed decision support system for location selection using fuzzy quality function deployment. In *Decision Support Systems, Advances in*; Ger, D., Ed.; INTECH: Rijeka, Croatia, 2010; p. 342.

21. Hauser, J.R.; Clausing, D. The house of quality. *Harvard Bus. Rev.* **1988**, *66*, 63–73.

22. Hashemi, S.B. Construction cost of underground infrastructure renewal: A comparison of traditional open-cut and pipe bursting technology. Master Thesis of Science in Civil Engineering, The University of Texas, Arlington, TX, USA, 2008.

23. Federal Highway Administration (FHWA). Vehicle Classification, US Department of Transportation. Available online: https://www.fhwa.dot.gov/policyinformation/tmguide/tmg_2013/vehicle-types.cfm (accessed on 25 February 2020).

24. Opus Central Laboratories in association with Transport Research Laboratory. *Review of VOC-Pavement Roughness Relationships Contained in Transfund's Project Evaluation Manual*; Opus Central Laboratories: Lower Hutt, New Zealand, 1999.

25. Labi, S.; Sinha, K.C. Life-cycle evaluation of flexible pavement preventive maintenance. *J. Transp. Eng.* **2005**, *131*, 744–751. [CrossRef]

26. Reizgevičius, M.; Ustinovičius, L.; Cibulskienė, D.; Kutut, V.; Nazarko, L. Promoting sustainability through investment in Building Information Modeling (BIM) technologies: A design company perspective. *Sustainability* **2018**, *10*, 600. [CrossRef]

27. Feili, H.; Molaee-Aghaee, E.; Jahed-Khaniki, G.; Rezaie, S.; Kohkheil, M. Applying fuzzy quality function deployment and fuzzy analytical hierarchy process approach in industrial bread production. *J. Food Saf. Hyg.* **2015**, *1*, 53–58.

Article

Geoadditive Quantile Regression Model for Sewer Pipes Deterioration Using Boosting Optimization Algorithm

Ngandu Balekelayi and Solomon Tesfamariam *

School of Engineering, University of British Columbia, Kelowna, BC V1V 1V7, Canada; ngbalek@mail.ubc.ca
* Correspondence: Solomon.Tesfamariam@ubc.ca

Received: 14 September 2020; Accepted: 17 October 2020; Published: 21 October 2020

Abstract: Proactive management of wastewater pipes requires the development of deterioration models that support maintenance and inspection prioritization. The complexity and the lack of understanding of the deterioration process make this task difficult. A semiparametric Bayesian geoadditive quantile regression approach is applied to estimate the deterioration of wastewater pipe from a set of covariates that are allowed to affect linearly and nonlinearly the response variable. Categorical covariates only affect linearly the response variable. In addition, geospatial information embedding the unknown and unobserved influential covariates is introduced as a surrogate covariate that capture global autocorrelations and local heterogeneities. Boosting optimization algorithm is formulated for variable selection and parameter estimation in the model. Three geoadditive quantile regression models (5%, 50% and 95%) are developed to evaluate the band of uncertainty in the prediction of the pipes scores. The proposed model is applied to the wastewater system of the city of Calgary. The results show that an optimal selection of covariates coupled with appropriate representation of the dependence between the covariates and the response increases the accuracy in the estimation of the uncertainty band of the response variable. The proposed modeling approach is useful for the prioritization of inspections and provides knowledge for future installations. In addition, decision makers will be informed of the probability of occurrence of extreme deterioration events when the identified causal factors, in the 5% and 95% quantiles, are observed on the field.

Keywords: Bayesian; geoadditive; quantile; P-splines; boosting optimization

---

1. Introduction

Wastewater systems collect and transport household, commercial and industrial used water through a complex underground sewer pipe network to treatment plant facilities [1,2]. The need to maintain the required level of service and protect the environment led municipalities to actively monitor and improve their systems based on actual and future performance indicators. Budget restrictions and large size of modern wastewater systems, however, limit the ability of municipalities to inspect and assess the condition of all pipes in their networks [3–6]. For example, in Canada, only 23% of large collection pipes have complete inspection data and the condition of the remaining proportion is estimated based on proxy information, such as age, soil types and expert opinion [7]. Existing deterioration models focus on relating the mean of the response variable to a given (linear, nonlinear) combination of the covariate effects [8,9]. The use of such models implies that the estimated effects describe the average condition of pipes. However, it is of interest to analyze the quantiles of the response, such as 5% and 95% quantiles, which provide information about the occurrence of extreme deterioration events (e.g., new pipes in worst structural condition or old pipes in excellent condition). Structured additive quantile regression coupled with the boosting algorithm (i.e., an optimization algorithm that aims at minimizing an expected loss criterion by stepwise updating of an estimator according to steepest gradient descent of the loss criterion) will simultaneously address the selection of relevant factors in the model and estimate model parameters in a single procedure [10].

Various mathematical models (physical, statistical, machine learning) have been proposed to model sewer pipes deterioration [11]. Lack of understanding of the complex physical process of pipe failure coupled with the limited data availabilities hinders use of physical models [12,13]. Artificial intelligence constitute an alternative approach that allow nonlinear modeling and fits well while the inherent bias and sparseness of sewer inspection data may prohibit the application of conventional statistical models [14]. The necessity of specification of internal model architecture, the need of large training sets and the sensitivity to imbalanced datasets limits their usage [15]. Furthermore, the causal relationship is not explicitly expressed. Statistical models are the most used techniques [16,17]. Regression models, a subcategory of statistical models, allow the identification of the causal relationship between the individual sewer attributes and its internal condition state [18]. Different regression approaches have been developed including linear regression [19,20] and nonlinear regression [9,21]. Despite efforts made by municipalities to collect information and build extensive and accurate databases of their wastewater systems, only physical characteristics of sewer pipes (e.g., length, diameter), are the most used attributes in statistical deterioration models for wastewater pipes [17,22–27]. However, various environmental and operational variables impact the structural safety of pipes [28–30]. Accuracy and reliability of existing regression based sewer deterioration models can be enhanced if the following shortfalls are addressed: (1) an oversimplification of the physical process (e.g., linear relationship between the covariates and the condition of pipes), (2) not all the required covariates are available nor included in deterioration models [30–33], (3) the spatial location of pipes that can inform about unknown and unobserved covariates is not considered [6,34,35].

Recent development of Gaussian structured additive regression models allows not only the inclusion of nonlinear effects of the covariates on the response variables but also the consideration of possible spatial autocorrelation to enhance predictive capabilities of wastewater deterioration models [9,13]. The geospatial location of pipes is introduced in the model to capture the information about unobserved covariates in the form of single surrogate covariate representing the district where the pipe is buried. A Gaussian Markov Random field (GMRF) approach allows pipes in neighborhood districts to have similar deterioration pattern, while a random effect parameter is applied to represent heterogeneities in pipes deterioration [34].

Davies et al. [30] listed 33 potential factors that might affect the deterioration of sewer pipes. However, all these factors are not always available in municipality databases to support the analysis of deterioration of sewer pipes. A parsimonious and interpretable model consisting of relevant factors represented at appropriate complexity is required. Statistically significant variables, in regression models, are identified based on heuristic techniques such as stepwise selection that are not efficient. In addition, in frequentist-based statistical models, the heuristic approach has a risk of overconfident estimation of the output variable [36,37]. Kabir et al. [38] applied Bayesian Model Averaging to identify significant covariates in a sewer system before applying a Bayesian logistic regression to predict the structural condition of pipes. Mohammadi et al. [39] considered thirteen variables including environmental factors to classify sanitary sewers. They proposed a gradient decision boosting algorithm to classify the pipes into two groups (good and critical pipes). The binary output of their these models limits its their use in the decision making. Laakso et al. [31] applied the Boruta algorithm to assess the importance of variable in a binary random forest sewer pipe deterioration model. Quantile regression aims at determining the influence of regressors on the conditional quantiles of the dependent variable which do not have necessarily a Gaussian distribution [40]. The advantage of the analysis of quantiles in sewer pipe deterioration is the assessment of the tails of the distribution of the response. For instance, the 5% quantile contains old pipes in good condition and the 95% quantile contains new pipes that have the high likelihood of failure. Both extreme cases influence the expected mean models. Furthermore, the pipes attributes that are responsible of the appearance of outliers may differ is each quantile. The identification of these factors through appropriate modeling of the deterioration process will allow decision makers to capture extreme deterioration patterns and reduce the uncertainty in the pipes useful life calculations. Boosting algorithm has been successfully applied for variable selection and model choice to estimate the time to failure of water pipes [41] and the classification of sanitary sewers [39]. Application of quantile regression coupled to boosting algorithm to simultaneously minimize an empirical risk defined as a loss function while estimating model parameters and iteratively select informative regressors are found in [10,42,43]. Balekelayi and Tesfamariam [42] applied quantile regression to capture the frequency of water pipes' failure in the city of Calgary. The boosting optimization algorithm has the advantage over other machine learning, such as random forests, artificial neural networks, that result in black-box predictions and do not allow quantification of partial influence of covariates on the response variable [37].

In this paper, a Structured Quantile Additive Regression (SQRM) using a boosting optimization algorithm for the selection of covariates is proposed. The approach presented in [9,42] is extended with the integration of linear components for continuous covariates, i.e., the continuous covariates are allowed to have two components including the linear and the nonlinear effects modeled following the ordinary least square and the penalized splines nonlinear smooth functions respectively. Only linear effect is considered for categorical variables. A Bayesian inference based on Markov Chain Monte Carlo simulation is used to capture uncertainty propagation in the definition of base learners' models. The response data, defined as a continuous variable (pipe scores obtained after grading observed defects following a given protocol as described in [44]), is iteratively estimated through an optimized process of learners selection. The Bayesian-based evaluation of the effect of individual factors coupled with the optimization-based selection of the most informative factors increases the accuracy in the quantile prediction and allows the decision makers to confidently quantify the uncertainty in the deterioration model.

This paper is organized as follow: the next section gives the mathematical formulation of the proposed methodology. The description of the case study, the available data and the proposed model follow. Results and discussions are presented before a conclusion is presented in the last section.

2. Mathematical Formulation

The proposed framework is presented in Figure 1. After the selection of the wastewater system, the steps to compute SQRM are:

- First, collection of the available data including inspection data (e.g., observed defects, their severity, their orientation and their location), pipe characteristics (e.g., length, diameter, material type, buried depth and slope) and maintenance data (e.g., number of previous flushes, number of previous reported backups). If available, the operational data (e.g., dry weather flow, wet weather flow, inflow infiltration) are also collected.
- Second, data processing where variables are classified into categorical (e.g., material) and continuous (e.g., diameter). The response variable is the internal condition of the pipe that is expressed in terms of score obtained after rating the observed defects following a grading protocol (e.g., Water Resource Center: WRc). The model equation is developed, and the base learners' priors are selected.
- Third, the last step is simulation and model performance evaluation. The number of iterations and step-length are set to the same values in both quantiles. The validated models are used for predictions and the effect of the selected covariates are examined.

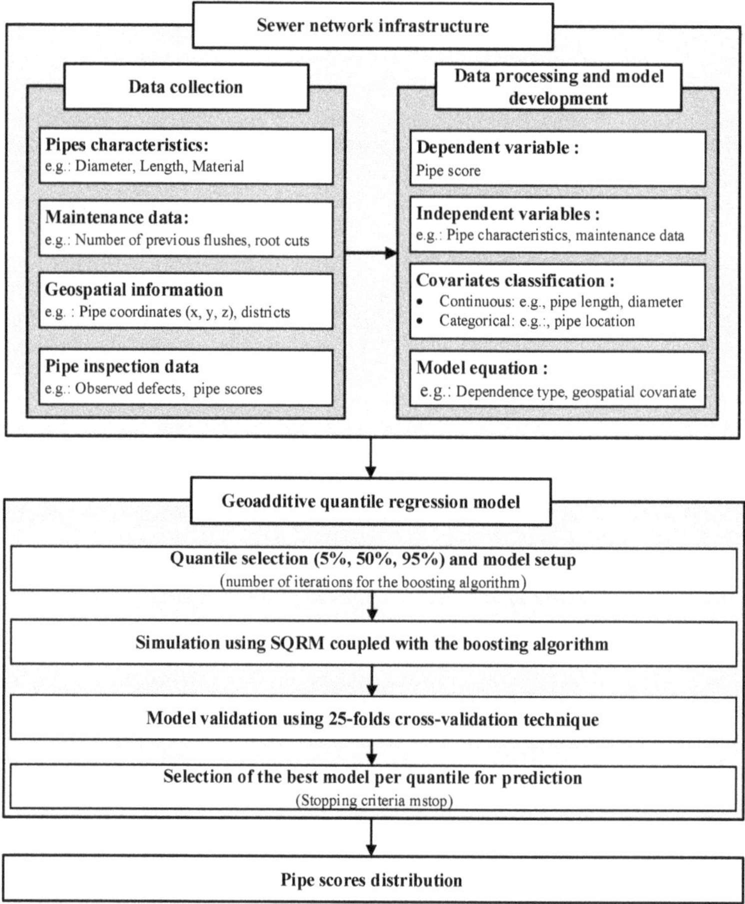

Figure 1. Proposed framework for geoadditive quantile regression model.

2.1. Geoadditive Quantile Regression

Flexible models better represent the complex nonlinear relationships between the regressors and the response variables in natural processes. These models allow the addition of noise that account for over dispersion caused by unobserved heterogeneities and possible autocorrelation. Assume the observations (Y_i, x_i, z_i), $i = 1, \ldots, n$ are elements of the inspection database where Y_i represents the pipe segment score (continuous), $x_i = (x_{i1}, \cdots, x_{iq})$ and $z_i = (z_{i1}, \cdots, z_{ip})$ are vectors of categorical (e.g., material) and continuous covariates (e.g., age), respectively. The relation between the observations Y_i and the predictions η_i is defined as:

$$Y_i = \eta_i + \epsilon_i, \quad \epsilon_i \sim F(\epsilon_i | \theta), i \in I \tag{1}$$

where i is a subset of $\{1, \ldots, n\}$; that is, not necessary all latent η_i are observed through the data Y_i. F is an unknown distribution function of error terms ϵ_i and may depend on some additional parameter θ.

Given a fixed and known quantile $\tau \in (0,100\%)$ and assuming the τth quantile of the error in Equation (1) is zero, i.e., $(F^{-1}(\tau|\theta) = 0)$, the general additive quantile regression model is expressed as [45].

$$Q_{(Y_i|x_i,z_i)}(\tau|x_i,z_i) = \eta_{\tau_i} = x_i^T \beta_{\tau_i} + \sum_{j=1}^{p} f_{\tau_j}(z_{ji}) + b_{\tau_i} \tag{2}$$

where $Q_{(Y_i|x_i,z_i)}(\tau|x_i,z_i)$ is a conditional quantile response, η_{τ_i} is the semi-parametric predictor, β_{τ_i} is the vector coefficient of the parametric component, f is the vector of smoothing functions for the nonparametric component and b_{τ_i} is an unstructured random effect accounting for overdispersion caused by unobserved heterogeneity or for correlation in longitudinal data [40].

In Bayesian framework, appropriate prior distributions (e.g., diffuse priors are considered for the linear effects (here applied to both discrete (categorical) and continuous covariates), first and second order random walk priors are used for continuous covariates and the structured spatial effect is modeled through GMRF specified as an intrinsic conditional autoregressive model) are defined and applied [9,42]. The parametric component in Equation (2) is estimated using the ordinary least squared algorithm. The nonlinear smooth functions f_{τ_j} (Equation (3)) are approximated through penalized polynomial splines (P-splines) of degree l_j represented as a linear combination of $m_j = h_j + l_j - 1$ B-splines basis functions $B_{j,k}$ evaluated at prespecified knots $z_{j,min} = \xi_{j1} < \xi_{j2} < \cdots < \xi_{j,h_j} = z_{j,max}$.

$$f_{\tau_j}(z_{ij}) = \sum_{k=1}^{m_j} \gamma_{\tau_{jk}} B_{jk} z_{ij} \quad i = 1 \cdots n; \quad j = 1 \cdots p \tag{3}$$

where the coefficients $\gamma_{\tau_{jk}}$ are the amplitude that scale the B-spline basis functions B_{jk} to fit the observed data. Smoothness of the functions and accurate fit to observed data are obtained through the definition of high number of knots as well as the imposition of penalties to adjacent B-splines [9,46]. The interactions between the covariates are defined as bivariate functions as explained in [47]. The conditional mean obtained for $\tau = 50\%$ in Equation (2) is the Structured Additive Regression model [43,48]. Information about the geospatial location of the wastewater pipes is added to Equation (2) to build the geoadditive SQRM as:

$$Q_{(Y_i|x_i,z_i)}(\tau|x_i,z_i) = \eta_{\tau_i} = x_i^T \beta_{\tau_i} + \sum_{j=1}^{p} f_{\tau_j}(z_{ji}) + f_{geo,\tau}(S_i) \tag{4}$$

where f_{geo} is spatial effect and has two components: the structured correlated effect f_{struct} and the unstructured effect $f_{unstruct}$, specific to the place where the pipe is buried (local heterogeneities), i.e., $f_{geo} = f_{struct} + f_{unstruct}$. For each district $s \in 1, \cdots, S$ in the area of interest (e.g., in a city, different communities may represent districts), a separate regression coefficient is estimated to represent the correlated spatial effect $f_{struct} = (f_{struct}(S_1) \cdots f_{struct}(S_n))' = Z_{struct}\gamma_{struct}$. The design matrix Z_{struct} is an incidence matrix whose entry in the ith row and kth column is equal to one if the observation i has been observed at location k and zero otherwise. Spatially adjacent regions are assumed to share similar effects on the deterioration of pipes. This is achieved by defining the adjacency matrix K as follows:

$$K[r,s] = \begin{cases} -1 & \text{if districts } s \text{ and } r \text{ are neighbors (i.e., } s \neq r, s \sim r) \\ |N(s)| & \text{if } s = r \\ 0 & \text{otherwise (i.e., } s \neq r, s \nsim r) \end{cases} \tag{5}$$

where $s \sim r$ denotes that the two regions s and r are neighbors and $N(s)$ is the total number of neighbors for region s. For example, in Figure 2, the selected district (Marlborough in the city of Calgary) has 8 neighbors

($N(s) = 8$). Further details about Markov Random Fields can be found in [49]. Information about the factors affecting the deterioration of sewer (e.g., sewerage type, load transfer, surface use, surface type, water main burst and leakage, ground disturbance, groundwater level, ground conditions, soil backfill type) are not systematically collected during inspection campaigns [9]. The addition of the geospatial location information allows to account for these factors [9,42]. The SQRM in Equation (4) is the final form of the equation used to represent the deterioration of sewer pipes in this research.

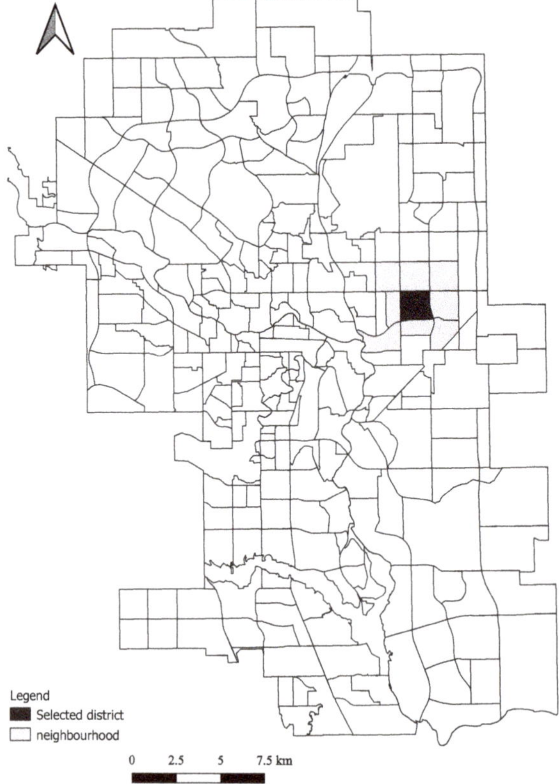

Figure 2. Example of neighborhood definition in Gaussian Markov Random field (GMRF).

For classic linear quantile, the regression coefficients are estimated by minimizing the asymmetrically weighted sum of absolute deviations [50]. The SQRM (Equation (4)) can alternatively be expressed as a minimization problem as [10]:

$$\underset{\eta_\tau}{\text{argmin}} \sum_{i=1}^{n} \rho_\tau(Y_i - \eta_{\tau i}) \tag{6}$$

where

$$\rho_\tau(y_i - \eta_{\tau i}) = \begin{cases} \tau |y_i - \eta_{\tau i}| & \text{if } (y_i - \eta_{\tau i}) \geq 0 \\ (\tau - 1)|y_i - \eta_{\tau i}| & \text{if } (y_i - \eta_{\tau i}) < 0 \end{cases}$$

is the "check function", i.e., a loss function for quantile regression problem that is appropriate for decision making perspectives [50]. The parametric component of the model is solved using the ordinary least square method.

2.2. Boosting Algorithm

The high number of covariates in the model (Equation (4)) led to the choice of boosting algorithm for variable and model selection [37,47]. The gradient descent (boosting) algorithm iteratively applies simple base learner procedures to residuals u obtained by differentiating the optimization criterion of interest (e.g., the loss function defined in Equation (6)) [47]. The stepwise increment of $\hat{f}_j^{[m]}$ (Algorithm 1) are small and the overall minimum is slowly approximated. At the same time, the additive structure of the resulting model fit is conserved since the final aggregation of the additive predictor and its single component are strictly additive. The algorithm selects the most informative base learners (model components) that minimize the empirical risk function. Since each base learner is composed of a single covariate, the variable selection is a very efficient process because the non-informative covariates will not be updated (i.e., they will never be the best fitting base learner in the optimization part of the algorithm) until the stopping criteria is reached. Thus, only influential base learners will appear in the final model. The step length factor v is of minor effect on the accuracy of the estimators if a sufficiently small number is selected [10]. In the developed approach, K-cross-validation technique is applied. The data are split into K subsets where a single subset is kept for the validation and the other (K-1) subsets are used for the training. The goal of boosting algorithm in structured additive quantile regression is to find a solution to the expected loss optimization problem given by the following equation [51]:

$$\eta^* = \underset{\eta}{\operatorname{argmin}} \, \mathbb{E}[L(\mathbf{Y}, \eta)] \tag{7}$$

where $\mathbf{Y}$ is the response and η is the predictor of a regression model. $L(\mathbf{Y}, \eta)$ corresponds to the loss function defined in Equation (6). From the observations $(\mathbf{Y}_i, \mathbf{x}_i, \mathbf{z}_i)$, with $i = 1, \cdots, n$ the number of observations; the expected loss in Equation (6) can be replaced by the following empirical risk:

$$\frac{1}{n}[L(\mathbf{Y_i}, \eta_i(\mathbf{x}_i, \mathbf{z}_i))] \tag{8}$$

where $\mathbf{x}_i, \mathbf{z}_i$ are the complete covariate vectors of observations.

Boosting algorithm advantage is able to combine the model fitting with the intrinsic variable selection and model choice through a functional gradient descent minimization framework [10,47,52,53]. In addition, in the presence of many predictors, the boosting algorithm shows high efficiency in the selection of variables compared to classical selection schemes [53]. The reduced number of parameters to be set by the user, including the number of iterations and the step-length factor (Algorithm 1), makes it a good tool to propose to municipalities for the modeling of the deterioration of wastewater pipes. Friedman [51] demonstrated that a small step-length is beneficial and never yields worse predictions using boosting algorithm. Thus, the step-length is set once. As explained in [51] and represented in Algorithm 1, the number of iterations will be changed after different runs until a better convergence is obtained. The optimal stopping criteria is determined using cross-validation and represents the optimal number of iterations needed to reach the minimum risk defined in Equation (8). The K-cross validation assists in avoiding the overfitting. The value of the stopping criteria (mstop) should be between 1 and the selected number of iterations.

Algorithm 1: Component-wise functional gradient descent boosting algorithm for structured additive quantile regression.

Initialization;

After selection the stopping criteria mstop, the definition and initialization of the predictor are the next steps to consider.;

From Equation (4), the type of interaction of each covariate is defined and the model built.;

Set the iterator m = 0 and the initialization of the response is done from suitable starting values such as the median of the response values $Y_1 \cdots Y_n$ as offset, i.e., $\eta_i^{[0]} = argmin_c \sum_{i=1}^{n} \rho_{0.5}(Y_i - c)$, and $\hat{f}_j^{[0]} = 0$ for $j = 1 \cdots p$;

Iteration;

while $m < mstop$ **do**

> Negative gradient;
>
> m = m + 1 (increment the iterator);
>
> Compute the negative gradient residuals of the loss function evaluated at the predictor values $\hat{\eta}_i^{[m-1]}$ (previous iteration estimated value);
>
> $$u_i^{[m]} = -\frac{\partial}{\partial \eta} L\left(Y_i, \eta\right) \Big|_{\eta = \hat{\eta}_i^{[m-1]}} \quad i = 1 \cdots n$$
>
> For quantile regression boosting;
>
> $$u_i^{[m]} = -\rho_\tau'\left(Y_i - \hat{\eta}_i^{[m-1]}\right) = \begin{cases} \tau & Y_i - \hat{\eta}_i^{[m-1]} \geq 0 \\ \tau - 1 & Y_i - \hat{\eta}_i^{[m-1]} < 0 \end{cases}$$
>
> Estimation;
>
> Fit all the base learners separately to negative gradient residuals and obtain for each base learner, the $\hat{g}_j^{[m]}$ estimators of $\hat{f}_j^{[m]}$.;
>
> Run an optimization process to estimate the best fitting base learner g_{j*} that minimizes the loss function below:
>
> $$j^* = argmin_j \left[\left(u^{[m]} - \hat{g}_j^{[m]}\right)^T \left(u^{[m]} - \hat{g}_j^{[m]}\right) \right]$$
>
> where $u^{[m]} = (u_1^{[m]}, \cdots, u_n^{[m]})^T$ is the vector of gradient residuals in the current iteration
>
> Update;
>
> Compute the update for the nest fitting base learner
>
> $$\hat{f}_{j*}^{[m]} = \hat{f}_{j*}^{[m-1]} + v\hat{g}_{j*}^{[m]}$$
>
> where $v \in [0, 1]$ is a given step length. Keep all the other effects constant, i.e., set $\hat{f}_j^{[m]} = \hat{f}_j^{[m-1]}$ for $d \neq d^*$ Compute the $\hat{\eta}_i^{[m]}$ for $i = 1 \cdots n$

end

---

3. Case Study

The city of Calgary is situated in Southern Alberta, Canada and has a population of 1.2 million people. Its wastewater system is composed of 76,380 sewer pipes representing a total length of 5545.9 km (Figure 3). These pipes connect three wastewater treatment plants (Fish Creek, Pine Creek and Bonnybrook) to residential homes, commercial, industrial and public buildings. The database of the city indicates 5.24% (299.5 km) of pipes have their age above 65 years with 3.4% (188.6 km) above 100 years old. Different pipe material has been used to build the system: asbestos cement (AC), concrete (CON), brick (BR), vitrified clay (VCT), steel (ST), cast iron (CI), corrugated metal pipes (CMP); polyvinyl chloride (PVC), polyethylene (PE) sewer pipes. Before 1945, most of the pipes were clay or metallic. From 1945 to date, a predominance of plastic and cementitious pipes is observed in the data. In this study, pipe materials are grouped into: (1) Cementitious: CON, AC; (2) clay: BRICK, VCT; (3) metallic pipes: ST, CI, CMP; and (4) plastic: PE, PVC. The plastic and cementitious pipes represent 89.4% of the total pipes length in the system

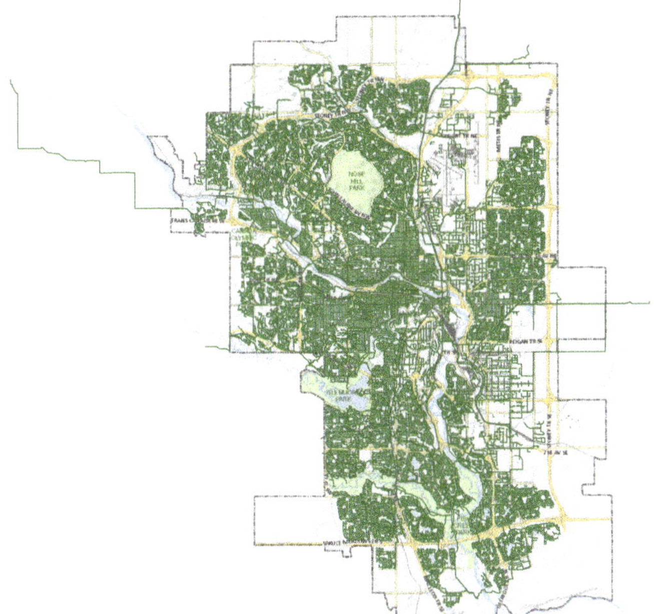

Figure 3. Calgary sewer pipe network.

3.1. Inspection Data Collection and Processing

Since early 1990, an inspection program using closed circuit television (CCTV) technique has been implemented in order to prioritize further inspection, maintenance and replacement of pipes. The selection of pipes for inspection is based on their age, material and the expert knowledge (e.g., out of 76,380 pipes in the city, 12,662 (1188.2 km i.e., 16.6%) have been inspected among which 68.2% are cementitious, 17.2% are clay, 10.1% are metallic and 4.5% are plastic pipes. Starting in 1940, the replacement of Ac and metallic pipes with concrete and plastic has been intensified in 1960. Figure 4 shows the age distribution of cementitious pipes that is varying between 0 and 55 and for plastic between 0 and 45 years. The installation period has been reported as important parameter that affect the sewer condition. It integrates the age, the quality of the material used and the workmanship [54,55].

The observed defects data, for the city of Calgary in Canada, have been graded following the WRc protocol [44]. In this protocol, defects are categorized and assigned a score describing their severity and orientation. The pipe segment grade is given based on its highest defect score that is converted into a crisp number varying from 1 (best) to 5 (likely to fail) representing the Internal Condition Grading (ICG).

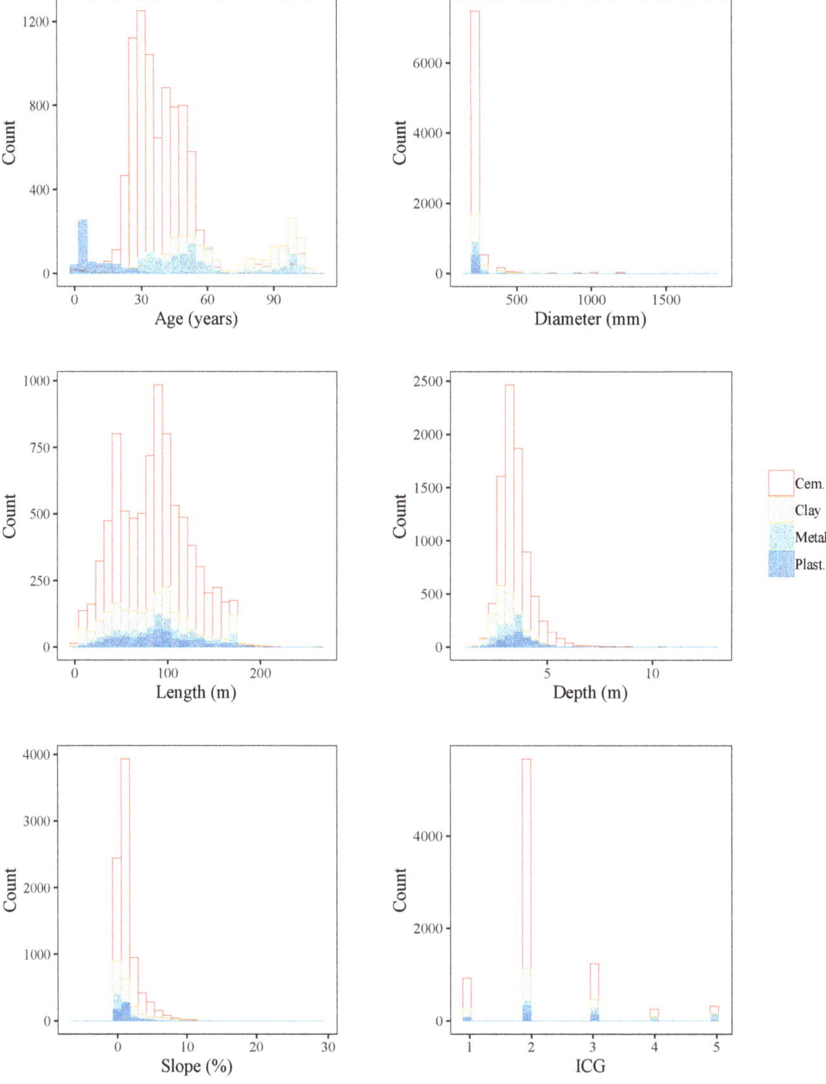

Figure 4. Paired characteristics of inspected pipes (City of Calgary).

3.2. Factors Selection and Model Development

Table 1 gives a brief description of the factors used in this study and assumed to influence the sewer pipe condition deterioration in the city of Calgary. These factors have been selected based on their availability from the database. Maintenance operations increase the pipes service life. However, some maintenance techniques (e.g., flushing, cleaning) may initiate erosion in the sewer material that will evolve with time. Maintenance factors are considered in the presented model.

Table 1. Covariates' description.

Covariates (Physical)	Description
Material	Designed material of pipes (categorical: 1 = cementitious, 2 = clay, 3 = metallic, 4 = plastic)
Age	The difference between the inspection date and the installation date (continuous)
Length	The manhole to manhole distance inspected (continuous)
Diameter	size of the pipes (continuous)
Depth	mean distance from the crown of the pipe to the ground surface (continuous)
Slope	Angle between the pipes axis with the horizontal (continuous)
rservs	Number of residential connections to the pipe (continuous)
cservs	Number of commercial buildings connected to the pipe (continuous)
Covariates (Maintenance)	
Flushes	Number of flushes done previously on the sewer pipe (continuous)
repairs	Number of previous repairs on the sewer pipe (continuous)
Rcuts	Number of times, the roots have been cut from inside the pipe (continuous)
Backups	Number of times, there have been water backups in the sewers (continuous)
Degrease	Number of time, the maintenance team have degreased the sewer pipe (continuous)
Cleaning	Number of time, the sewer pipes have been cleaned (continuous)
Covariates (Environmental)	
Geospatial location	The geographic location of the sewer pipe. While it is widely accepted that pipes having the same characteristics deteriorates at the same pace, there is a variation in the deterioration across pipes in the same cohort but affected by different environmental covariates such as the nature of the soil they are buried in, the groundwater fluctuation, soil compaction, traffic on the ground, activities in the vicinity. These unobserved and unknown data have correlated and uncorrelated effects on the structural response

Penalized splines with second order difference penalty was assumed as prior distribution for the continuous variables (e.g., [10,47]) and ordinary least squares approach is applied for the linear effects (e.g., [10,47,56]). Only the material covariate is considered as a categorical variable in the modeling and its fixed effect is assessed for each material type. The number of knots is set to 40 for the following covariates: diameter, age, spatial effect, flushes, cleaning, repairs, degrease and 20 for the others.

The geospatial information is captured through the location of pipes in districts. All the pipes of a district are assigned the same identification number, which means they are affected by the same unobserved covariates. The city of Calgary has 301 communities representing the districts (id) (Figure 2).

3.3. Response Variable

The response variable (score of pipes) is a continuous variable (following WRc protocol) that has been observed on the pipe segment during inspection. The geoadditive SQRM (Equation (4)) is represented in the following form:

$$\begin{aligned}
\eta_\tau = {} & a_1 \times (Age) + f(Age) + a_2 \times (rservs) + f(rservs) + a_3 \times (cservs) + \\
& f(cservs) + a_4 \times (Length) + f(Length) + a_5 \times (Diameter) + f(Diameter) + \\
& a_6 \times (Material) + a_7 \times (Flushes) + f(Flushes) + a_8 \times (Cleaning) + \\
& f(Cleaning) + a_9 \times (Degrease) + f(Degrease) + a_{10} \times (Backups) + \\
& f(Backups) + a_{11} \times (Roots) + f(Roots) + a_{12} \times (repairs) + f(repairs) + \\
& a_{13} \times (Replaced) + f(Replaced) + a_{14} \times (Slope) + f(Slope) + \\
& a_{15} \times (Depth) + f(Depth) + map(id, bnd = clgr) + rd(id, bnd = clgr)
\end{aligned} \tag{9}$$

where a_i is a fixed parameter coefficient estimated for linear effect, f is smooth nonlinear function expressing nonlinearity in the relationships, map is the structured effect of the geospatial information, rd is the unstructured geospatial effect. The boosting algorithm iteratively builds each quantile model by selecting and updating the more informative covariates. Model parameters are selected as follow: initial number of iterations mstop = 30,000 and the step length = 0.25. The cross-validation of the models is done using 25-folds cross-validation technique. The optimal model used for prediction correspond to the one obtained at the optimal stopping criteria. The selected covariates at this point are the most influential parameters for the quantile under analysis. The effect of the interaction between factors is included in the model as a smooth function f.

4. Results and Discussion

4.1. Model Selected Covariates

Several reasons may explain the appearance of extreme values (e.g., human errors in the interpretation of the inspection data, not updating the database after replacement, maintenance and repairs, other reasons that need to be investigated on the field). In the presented results, human causes of errors are excluded, and the assigned sewer pipes condition grades are assumed to reflect the real actual condition of pipes.

Three quantiles regression models (5%, 50% and 95%). are developed and evaluated in this study. For each model, the score of pipes is predicted and the influential factors identified. The 5% and 95% quantile models are important in the schedule of further inspections and maintenance planning.

The optimal number of iterations obtained after the cross validation is 3675 for the 5% quantile, 18,656 for the 50% quantile and 20,000 for the 95% quantile. At this value (e.g., 3675 iterations as shown in Figure 5), the variables that have been incorporated in the deterioration model are the most influential, i.e., the most informative (Table 2). Both types of relationships are iteratively included in the model. Thus, the final optimal model may have linear, nonlinear and geospatial effects of the covariates on the output ICG response as presented in Table 2.

Interactions between the covariates have been considered in the developed model (Equation (9)). However, no interaction has been found informative by the applied boosting algorithm. Different covariates have been found to have mixed (linear and a nonlinear) effect to the pipe score. Table 2 shows the identified types of effects for all the included covariates in the model. Several studies, where the relationship between the pipe condition and the diameter is expressed in a linear form, found the diameter as an influential factor that affects the deterioration of sewer pipes [25,26,54,57–60]. However, none of these studies gave details on how this covariate acts upon the sewer pipe scores. It is shown here that the diameter of the pipe influences linearly the observation of extreme high score values (95% quantile) of sewer pipes deterioration only (Figure 6a). The diameter does not influence the 5% quantile and 50% quantile. The length covariate is the only characteristic of sewer pipe segments that influences nonlinearly the 5% quantile (Figure 6b) and does not affect the 50% and 95% quantiles. It has been reported in previous studies that this covariate has an influence on the ICG. However, the dependence type and the affected quantile are not described [15,57].

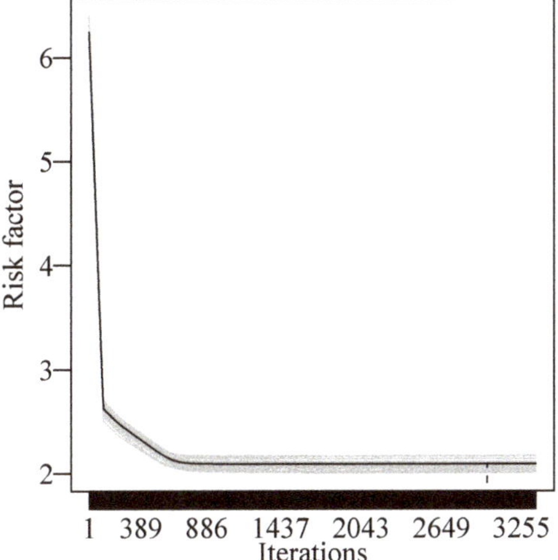

Figure 5. Optimal number of function evaluation to avoid overfitting in the 5% quantile.

Table 2. Influential covariates for each selected percentile.

Covariates	5th Percentile		50th Percentile		95th Percentile	
	Linear	Nonlinear	Linear	Nonlinear	Linear	Nonlinear
Age	-	-	x	x	x	x
rservs	-	x	x	x		x
cservs	-	x	x	-	x	-
Length	-	x	-	-	-	-
Diameter	-	-	-	-	x	-
Material	x	-	x	-	x	-
Flushes	-	x	x	x	-	x
Cleaning	-	x	-	x	-	-
Degrease	-	x	-	x	-	-
Backups	-	x	-	x	-	-
Roots	-	x	-	x	-	-
repairs	x	x	x	x	x	x
Replaced	-	-	X	-	x	-
Slope	-	x	-	x	-	-
Depth	-	x	-	x	-	-

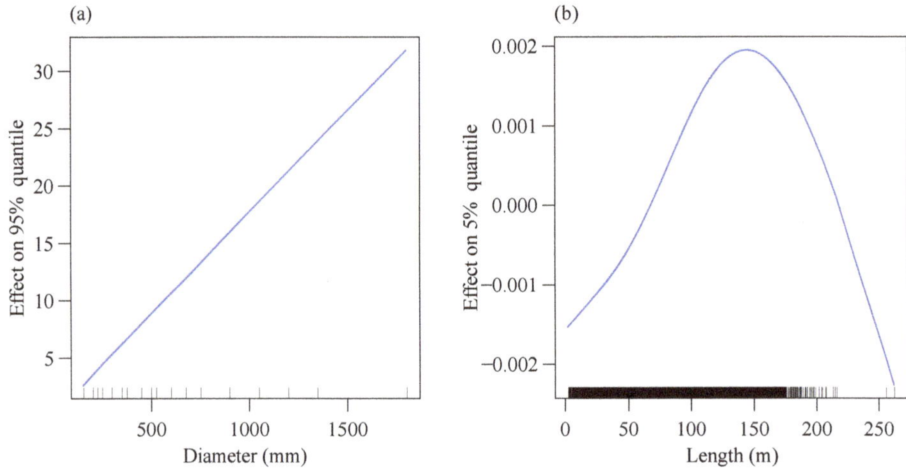

Figure 6. (**a**) Linear effect of the diameter on the 95% quantile; (**b**) nonlinear effects of the length covariate on the 5% quantile.

The age covariate only affects the 50% and 95% quantiles. Municipalities mostly use age covariate to support their decision making for inspections and budget planning. The following studies highlighted the influence of the age variable on the condition of sewer pipes [25,26,57,58,61,62]. When analyzing sewer pipes in the 5% quantile, the age covariate may be misleading as decision variable because it does not affect the 5% quantile. However, the 50% and 95% quantiles are both affected linearly and nonlinearly by the age covariate. Figure 7a,b shows both effect of the age covariates on the 95% quantile.

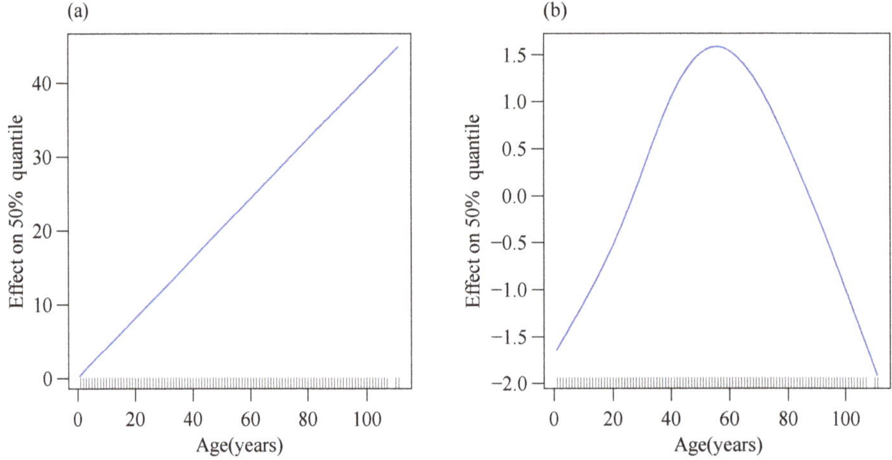

Figure 7. (**a**) Linear and (**b**) nonlinear effect of the age covariate on the 95% quantile.

The 5% quantile is affected by most of the independent variables except the following: Age, diameter and number of previous replacements. However, the 50% quantile (expected mean regression) is not affected by the diameter and length variables. Finally, only three maintenance covariates influence the 95% quantile predicted ICG response including the number of previous flushes flushes that have a nonlinear effect, the number of previous repairs with both types (linear and nonlinear) of relationships and the linear effect of the number of previous replacements.

The values of the covariate's effect in the 5% quantile are less than one. Both covariates have either negative influence (meaning the pipes are in the resistance period, i.e., a newly correctly installed pipe is in the flat part of the bath-tube shape deterioration pattern [63] or positive but below one demonstrating a low influence of the covariate to the deterioration process. In the 50% and 95% quantiles, however, the effect of some covariates goes above 250 (e.g., effect of number of previous repairs in the 95% quantile).

The response variable is an additive combination of simple estimators (constant and varying parameters) representing the partial effect of the selected covariates to the deterioration process. In the case, both types of effects are identified as influential for a covariate, the resulting effect is a mixture of the individual effects that are not easy to be captured with a single type of relationship between the predictor and the output ICG variable. An example is given in Figure 8a,b, where both relationships (linear and nonlinear) for the age and repairs covariates have been selected by the boosting algorithm as influential factors. The resulting mixed effect ("combined effect") shows that the age effect (Figure 8a) in the 50% quantile is mostly controlled by the linear part. However, the nonlinear part should be included to get a precise representation of the overall covariate effect. The repairs covariate effect Figure 8b is also a mixture of both linear and nonlinear effects when the number of previous repairs is below 4. However, after 4 repairs of a pipe, the repairs' effect is led by the nonlinear component.

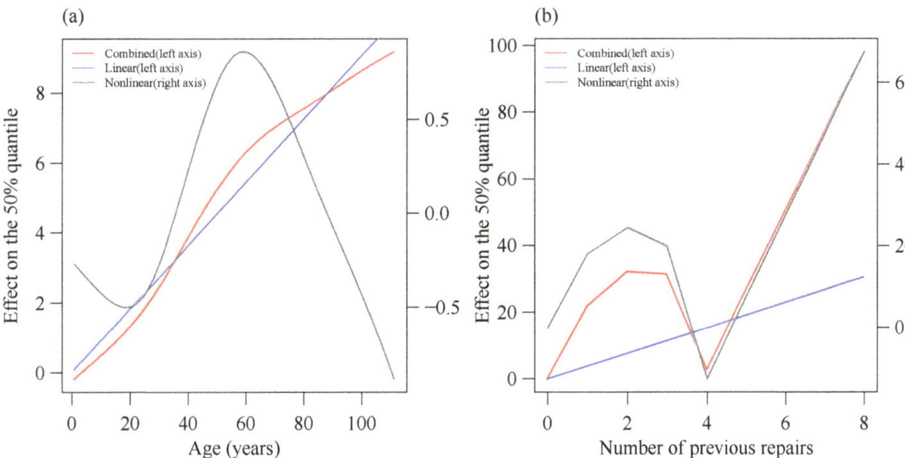

Figure 8. Combined effects of (**a**) age and (**b**) repairs covariates on the 50% quantile.

The inclusion in the model of either linear or nonlinear effect is in some cases the source of errors in the predictions of the output ICG.

4.2. Covariates' Effects on the Response Variable

Both quantile responses are influenced linearly and nonlinearly (Figures 6–8). The 5% quantile has the highest number of nonlinear factors' effects compared to the 50% and 95% quantiles. The type of dependence (linear and nonlinear) between the response variable and the predictors added to their effect's ranges describe the complexity of the wastewater deterioration process. Only the repairs covariate has both types of relationships (linear and nonlinear) with the pipes scores within all the selected quantile. Figure 8b shows both effects for the 50% quantile model. The 50% quantile, i.e., the normal (usual) expected mean regression model, gives a balance between the linear and the nonlinear relationships. Finally, the extremes observed high scores (95% quantile) are caused by only 8 factors (age, the number of residential connections, the number of commercial connections, diameter, material, the number of previous flushes, repairs and replacement) that influence linearly and/or nonlinearly the 95% quantile (Table 2).

Material, flushes and repairs covariates are the only factors that influence all the three selected quantile regression models (Table 2). The partial effects of the material (Figure 9) agree with the age of the pipes in different types of material as represented in Figure 4. Metallic and clay pipes are the oldest types of material in the network and show the highest partial effects compared to the newly installed types (cementitious and plastic pipes). In the 50% quantile, plastic pipes show a high partial effect compared to cementitious pipes while the revers is observed in the 95 % quantile. In additive models, each covariate is fitted individually to the response score (Figures 10 and 11). The model minimizes the differences between the output score and its predicted value from each single covariate (see Equation (6)). Once a good fit is obtained for each covariate, the global output is a weighted sum of the fit of the covariates.

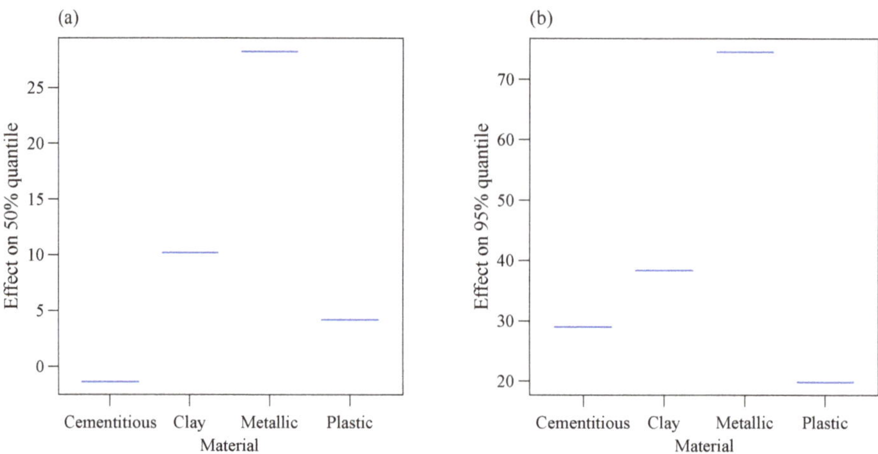

Figure 9. Material linear effect for (**a**) 50% quantile and (**b**) 95% quantile.

It can be seen from Figures 10 and 11 that the 5% quantile is not supported by much data. Thus, the decision maker should, in accordance with the map effect provided in the next section, investigate the delineated high effects areas to get more evidence (investigate new factors) that can support his decision-making process. However, for the 50% and 95% quantiles, Figures 10 and 11 show a good fit of the predictions over the observations. The developed quantile regression tool allow to capture a wide range of dependent variable.

The identified range is the band of uncertainty that the decision maker should consider when taking decisions about replacement, repairs or maintenance. The uncertainty range is skewed to the left with a long tail to the right (95% quantile). Thus, the decision maker can focus only on the 5% and 95% quantiles. In addition, the 95% quantile is the subset that has the highly risky pipes that should be prioritized for inspection and maintenance.

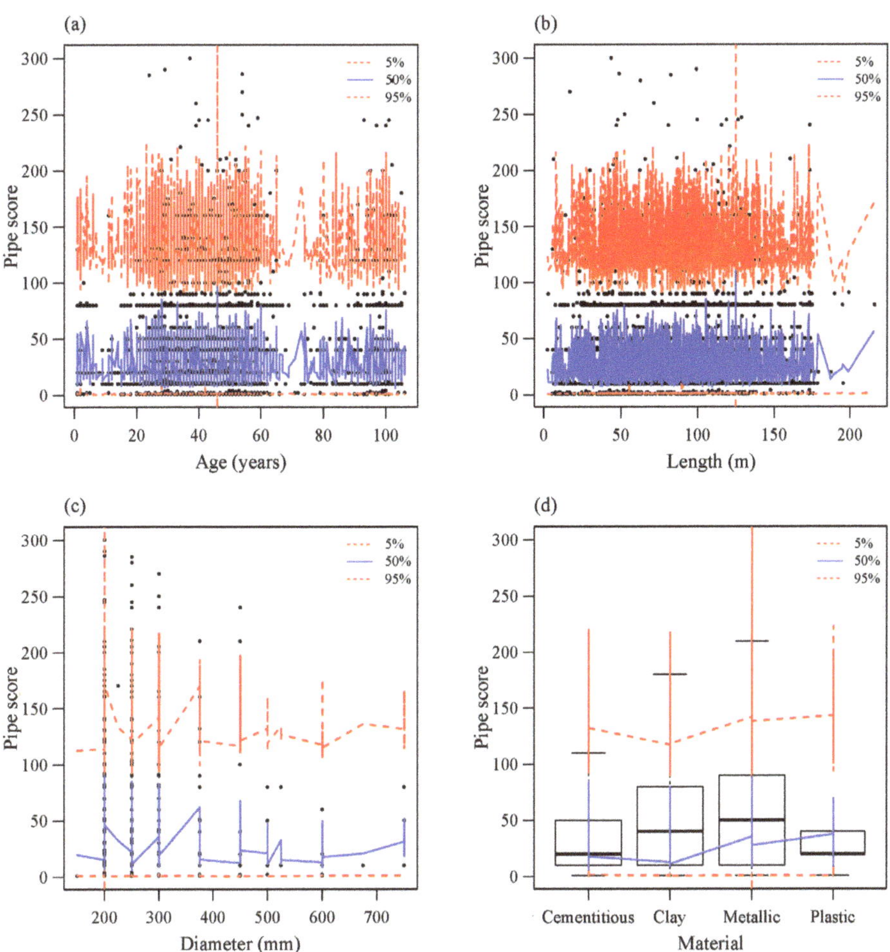

Figure 10. Physical characteristics covariates fitting to the response variables (**a**) age, (**b**) length, (**c**) diameter, (**d**) material.

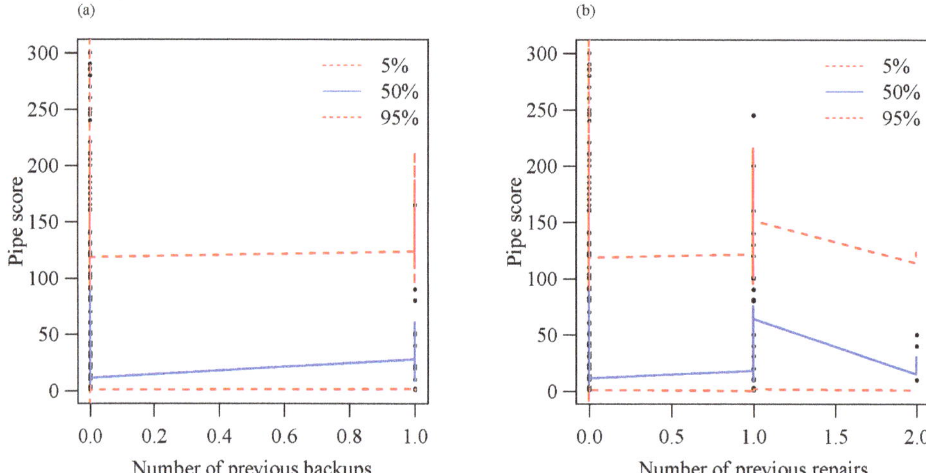

Figure 11. Example of maintenance covariates fitting to the output response (**a**) number of previous backups, (**b**) number of previous repairs.

4.3. Spatial Effects

Maps in Figure 12 are an important source of information to support early decision making for inspection needs and search of more explanatory variables that could better explain the undergoing deterioration of sewer pipes in each district of the city of Calgary. Furthermore, the geospatial effect represents the range of the rate of deterioration in each area of the city. The area around the airport shows the highest effect on the observed pipes condition scores (in the 50% quantile and 95% quantiles). Pipes in this area are cementitious. The causes of high deterioration rates in the airport area can be: (1) rigid pipes subject to intermittent surface loading due to frequent take-offs and landings of airplanes will have higher failure rate than plastic or other material types in the outer areas of the city, (2) the soil chemistry composition may have compound (e.g., H_2S) that reacts with the cementitious pipes and reduces their thickness. An in-depth investigation will likely reveal the main causes of the obtained high deterioration rate in the airport area. However, on the decision-making side, early decision making recommends an increase of inspection frequency in this area. The south west districts, where there are a mix of cementitious and plastic pipes, present also high deterioration rates for both 50% and 95% quantiles (Figure 12b,c).

The high deterioration rates in the north-east and south-west explain why city managers prioritized the replacement of metallic and clay pipes in the outer part of the city. The remaining metallic and clay pipes are buried in the center of the city of Calgary where the deterioration rate is not high. They are the oldest material type in the network (See Figure 4). The 50% quantile regression model (Figure 12b) highlights the same area of high influence (with a focus on the center of the network where more metallic and clay pipes are installed). The critical area (near the airport) is captured in the 95% quantile model. The analysis of quantiles allows to discover other causes (human activities) that may influence the deterioration pattern of sewer pipes in the system and delineate critical areas. Furthermore, the spatial effect maps show the existence of autocorrelation between the districts. The range values of the geospatial effects indicate that high geospatial effects are in the 95% quantile.

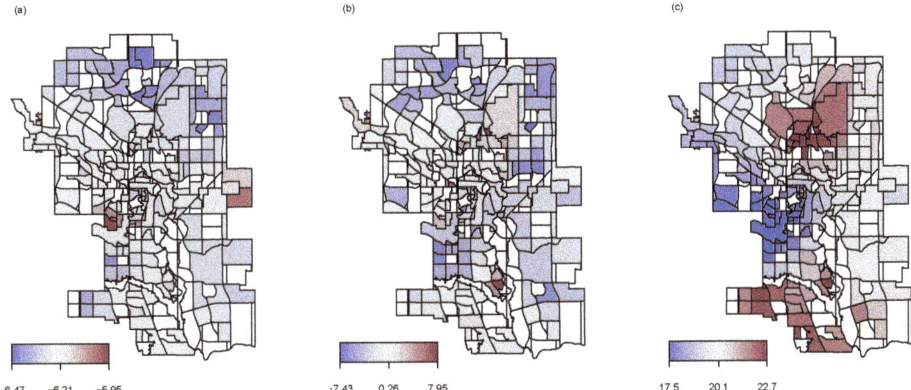

Figure 12. Spatial effects for (**a**) 5% quantile model, (**b**) 50% quantile and (**c**) 95% quantile.

4.4. Discussion

Quantile regression modeling is an effective way of addressing deterioration of sewer pipes. The proposed model is robust against outliers which are a frequent issue in the management of wastewater systems. It allows to elucidate factors that make extreme values to be observed. Moreover, it gives a credible interval in which the observed values are identified. This interval quantifies the band of uncertainty in the output variable for each independent variable (Figures 10 and 11). The band of uncertainty is key component of the decision-making process. Accuracy in the predictions is gained not only through acquisition of unknown information about contributing factors but also through a realistic representation of the complex dependence (linear, nonlinear) of the response (internal condition score) to the covariates. The obtained results agree with results obtained in previous studies [9]. For instance, the selected influential factors (see Table 2) have been found significant in many previous studies [62,64].

Kabir et al. [38] found in their application on the same case study that the age, the length and the rim elevation are the most significant factors for the cementitious pipes. For the clay pipes, their results show the importance of the age and the length. Finally, the plastic and metallic pipe deteriorations are influenced by the age and diameter. The analysis of the results in Table 1 show that the selected factors influence the deterioration of pipes but at different quantiles. For instance, the age is significant both (linearly and nonlinearly) for the 50% and 95% quantiles, not for the 5% quantile for both types of material. However, the diameter only influences the 95% quantile. The length is significant in the 5% quantile. The quantile approach highlights the effects of factors on different quantiles for both material types in the system. In addition, Balekelayi and Tesfamariam [42] applied nonlinear P-splines smooth functions to all the continuous factors and the only parametric factor was the material covariate. Their model found a root mean squared error (RMSE) equal to 17.33 while the RMSE for the 50% quantile in this study is equal to 17.46. In this study, the optimal 50% quantile model has a reduced number of covariates compared to the model developed in [9]. This study shows that it is efficient to consider only the most informative covariates in the quantile of interest. These covariates may differ from one quantile to another. Furthermore, the output maps highlight areas where further search of causing factors and pipes inspection should be prioritized. For inspection prioritization, the areas where no information have been recorded (district where only local heterogeneities are found) and districts with high geospatial effects are the priority areas for future inspection campaigns. The new information will be integrated in the developed model and the model will be updated. Regarding the quantiles, the 95% quantile is more informative on pipes having a high deterioration rate and gives a better representation of critical zones than the 5% and 50% quantiles. From the analysis of the predicted ICG (Figures 10 and 11), it is observed that the distribution of the predicted response is skewed to the left and considering only the 50% and the 95% quantiles will be sufficient to give insights on the range of possible output responses.

5. Conclusions

In this paper, an individual deterioration model is developed using a Bayesian quantile inference with boosting algorithm for variable and model selection. The model considered the effects of physical, maintenance and environmental factors on the structural condition of sewer pipes using linear, nonlinear and geospatial effects on different quantiles (5%, 50% and 95%). The operator does not decide on the type of relationships between the model factors and the response. All possible combinations are incorporated in the model and an unsupervised selection of factors and model complexity is performed during the simulation by the model. Modeling quantiles allows the analysis and the identification of pipes in the tails of the condition distributions. Thus, different decisions should be made for pipes located in the same district and having the same characteristics. On the contrary, single decision (as suggested by several deterioration models) for both pipes (on the entire distribution) lead to premature replacement or the existence of high-risk pipes in the network. At a practical point of view, this research presents a powerful tool to orient and guide the prioritization of inspections in the network area. A map is a universal language tool, accessible to people from different personal and professional backgrounds. The geospatial effect maps on the deterioration of sewer pipes provide a sense of scale and distance of critical areas. The decision makers will have to decide on different priorities focusing on the 5% quantile and 95% quantiles for inspections. The rehabilitation plan may be based on the 50% quantile. Finally, this research showed that the pipe characteristic data representing the physical data (e.g., length, material, age, depth, slope), the maintenance data and the geospatial location of the pipes can efficiently represent a realistic deterioration model for all (inspected and uninspected) the pipes in the network. The proposed modeling approach allows the identification of factors that explain the difference occurring in the deterioration of pipes having the same characteristics and exposed to the same environmental factors. The defined uncertainty band is an important tool that should be efficiently used in maintenance planning and inspection prioritization. The presented methodology provides an accurate and effective tool that is based on reduced number of factors and through a proper representation of their effects to the response variable allows decision makers to analyze the entire distribution of the condition of pipes. Future research direction should include fine resolution in the geospatial location information. This information will allow to differentiate the effect of the environment on the pipes buried in the same district.

6. Future Research Needs

To support the proactive management of water utilities, there is a need of accurate deterioration models at reduced cost of data acquisition. The use of complex mathematical modeling to explain the deterioration process with available data is more than needed. Future research directions will use the proposed methodology in this study and reduced granularity for the geospatial location of the pipes (sub-district, roads). The objective is to have the geospatial location of pipes expressed in terms of its geographical coordinates.

Author Contributions: Conceptualization, methodology, writing—Original draft preparation and formal analysis, N.B.; review and editing, S.T. All authors have read and agreed to the published version of the manuscript.

Funding: This research received no external funding.

Acknowledgments: The authors acknowledge the financial support through the Natural Sciences an Engineering Research Council of Canada (RGPIN-2019-05584) under the Discovery Grant programs.

Conflicts of Interest: The authors declare no conflict of interest.

Abbreviations

SQRM	Structured Quantile Regression model
GMRF	Gaussian Markov Random Filed
ICG	Internal Condition Grading
EPA	Environmental Protection Agency
AC	Asbestos Cement
CON	Concrete
BR	Brick
VCT	Vitrified clay
ST	Steel
CI	Cast iron
CMP	Corrugated metal pipes
PVC	Polyvinyl chloride
PE	Polyethylene

References

1. US-EPA. *Rehabilitation of Wastewater Collection and Water Distribution Systems, State of the Technology Review Report;* US Environmental Protection Agency: Edison, NJ, USA, 2009.
2. Vahidi, E.; Jin, E.; Das, M.; Singh, M.; Zhao, F. Environmental life cycle analysis of pipe materials for sewer systems. *Sustain. Cities Soc.* **2016**, *27*, 167–174. [CrossRef]
3. Fenner, R.A. Approaches to sewer maintenance: A review. *Urban Water* **2000**, *2*, 343–356. [CrossRef]
4. Korving, H.; Van Noortwijk, J.M.; Van Gelder, P.H.; Clemens, F.H. Risk-based design of sewer system rehabilitation. *Struct. Infrastruct. Eng.* **2009**, *5*, 215–227. [CrossRef]
5. Mancuso, A.; Compare, M.; Salo, A.; Zio, E.; Laakso, T. Risk-based optimization of pipe inspections in large underground networks with imprecise information. *Reliab. Eng. Syst. Saf.* **2016**, *152*, 228–238. [CrossRef]
6. Syachrani, S.; Jeong, H.S.; Chung, C.S. Dynamic deterioration models for sewer pipe network. *J. Pipeline Syst. Eng. Pract.* **2011**, *2*, 123–131. [CrossRef]
7. Infrastructure Canada. Canadian Infrastructure Report Card-Informing the Future. 2016. Available online : http://canadianinfrastructure.ca/en/index.html (accessed on 20 October 2020).
8. Malek Mohammadi, M.; Najafi, M.; Kaushal, V.; Serajiantehrani, R.; Salehabadi, N.; Ashoori, T. Sewer Pipes Condition Prediction Models: A State-of-the-Art Review. *Infrastructures* **2019**, *4*, 64. [CrossRef]
9. Balekelayi, N.; Tesfamariam, S. Statistical inference of sewer pipe deterioration using Bayesian geoadditive regression model. *J. Infrastruct. Syst.* **2019**, *25*, 04019021. [CrossRef]
10. Fenske, N.; Kneib, T.; Hothorn, T. Identifying risk factors for severe childhood malnutrition by boosting additive quantile regression. *J. Am. Stat. Assoc.* **2011**, *106*, 494–510. [CrossRef]
11. Ana, E.; Bauwens, W. Modeling the structural deterioration of urban drainage pipes: The state-of-the-art in statistical methods. *Urban Water J.* **2010**, *7*, 47–59. [CrossRef]
12. Rajani, B.; Tesfamariam, S. Uncoupled axial, flexural, and circumferential pipe soil interaction analyses of partially supported jointed water mains. *Can. Geotech. J.* **2004**, *41*, 997–1010. [CrossRef]
13. Kleiner, Y.; Rajani, B. Comprehensive review of structural deterioration of water mains: Statistical models. *Urban Water* **2001**, *3*, 131–150. [CrossRef]
14. Mashford, J.; Marlow, D.; Tran, D.; May, R. Prediction of sewer condition grade using support vector machines. *J. Comput. Civ. Eng.* **2010**, *25*, 283–290. [CrossRef]
15. Kabir, E.; Guikema, S.; Kane, B. Statistical modeling of tree failures during storms. *Reliab. Eng. Syst. Saf.* **2018**, *177*, 68–79. [CrossRef]

16. Balekelayi, N. Advanced Deterioration Models for Wastewater Inspection Prioritization. Ph.D. Thesis, University of British Columbia, Vancouver, BC, Canada, 2019.

17. Duchesne, S.; Beardsell, G.; Villeneuve, J.P.; Toumbou, B.; Bouchard, K. A survival analysis model for sewer pipe structural deterioration. *Comput. Aided Civ. Infrastruct. Eng.* **2013**, *28*, 146–160. [CrossRef]

18. Baah, K.; Dubey, B.; Harvey, R.; McBean, E. A risk-based approach to sanitary sewer pipe asset management. *Sci. Total Environ.* **2015**, *505*, 1011–1017. [CrossRef]

19. Gedam, A.; Mangulkar, S.; Gandhi, B. Prediction of sewer pipe main condition using the linear regression approach. *J. Geosci. Environ. Prot.* **2016**, *4*, 100–105. [CrossRef]

20. Coppi, R. Management of uncertainty in statistical reasoning: The case of regression analysis. *Int. J. Approx. Reason.* **2008**, *47*, 284–305. [CrossRef]

21. Su, Z.G.; Wang, Y.F.; Wang, P.H. Parametric regression analysis of imprecise and uncertain data in the fuzzy belief function framework. *Int. J. Approx. Reason.* **2013**, *54*, 1217–1242. [CrossRef]

22. Caradot, N.; Riechel, M.; Fesneau, M.; Hernandez, N.; Torres, A.; Sonnenberg, H.; Eckert, E.; Lengemann, N.; Waschnewski, J.; Rouault, P. Practical benchmarking of statistical and machine learning models for predicting the condition of sewer pipes in Berlin, Germany. *J. Hydroinform.* **2018**, *20*, 1131–1147. [CrossRef]

23. Kaddoura, K.; Zayed, T. An integrated assessment approach to prevent risk of sewer exfiltration. *Sustain. Cities Soc.* **2018**, *41*, 576–586. [CrossRef]

24. Salman, B. Infrastructure Management and Deterioration Risk Assessment of Wastewater Collection Systems. Ph.D. Thesis, University of Cincinnati, Cincinnati, OH, USA, 2010.

25. Baik, H.S.; Jeong, H.S.; Abraham, D.M. Estimating transition probabilities in Markov chain-based deterioration models for management of wastewater systems. *J. Water Resour. Plan. Manag.* **2006**, *132*, 15–24. [CrossRef]

26. Ariaratnam, S.T.; El-Assaly, A.; Yang, Y. Assessment of infrastructure inspection needs using logistic models. *J. Infrastruct. Syst.* **2001**, *7*, 160–165. [CrossRef]

27. Wirahadikusumah, R.; Abraham, D.; Iseley, T. Challenging issues in modeling deterioration of combined sewers. *J. Infrastruct. Syst.* **2001**, *7*, 77–84. [CrossRef]

28. Wang, C.; Zhang, H.; Li, Q. Reliability assessment of aging structures subjected to gradual and shock deteriorations. *Reliab. Eng. Syst. Saf.* **2017**, *161*, 78–86. [CrossRef]

29. Dirksen, J.; Clemens, F.; Korving, H.; Cherqui, F.; Le Gauffre, P.; Ertl, T.; Plihal, H.; Müller, K.; Snaterse, C. The consistency of visual sewer inspection data. *Struct. Infrastruct. Eng.* **2013**, *9*, 214–228. [CrossRef]

30. Davies, J.; Clarke, B.; Whiter, J.; Cunningham, R. Factors influencing the structural deterioration and collapse of rigid sewer pipes. *Urban Water* **2001**, *3*, 73–89. [CrossRef]

31. Laakso, T.; Kokkonen, T.; Mellin, I.; Vahala, R. Sewer condition prediction and analysis of explanatory factors. *Water* **2018**, *10*, 1239. [CrossRef]

32. Syachrani, S.; Jeong, H.S.D.; Chung, C.S. Decision tree–based deterioration model for buried wastewater pipelines. *J. Perform. Constr. Facil.* **2013**, *27*, 633–645. [CrossRef]

33. Kneib, T.; Fahrmeir, L. A mixed model approach for geoadditive hazard regression. *Scand. J. Stat.* **2007**, *34*, 207–228. [CrossRef]

34. Balekelayi, N.; Tesfamariam, S. Geoadditive Bayesian regression models for water mains failure rate prediction. In Proceedings of the 13th International Conference on Applications of Statistics and Probability in Civil Engineering, ICASP13, Seoul, Korea, 26–30 May 2019; Volume 1, pp. 1–8.

35. Post, J.; Langeveld, J.; Clemens, F. Analysing spatial patterns in lateral house connection blockages to support management strategies. *Struct. Infrastruct. Eng.* **2017**, *13*, 1146–1156. [CrossRef]

36. Fragoso, T.M.; Bertoli, W.; Louzada, F. Bayesian model averaging: A systematic review and conceptual classification. *Int. Stat. Rev.* **2018**, *86*, 1–28. [CrossRef]

37. Hofner, B.; Mayr, A.; Robinzonov, N.; Schmid, M. Model-based boosting in R: A hands-on tutorial using the R package mboost. *Comput. Stat.* **2014**, *29*, 3–35. [CrossRef]

38. Kabir, G.; Balekelayi, N.C.; Tesfamariam, S. Sewer structural condition prediction integrating Bayesian model averaging with logistic regression. *J. Perform. Constr. Facil.* **2018**, *32*, 04018019. [CrossRef]

39. Malek Mohammadi, M.; Najafi, M.; Salehabadi, N.; Serajiantehrani, R.; Kaushal, V. Predicting Condition of Sanitary Sewer Pipes with Gradient Boosting Tree. In *Pipelines 2020*; American Society of Civil Engineers Reston: Reston, VA, USA, 2020; pp. 80–89.
40. Yue, Y.R.; Rue, H. Bayesian inference for additive mixed quantile regression models. *Comput. Stat. Data Anal.* **2011**, *55*, 84–96. [CrossRef]
41. Snider, B.; McBean, E.A. Improving time to failure predictions for water distribution systems using extreme gradient boosting algorithm. In Proceedings of the WDSA/CCWI Joint Conference, Kingston, ON, Canada, 23–25 July 2018; Volume 1.
42. Balekelayi, N.; Tesfamariam, S. Advanced Water Main Deterioration Model Using Bayesian Geoadditive Quantile Regression. In *Encyclopedia of Water: Science, Technology, and Society*; John Wiley & Sons Ltd.: Hoboken, NJ, USA, 2019; pp. 1–14. Available online: https://onlinelibrary.wiley.com/doi/book/10.1002/9781119300762 (accessed on 19 October 2020).
43. März, A.; Klein, N.; Kneib, T.; Musshoff, O. Analysing farmland rental rates using Bayesian geoadditive quantile regression. *Eur. Rev. Agric. Econ.* **2016**, *43*, 663–698. [CrossRef]
44. Marlow, D.; Heart, S.; Burn, S.; Urquhart, A.; Gould, S.; Anderson, M.; Cook, S.; Ambrose, M.; Madin, B.; Fitzgerald, A. *Condition Assessment Strategies and Protocols for Water and Wastewater Utility Assets*; Project Ref. 03-CTS-20CO; Water Environment Research Foundation: Alexandria, VA, USA, 2007.
45. Waldmann, E.; Kneib, T.; Yue, Y.R.; Lang, S.; Flexeder, C. Bayesian semiparametric additive quantile regression. *Stat. Model.* **2013**, *13*, 223–252. [CrossRef]
46. Scheipl, F.; Kneib, T.; Fahrmeir, L. Penalized likelihood and Bayesian function selection in regression models. *AStA Adv. Stat. Anal.* **2013**, *97*, 349–385. [CrossRef]
47. Sobotka, F.; Kneib, T. Geoadditive expectile regression. *Comput. Stat. Data Anal.* **2012**, *56*, 755–767. [CrossRef]
48. Kneib, T.; Hothorn, T.; Tutz, G. Variable selection and model choice in geoadditive regression models. *Biometrics* **2009**, *65*, 626–634. [CrossRef]
49. Rue, H.; Held, L. *Gaussian Markov Random Fields: Theory and Applications*; CRC Press: Boca Raton, FL, USA, 2005.
50. Koenker, R.; Bassett, G., Jr. Regression quantiles. *Econom. J. Econom. Soc.* **1978**, *46*, 33–50. [CrossRef]
51. Friedman, J.H. Greedy function approximation: A gradient boosting machine. *Ann. Stat.* **2001**, *29*, 1189–1232. [CrossRef]
52. Winkler, D.; Haltmeier, M.; Kleidorfer, M.; Rauch, W.; Tscheikner-Gratl, F. Pipe failure modelling for water distribution networks using boosted decision trees. *Struct. Infrastruct. Eng.* **2018**, *14*, 1402–1411. [CrossRef]
53. Bühlmann, P.; Hothorn, T. Boosting algorithms: Regularization, prediction and model fitting. *Stat. Sci.* **2007**, *22*, 477–505. [CrossRef]
54. Baur, R.; Herz, R. Selective inspection planning with ageing forecast for sewer types. *Water Sci. Technol.* **2002**, *46*, 389–396. [CrossRef]
55. Davies, J.; Clarke, B.; Whiter, J.; Cunningham, R.; Leidi, A. The structural condition of rigid sewer pipes: A statistical investigation. *Urban Water* **2001**, *3*, 277–286. [CrossRef]
56. Langrock, R.; Kneib, T.; Sohn, A.; DeRuiter, S.L. Nonparametric inference in hidden Markov models using P-splines. *Biometrics* **2015**, *71*, 520–528. [CrossRef]
57. Salman, B.; Salem, O. Risk assessment of wastewater collection lines using failure models and criticality ratings. *J. Pipeline Syst. Eng. Pract.* **2012**, *3*, 68–76. [CrossRef]
58. Chughtai, F.; Zayed, T. Infrastructure condition prediction models for sustainable sewer pipelines. *J. Perform. Constr. Facil.* **2008**, *22*, 333–341. [CrossRef]
59. Tran, D.; Ng, A.; Perera, B.; Burn, S.; Davis, P. Application of probabilistic neural networks in modelling structural deterioration of stormwater pipes. *Urban Water J.* **2006**, *3*, 175–184. [CrossRef]
60. Micevski, T.; Kuczera, G.; Coombes, P. Markov model for storm water pipe deterioration. *J. Infrastruct. Syst.* **2002**, *8*, 49–56. [CrossRef]
61. Younis, R.; Knight, M.A. A probability model for investigating the trend of structural deterioration of wastewater pipelines. *Tunn. Undergr. Space Technol.* **2010**, *25*, 670–680. [CrossRef]

62. Ana, E.; Bauwens, W.; Pessemier, M.; Thoeye, C.; Smolders, S.; Boonen, I.; De Gueldre, G. An investigation of the factors influencing sewer structural deterioration. *Urban Water J.* **2009**, *6*, 303–312. [CrossRef]
63. Røstum, J. Statistical Modelling of Pipe Failures in Water Networks. Ph.D. Thesis, Norwegian University of Science and Technology NTNU, Trondheim, Norway, 2000.
64. Kley, G.; Caradot, N. *Review of Sewer Deterioration Models*; KWB Project SEMA, Report 1; Cicerostr: Berlin, Germany, 2013.

Publisher's Note: MDPI stays neutral with regard to jurisdictional claims in published maps and institutional affiliations.

sustainability

Review

Shipping Bunker Cost Risk Assessment and Management during the Coronavirus Oil Shock

Tzeu-Chen Han [1] and Chih-Min Wang [2,*]

[1] Department of Shipping and Transportation Management, National Penghu University of Science and Technology, Penghu 880011, Taiwan; tchan@gms.npu.edu.tw
[2] Department of Aviation and Maritime Transportation Management, Chang Jung Christian University, Tainan 71101, Taiwan
*   Correspondence: wang8200@mail.cjcu.edu.tw

Abstract: This research explores ways to develop a risk management strategy that enables shipping companies to reduce unnecessary fuel cost risks, fuel price fluctuations and improve financial management. Through the Monte Carlo method, the study makes use of the simulation of the conditional value-at-risk (CVaR) model. First, the VaR of various shipping-fuel-cost combination over a ten-year period is simulated. Then, through the most appropriate probability distribution test, it is found that most of the VaR of shipping fuel cost combination are in Beta–Arcsine distribution. In other words, the high-frequency data are concentrated at both tails (minimum and maximum) with high volatility. Therefore, the best strategy is to install scrubbers on existing ships to purify their exhaust gas and choose natural gas-based marine fuel for new ships. This will benefit the shipping companies significantly more compared to the use of low-sulfur fuel and choosing forward bunker agreements. Bunker swaps and options of bunker prices to hedging the risk of bunker cost raised in the end of Coronavirus oil shock, the strategy could help achieve the goal of risk management in the sustainable supply chain.

Keywords: shipping cost; risk assessment and management; conditional value-at-risk

Citation: Han, T.-C.; Wang, C.-M. Shipping Bunker Cost Risk Assessment and Management during the Coronavirus Oil Shock. *Sustainability* **2021**, *13*, 4998. https://doi.org/10.3390/su13094998

Academic Editors: Golam Kabir, Sanjoy Kumar Paul and Syed Mithun Ali

Received: 25 March 2021
Accepted: 27 April 2021
Published: 29 April 2021

Publisher's Note: MDPI stays neutral with regard to jurisdictional claims in published maps and institutional affiliations.

1. Introduction

On 1 January 2020, the International Maritime Organization (IMO) required ships to lower their sulphur emissions from 3.5% to 0.5%. A lot of ship operators' attention has now focused on alternative fuels and technologies to serve it as their solution. IMO regulations prompt the adoption of positive measures that cost ship operating companies an amount of money as they must choose alternative fuels and technologies. In other words, companies suffer economically when forced to switch to a more expensive fuel type [1]. The proposed strategies are as follows.

1.   Switch from high sulphur fuel oils (HSFO) to low sulphur fuel oils (LSFO).
2.   Install exhaust gas cleaning systems (EGSC or scrubbers) and use heavy fuel oil (HFO).
3.   Change the use of fuel altogether, such as liquefied natural gas (LNG) or other fuels.

A combination of weaker demand due to the coronavirus pandemic and a warm winter [2] has led to a price war, resulting in oil prices falling over 70% from January to April in 2020. On 20 April 2020, the May contract for West Texas intermediate crude touched negative USD 40.32, triggering shock waves among global oil companies and related sectors. Oil price declines are a boon for shipping bunker cost, which is 50–60% of a ship's total operating costs. Generally, low transportation costs would encourage the growth of shipping transportation demand. However, the rapid spread of coronavirus has forced countries to impose personal travel restrictions. These restrictions make it impossible to handle cargos at ports and the inactivity induced shipping lines. Moreover, several ports remain closed in many countries due to the cancellation of sailing. It is indicated that the loading and unloading of the world's major ports have decreased more than 30% between January and June this year during the coronavirus pandemic. Carriers had announced blank sailings and the minimum number of vessels to manage capacity while potential and sustained outbreaks of coronavirus might cause a prolonged downturn in transportation demand. Meanwhile, the reduction of sailings also decreases ship fuel consumption and pollution emissions.

According to Ship&Bunker website, by the end of April 2020, the difference between the price of low-sulfur oil and heavy fuel oil fell from US$291.75/ton to US$94.25/ton, or down 67.7% compared to the beginning of the year. This has posed a huge question on the scrubber installation/switch to low-sulphur fuel for ship operators.

During this current coronavirus oil shock, the current paper focuses on the following steps.

1. Identify the risk of a cost difference between alternative fuels and technologies.
2. Assess the vulnerability of alternative fuels and technologies.
3. Determine the risk (i.e., the likelihood and consequences of specific types of risk on fuel cost).
4. Identify ways to reduce those risks.
5. Prioritize risk reduction measures.

In this study, we aim to minimize a shipping company's expected total bunker costs based on its alternative fuels and technologies and risk attitude.

2. Methods

Alizadeh and Nomikos [3] invest in the VaR to a particular amount of bunker cost, which is likely to be a lost due to the changes in the market over a certain period of time and given some probability. However, CVaR is used to measure the risk-aversion attitude in shipping lines' pricing strategies that can significantly get the best effect to exceed VaR [4]. Therefore, in this paper we adopt the CVaR model to measure the fuel costs of alternative fuel as they are more focus on price fluctuations and financial management. The following equations are used to estimate CvaR of fuel portfolio type cost.

2.1. Estimation of Fuel Portfolio Types' Cost Distribution

Gu et al. [5] show that fuel cost has high volatility, and in order to control the risk of bunker cost, one can apply CVaR from the field of financial portfolio management. This study thus applies CvaR as an accurate fuel portfolio type of cost risk management to line shipping financials, which can be described into three steps process shown in Figure 1. The first step is to analyze the cost data of fuel portfolio types and their degree of volatility by the coefficient of variation. The second step is to determine a specific distribution by a goodness of fit test on the historical data distribution. When the distribution and further details have been selected, the third step is to use Monte Carlo simulation and evaluate fuel portfolio type cost valuations that simulations worst-case, best-case in risk analysis. Additionally, the methods amount to a new distribution of fuel portfolio type cost for evaluating CVaR at an optimized confidence level. It is also a measure of risk based on the idea of VaR quantile [6] and mean value [7].

The cost fuel portfolio types exhibit a risk distribution. As illustrated in Figure 2, the frequency distribution allows one to estimate a voyage fuel portfolio type cost CvaR, including its minimum and maximum. They might be expected to hold over a given period at the 5% and 95% levels of confidence (probability) [8].

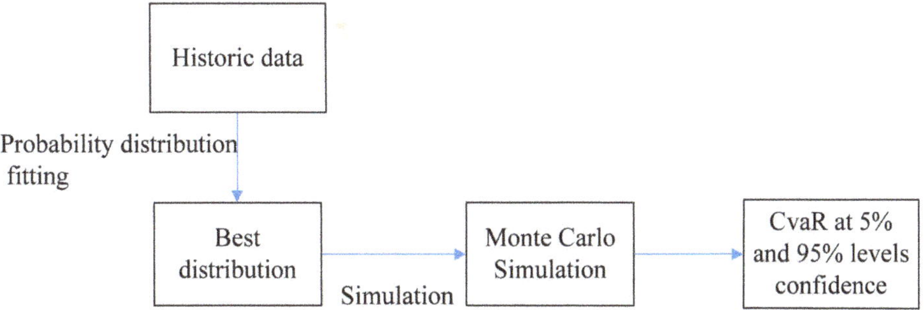

Figure 1. The risk measuring process of fuel portfolio type of cost risk.

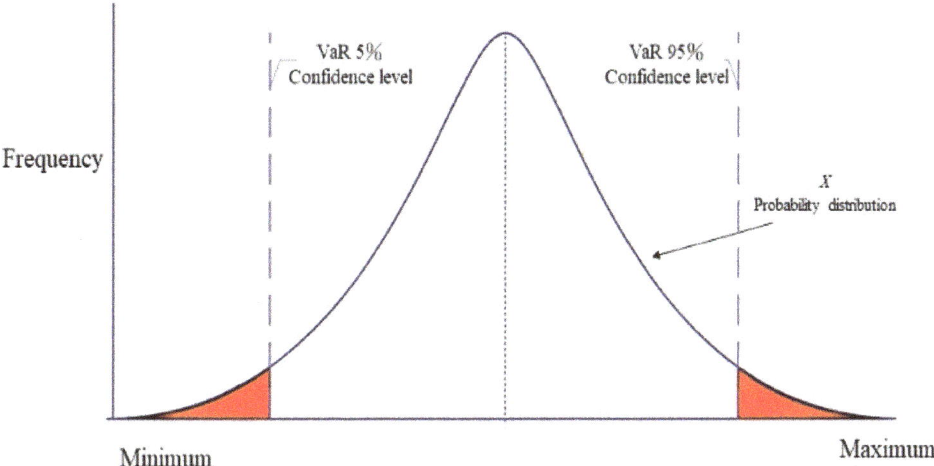

Figure 2. Fuel portfolio types' cost distribution at 5% and 95% levels of confidence. [8].

2.2. Trip Fuel Consumption Calculation

The different phase of trip fuel consumption for each ship can be calculated by ship engine type, fuel types, and time spent in the different navigation phases [9]. Based on design ship speed and daily fuel consumption, trip fuel consumption can be calculated from a detailed knowledge of the installed main and auxiliary engine power, and load factor, as expressed in the following equation.

$$FC_{Trip,p,j,n} = \sum \left[T_p \sum (FC_{os} + FC_{sec\,a} + FC_h + FC_m)_{j,n} \right] \tag{1}$$

where:

FC_{Trip} = fuel consumption over a complete trip (tonnes),

j = engine category (main, auxiliary),

n = fuel portfolio type (LSFO+ marine diesel oil (MGO), HFO with EGSC (capital cost (K) and maintain cost (MC) + MGO, MGO, LNG + MGO, HFO with EGSC (MC) + MGO)

p = different phase of trip (cruise at open sea (os), emission control area (seca), hoteling (h), maneuvering (m)).

T_p = different phase of trip time.

The different phase of trip time can be calculated as:

$$T_p = \sum_{p=1}^{n} \frac{dist_p}{Average_p} \tag{2}$$

2.3. Estimating the Cost of Fuel for a Trip

To estimate the fuel cost of a trip, one must know the trip time, fuel consumption, and fuel price, as in following equation [10].

$$F_{cost,Trip,p,j,n} = \sum (FC \times FP_t)_{Trip,j,n} + D(EGSC_{K\&MC}) \tag{3}$$

where:

F_{cost} = cost of fuel portfolio type (USD)
FP_t = *current* price of fuel at time t.
D = dummy variable, d = 0 it is not installation EGSC, d = 1 it is installation EGSC
EGSC = the expenses installation of EGSC in ship
K = capital cost (existing ship)
MC = maintain cost (newbuilding ship)

2.4. Fuel Cost Risk Assessment of Selected Alternative Fuels and Technologies

Fuel cost risk assessment is based on the CVaR. The volatility of fuel costs can be observed from the distribution of the volatility in expenses over time. CVaR can be calculated based on the calculation of VaR. According to Sarykalin et al. [11], CVaR of x with confidence level α is the mean of the generalized α-tail distribution, as expressed in the following equation

$$CVaR = \frac{1}{1-c} \int_{-1}^{VaR} xp(x)dx \tag{4}$$

where:

$CVaR$ = conditional value at risk
$P(x)\ dx$ = a probability distribution for the possible of fuel cost with value "x"
c = the cut-off point on the distribution where the analyst sets the *VaR* breakpoint
VaR = the agreed-upon *VaR* level.

3. Data and Descriptive Statistics

In this section we briefly describe the test case in Section 1, while the scenario generation process is discussed in Section 2.

3.1. Data Sources

Table 1 lists the data of New Panamax container route from East Asia to Europe and the overall data.

Table 1. New Panamax container and voyage data.

Item	Data	Item	Data
Dead Weight	156,907 DWT	Design speed	23.5 knots
Capacity	14,078 TEU	fuel consumption	1660 gallons/h
Route	Far East-North Europe	Route distance	11,944 miles
Port time	10.94 day	Ports of call	9 port
Roundtrip	71 day	Actual speed	15.0 knots

Source: Supply chain digest, 2013. [12].

Figure 3 below shows the monthly price of ship bunker and LNG gas data between January 2011 and February 2021, which is based on fuel selection and indicated the cost of a one-way ship. The fuel costs are based on historical HFO, LSFO, MGO, and LNG prices. They depict the impact of fuel price changes on fuel costs.

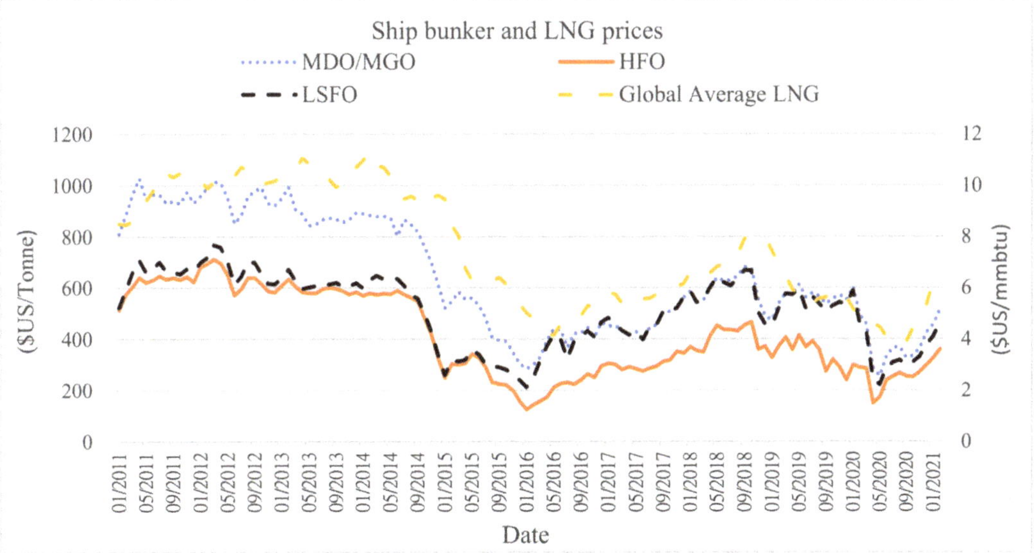

Figure 3. Ship bunker and LNG gas price. Source: Clarkson Research Services and Bluegold research.

3.2. Estimating the Shipping Bunker Cost for a Trip

We now measure the dispersion of fuel portfolio types' cost data and their degree of volatility. Table 2 summarizes and identifies the HFO + MGO + EGSC fuel portfolio types' costs that have a high degree of volatility that is more than other fuel portfolio types' cost via the coefficient of variation (CV).

Table 2. Estimation of fuel portfolio types' cost.

	Existing Ship			Newbuilding Ship	
	HFO + MGO + EGSC (K&MC)	LSFO + MGO	MGO	LNG + MGO	HFO + MGO + EGSC
Mean	2,254,037	2,288,117	2,637,784	967,867	2,060,037
Median	1,991,115	2,261,808	2,361,331	865,161	1,797,115
Maximum	3,575,312	3,504,137	4,203,282	1,535,757	3,381,312
Minimum	964,181	960,496	1,050,713	388,903	770,181
Std. Dev.	755,604	682,661	926,107	338,637	755,604
Skewness	0.19	−0.10	0.19	0.19	0.19
Kurtosis	1.60	1.81	1.61	1.60	1.60
Coefficient of Variation	0.34	0.30	0.35	0.35	0.37
Jarque-Bera	10.77	7.46	10.58	10.66	10.77
Probability	0.00	0.02	0.01	0.00	0.00

Equation (3) uses the estimated fuel portfolio types' cost in Table 2. From the existing ships' average fuel cost, we see that the average cost of HFO + MGO + EGSC (K&MC) (USD 2,254,037) per trip is cheaper than LSFO (USD 2,288,117) and MGO (USD 2,637,784). For a newly built ship, the average cost of LNG per trip (USD 967,867) is cheaper than HFO + MGO + EGSC (USD 2,064,037). These costs are only one component of fuel usage and do not include installation expenses, because the price of a newly built ship covers main engine equipment. Comparing the per trip fuel cost on fuel portfolio types can help one get a better handle on ship expenses and overall budget.

Figure 4 shows the historical fuel portfolio types' cost. We see the highest cost in April 2011 and the lowest in April 2020. Thus, fuel costs are closely linked to fuel price and crude oil prices. For all shipping companies, the cost of fuel is the largest expenditure, and a sudden sharp rise in crude oil prices will severely impact profits. Choosing low-cost ship fuel will reduce operational risk.

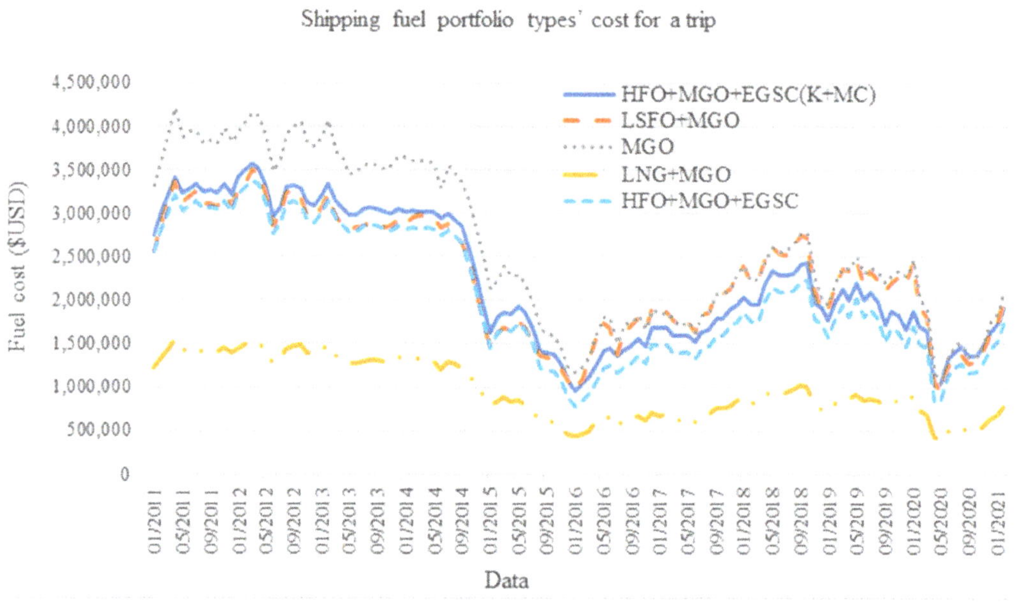

Figure 4. Shipping fuel portfolio types' cost for a trip. Source: Authors' elaboration based on data from Equation (3).

3.3. Situational Analysis of Shipping Bunker Cost Risk

In order to choose the correct distribution for shipping bunker cost, we need to understand each of the possible distributions for fuel portfolio types' cost. First, it is determined that a specific distribution by the historical data distribution goodness of fit test, Stephens [13] propose three indicates has been employed to assess the most appropriate probability distribution. The first choice is Kolmogorov–Smirnov (K–S) test, which is performed to measure the fitness, the second choice is the Chi-squared test (X^2), and the third choice is Anderson–Darling test (A–D) which can gain the best tail distribution simulation. Table 3 show the goodness of fit of a probability distribution to each cost of fuel portfolio series of data.

Since the fuel portfolio types' cost critical values of K–S test statistic is less than the p-value 0.95 critical values (0.12) [13], Chi-squared test (X^2) tested statistic less than the p-value 0.95 critical values ($X^2_{0.05,121}$ = 146.57) with 121 degrees of freedom and significance level 0.95, obtained from X^2 distribution threshold table for n = 121. A–D test statistic is greater than the p-value 0.95 critical values (1.5) [14] which rejected the null hypothesis in any distribution fits. The result shows the K–S and X^2 test fails to reject the null hypothesis for the beta distribution data set and goodness of fit more than the other distribution [15]. Additionally, it was found that a close fit given by the beta distribution can help lead to a good prediction in all fuel portfolio types.

Table 3. Goodness of fit distribution of fuel portfolio types' cost.

Test Value Distribution	Existing Ship			Newbuilding Ship	
	HFO + MGO + EGSC (K&MC)	LSFO + MGO	MGO	LNG + MGO	HFO + MGO +EGSC
	K–S X² A–D	K–S X² A–D	K–S X² A–D	K–S X² A–D	K–S X² A–D
Beta	0.09 38.1 19.9	0.07 7.44 2.37	0.08 30.9 22.8	0.08 31.3 22.7	0.09 38.1 19.9
Uniform	0.12 56.6 2.02	0.10 19.8 1.55	0.10 41.1 1.73	0.11 41.1 1.78	0.12 56.6 2.02
Triangular	0.24 57.6 7.62	0.16 22.9 4.53	0.23 65.9 7.94	0.23 67.8 8.23	0.24 57.6 7.62
Normal	0.17 63.1 4.41	0.13 24.8 1.86	0.15 55.1 4.02	0.16 57.4 4.09	0.17 63.1 4.41

Note: goodness-of-fit test for K–S critical value is ≤ 0.12, $X^2 \leq X^2_{0.95,(121)} = 146.57$, A–D ≤ 1.5. Source: Evans and Olson (1998). [14].

4. Empirical Results and Discuss

4.1. Simulation of CVaR of Fuel Portfolio Types' Cost

The distributions of the Monte Carlo simulation of fuel portfolio types' cost data are expressed graphically by using a histogram and the results of simulation for 1000 times as shown in Figures 5–9. Overall, fuel portfolio types' cost are known as the beta family distribution and we found Figures 5 and 7–9 proposed U-shaped and bimodal as a beta–arcsine distribution probability density; it represents uncertainty for a Bernoulli or a binomial distribution in Bayesian inference [14]. Figure 6 is LSFO + MGO fuel portfolio types' costs beta distribution which was described to show a peak at the right tail only.

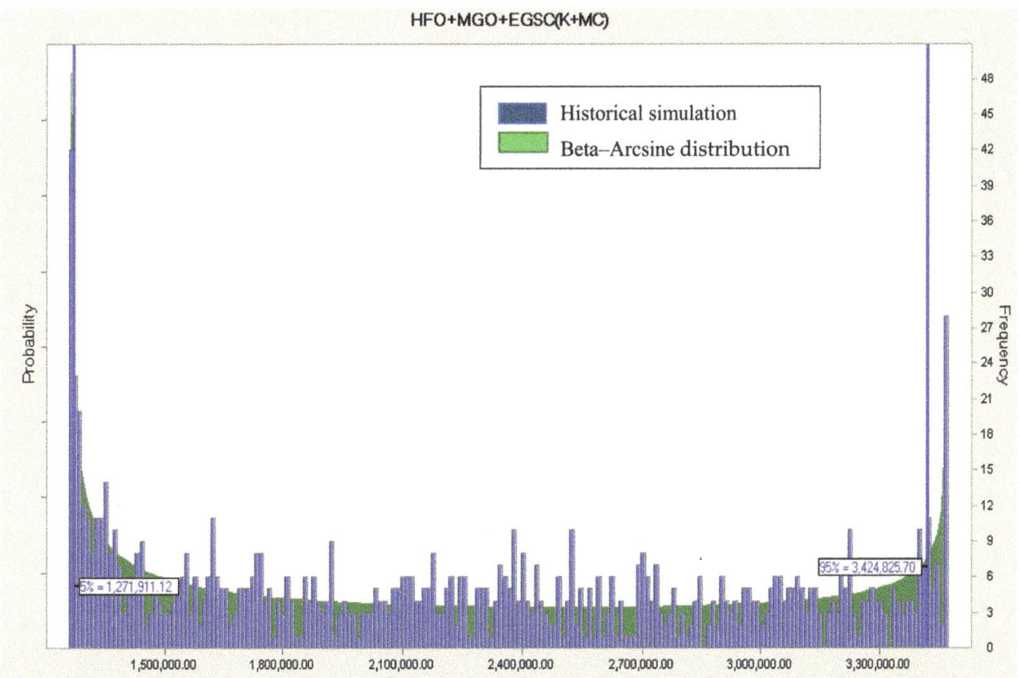

Figure 5. Historical simulation and Beta–Arcsine distribution of Monte Carlo of HFO + EGSC (K&MC) fuel portfolio types' cost for existing ships.

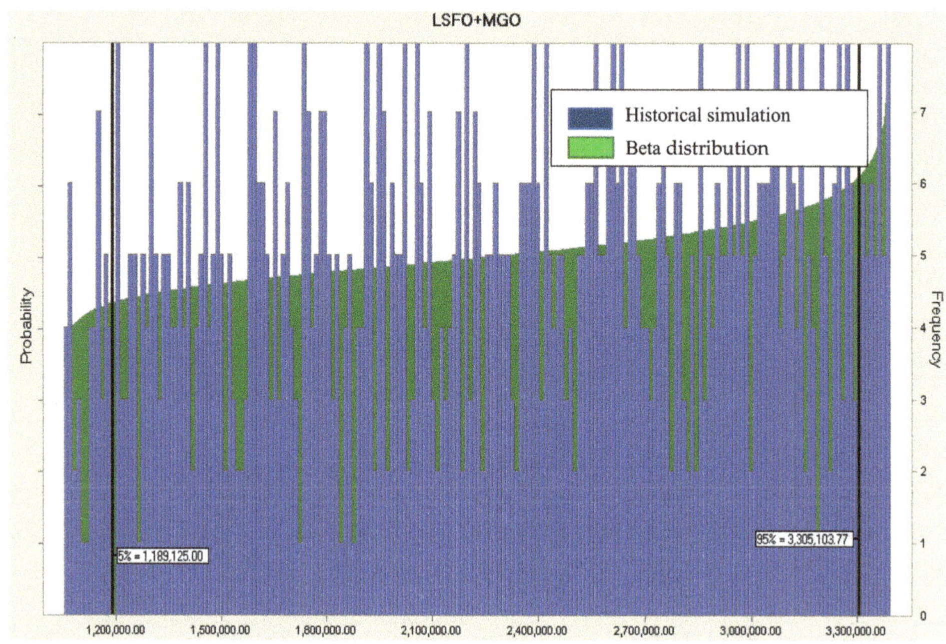

Figure 6. Historical simulation and Beta distribution of Monte Carlo of LSFO + MGO fuel portfolio types' cost for existing ships.

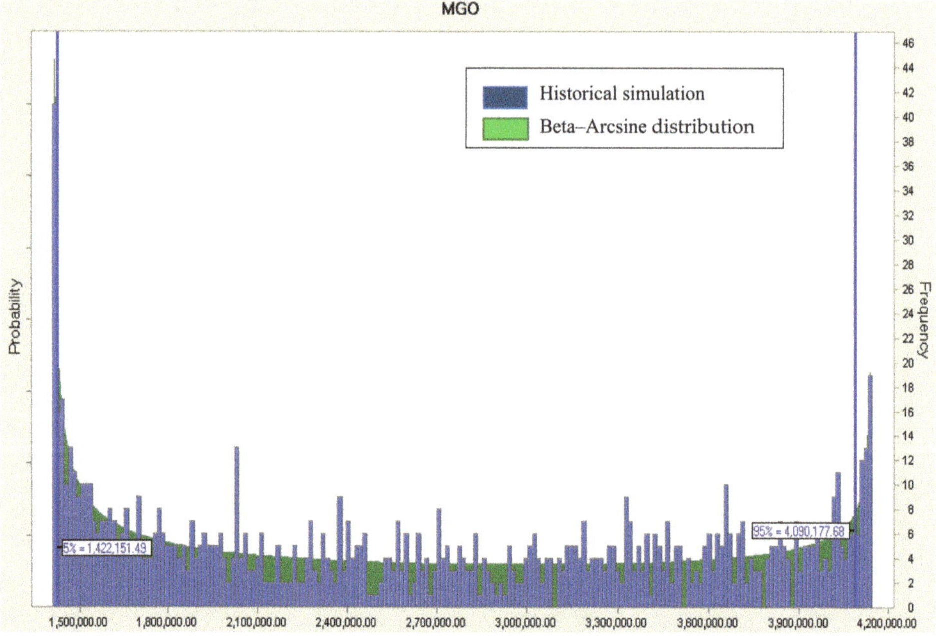

Figure 7. Historical simulation and Beta–Arcsine distribution of Monte Carlo of MGO fuel portfolio types' cost for existing ships.

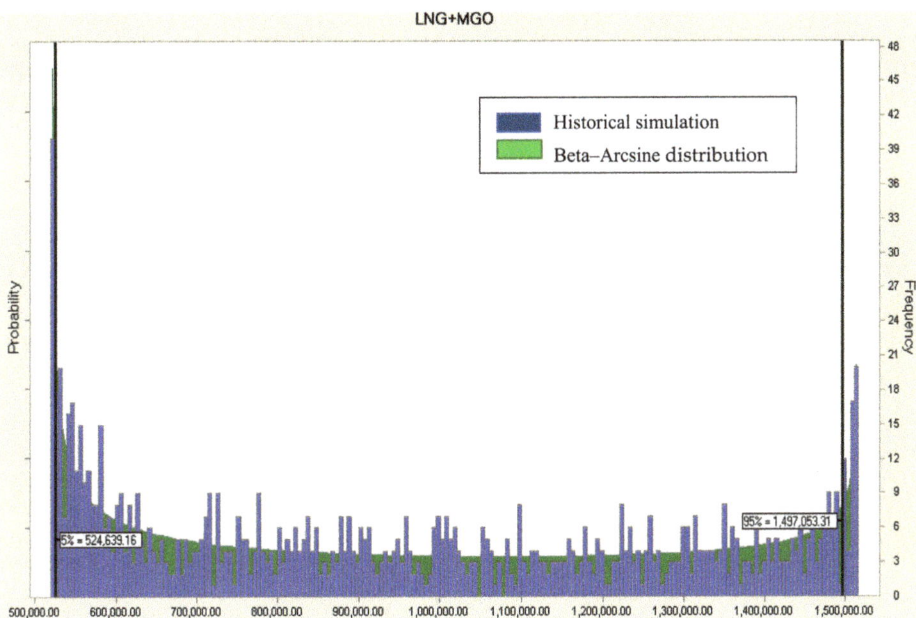

Figure 8. Historical simulation and Beta–Arcsine distribution of Monte Carlo of LNG fuel portfolio types' cost for newly built ships.

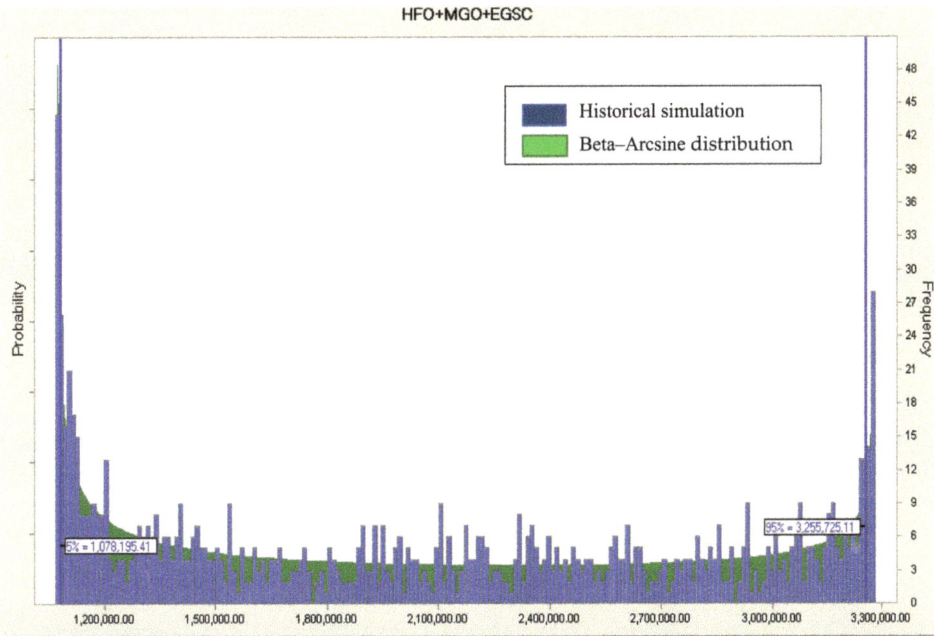

Figure 9. Historical simulation and Beta–Arcsine distribution of Monte Carlo of HFO + MGO + EGSC Fuel portfolio types' cost for newly built ships.

Figure 5 illustrates fuel portfolio types' cost with high-frequency observations on both tails. The low cost at the 5% confidence level is USD 1,271,911, and the high cost at the 95% confidence level is USD 3,424,825. The fuel cost increase from 0.05% to 0.95% confidence level reaches about 2.69 times. Thus, as expected, Installation is a significant investment cost but passing the payback period, the use of HFO with scrubber fuel will cost down from USD 889,503 (0.05%) to USD 3,036,125 (0.95%) saving more expenses.

Figure 6 shows fuel portfolio types' cost and their high-frequency observations on right tails. The low cost at the 5% confidence level is USD 1,302,603, and the high cost at the 95% confidence level is USD 3,323,054. The fuel cost increase from 0.05% to 0.95% confidence level reaches about 2.55 times. In the past few years, the price of LSFO about 30 to 50% more expensive than HFO. Thus, it costs more to switch from HFO to LSFO.

Figure 7 presents MGO fuel portfolio types' cost and their high-frequency observations on both tails. The low cost at the 5% confidence level is USD 1,506,065, and the high cost at the 95% confidence level is USD 4,091,971. The fuel cost increase from 0.05% to 0.95% confidence level reaches about 2.71 times. MGO price is expensive relative to its fuel cost higher than using other fuels.

Figure 8 illustrates LNG portfolio types' cost and their high-frequency observations on both tails. The low cost at the 5% confidence level is USD 555,726, and the high cost at the 95% confidence level is USD 1,503,463. The fuel cost increase from 0.05% to 0.95% confidence level reaches about 2.71 times. LNG is expected to be less costly than all marine fuel, but ship-owners always interested in new buildings ship only, the reason is saving relatively high investment costs for the power system.

Figure 9 shows HFO + MGO + EGSC portfolio types' cost and the high-frequency observations on both tails. The low cost at the 5% confidence level is USD 1,086,639, and the high cost at the 95% confidence level is USD 3,240,719. The fuel cost increase from 0.05% to 0.95% confidence level reaches about 2.98 times. Scrubber capital requirement is not installed in newbuilding ship. Thus, compare the fuel cost of installed scrubber in the existing ship and used scrubber with HFO in newbuilding ship, which is found newbuilding ship fuel cost could down about 5 to 15% than the existing ship.

4.2. Summarizes and Discussion

Table 4 summarizes the CVaR of fuel portfolio types' cost at the 5% and 95% probability levels to measure possible costs of different portfolio types. For an existing ship, we find CVaR at 5% HFO + EGSC (K&MC) is the lowest fuel cost and CVaR at 95% MGO is the highest fuel cost. It means a change in fuel cost to MGO has high risk and HFO with installed EGSC has low risk. However, in terms of the variance of fuel cost, LSFO has the lowest fluctuation in fuel cost.

Table 4. CVaR of fuel portfolio types' cost at 5% and 95% probability levels.

Probability Level	Existing Ship			Newbuilding Ship	
	HFO + MGO + EGSC (K&MC)	LSFO + MGO	MGO	LNG + MGO	HFO + MGO + EGSC
5% (Minimum cost)	1,271,911	1,302,603	1,506,065	555,726	1,086,639
95%(Maximum cost)	3,424,825	3,323,054	4,091,971	1,503,463	3,240,719

For a newly built ship with CVaR at 5% and 95%, LNG has the lowest fuel cost versus HFO + EGSC. It means shipowners should choose LNG fuel to reduce fuel cost and lower risk.

4.3. Risk Management Strategies in Bunker Cost

Our CVaR analysis indicates that bunker price volatility significantly increases bunker cost in the shipping, and it could reduce the risk through risk management strategy. The reliability practices on the risk management strategies in shipping are listed as follows:

(1) Suitable alternative fuel selection

A selection that costs the lowest and is the most efficient and eco-friendly fuel choice. For instance, the newbuilding LNG fueled ships have not only economic benefits but also practical and environmental advantages, simply by having current ships replace fuel to LSFO or HFO with EGSC. In addition, there are which fuels may be more beneficial for a shipping company in terms of many criteria, such as fleet size, ship type and bunking port limited in a variety of routes.

(2) Hedging bunker cost using alternatives

The variety of fuel prices have seriously affected the profitability of the shipping business. Alizadeh and Nomikos [16] suggested the strategy of hedging bunker cost, they claim to choose the OTC bunker derivatives products, such as forward bunker agreements, bunker swaps and alternatives on bunker prices to hedging bunker cost.

(3) Passing the bunker surcharges on to customers

Ocean carriers pass some of that operational cost burden onto their customers such as Bunker Adjustment Factor (BAF), Emergency Bunker Surcharges (EBS) or Low Sulphur Fuel Surcharge (LSF), but when long-term ocean freight rates decline on poor economic will limit the carriers' ability to reduce their loss of fuel costs.

5. Conclusions

Coronavirus oil shock and IMO low sulphur fuel limitation make the management of shipping bunker complex and crucial. It involves decision-making in both existing ships and newbuilding ships. That is why these two types of grades levels are discussed separately in this paper.

1. Risk management of fuel portfolio types' cost for existing ships

More than 5.9% of the total number of container ships or 11.8% of total fleet capacity fitted with scrubbers by January 2020 when IMO's global 0.5% fuel sulphur cap comes into effect [3]. Ship operators are vigorously pursuing scrubber installation, but they also face a risk in which there is no guarantee that they will not be targeted by IMO again in the foreseeable future. Thus, strategies overcoming the negative impact of regulatory uncertainties should focus on reducing the negative impact or likelihood of threats. One can consider changing to low-sulfur oil or scrapping and dismantling ships to reduce and eliminate threats to existing ships.

2. Risk management of fuel portfolio types' cost for newly built ship

As suggested by Hansson et al. [17], LNG is best in terms of fuel price and available infrastructure, because it is cheap, clean fuel. However, the problem is that the necessary equipment and heavy-weight metal tanks for storing LNG may add more volume to the LNG-powered vessels than it could carry. The possible solution is to build a carbon-fiber tank that is up to 90% lighter than metal tanks, so that a long-term natural gas supply contract could be executed to stabilize fuel cost as the expected refueling cost could be minimized by the shipping company [18].

The fact is that the price of oil should recover to near pre-coronavirus levels in the future as economies recover. Even as oil prices are battered down to 18-year lows, one energy fund thinks USD 100 a barrel is still achievable in the future [19]. Ship operators will still face high-fuel prices and high expenditures from fuel costs. Therefore, alternative fuels and technologies must focus on long-term plans that entail higher expected return with lower risk compared to high sulphur fuel oils. For risk management, we suggest that operators install scrubbers to clean the exhaust gas and use HFO for existing ships. For newly built ships, operators can change the use of fuel into LNG, which could help save on capital expenses.

Author Contributions: Conceptualization, T.-C.H. and C.-M.W.; Literature review and information search, T.-C.H. and C.-M.W.; Draft writing, T.-C.H. and C.-M.W.; Reviewing and editing: T.-C.H. and C.-M.W., Administration, T.-C.H. All authors have read and agreed to the published version of the manuscript.

Funding: This research received no external funding.

Informed Consent Statement: Informed consent was obtained from all subjects involved in the study.

Data Availability Statement: The data presented in this study are available on request from the corresponding author. The data are not publicly available as it contains information that could compromise the privacy of research participants.

Acknowledgments: We would like to thank three anonymous (unknown) reviewers and the editor for their comments.

Conflicts of Interest: The authors declare no conflict of interest.

References

1. Kalli, J.; Repka, S.; Alhosalo, M. Estimating Costs and Benefits of Sulphur Content Limits in Ship Fuel. *Int. J. Sustain. Transp.* **2015**, *9*, 468–477. [CrossRef]
2. S&P Global. Coronavirus Impact Masking Post-IMO 2020 Risks for Marine Fuels. 24 February 2020. Available online: https://www.spglobal.com/platts/en/market-insights/latest-news/oil/022420-coronavirus-Impact-masking-post-imo-2020-risks-for-marine-fuels (accessed on 1 May 2020).
3. Alpha Liner. Weekly Newsletter. Volume 2019. Issue 50. Available online: https://files.constantcontact.com/8e1f5de2401/f102b5a0-9019-44b2-bbb3-97a6cf5bdc1f.pdf (accessed on 1 May 2020).
4. Zheng, W.; Li, B.; Song, D.P. Effects of risk-aversion on competing shipping lines' pricing strategies with uncertain demands. *Transp. Res. Part B* **2017**, *104*, 337–356. [CrossRef]
5. Gu, Y.; Wallace, S.W.; Wang, X. The Impact of Bunker Risk Management on CO2 Emissions in Maritime Transportation under ECA Regulation. Available online: https://openaccess.nhh.no/nhh-xmlui/bitstream/handle/11250/2421290/1716.pdf?sequence=1&isAllowed=y (accessed on 1 May 2020).
6. Madadi, A.; Kurz, M.E.; Mason, S.J.; Taaffe, K.M. A Metaheuristic Approach to Supply Chain Network Design Using CVaR. In Proceedings of the IIE Annual Conference Proceedings, Norcross, GA, USA, 1–9 November 2012; pp. 1–9.
7. Sun, R.; Ma, T.; Liu, S.; Sathye, M. Improved Covariance Matrix Estimation for Portfolio Risk Measurement: A Review. *J. Risk Financ. Manag.* **2019**, *12*, 48. [CrossRef]
8. Han, T.C.; Chou, C.C.; Wang, C.M. Cash flow at risk and risk management in bulk shipping company: Case of Capesize bulk carrier. *J. Eng. Marit. Environ.* **2016**, *230*, 13–21. [CrossRef]
9. Carlo, T. Emission estimate methodology for maritime navigation. In Proceedings of the US EPA 19th International Emissions Inventory Conference, San Antonio, TX, USA, 27–30 September 2010.
10. Stopford, M. *Maritime Economics 3e*; Routledge: London, UK, 2009; p. 233.
11. Sergey, S.; Gaia, S.; Stan, U. Value-at-Risk vs. Conditional Value-at-Risk in Risk Management and Optimization. In Proceedings of the INFORMS Annual Meeting, Washington, DC, USA, 5–12 October 2008.
12. Supply Chain Digest Global Supply Chain News: Maersk Triple Ecost Advantages are too great to Ignore. 2013. Available online: http://www.scdigest.com/ontarget/13-09-12-1.php?cid=7401&ctype=content (accessed on 20 September 2014).
13. Stephens, M.A. Goodness-of-fit for the extreme value distribution. *Biometrika* **1977**, *64*, 583–588. [CrossRef]
14. James, R.E.; David, L. *Introduction to Simulation and Risk Analysis*; Prentice Hall: Olson Upper Saddle River, NJ, USA, 1998.
15. U.S.D. of Commerce. NIST/SEMATECH e-Handbook of Statistical Methods. 2012. Available online: https://www.itl.nist.gov/div898/handbook/ (accessed on 20 September 2014).
16. Alizadeh, A.H.; Nomikos, N.K. *Shipping Derivatives and Risk Management*; Palgrave, Macmillan: New York, NY, USA, 2009.
17. Hansson, J.; Grahn, M.; Månsson, S. Assessment of the possibilities for Selected Alternative Fuels for the Maritime Sector. In Proceedings of the Conference Proceedings, Shipping in Changing Climates (SCC), London, UK, 4–5 September 2017.
18. Pedrielli, G.; Lee, L.H.; Ng, S.H. Optimal bunkering contract in a buyer seller supply chain under price and consumption uncertainty. *Transp. Res. Part E* **2015**, *77*, 77–94. [CrossRef]
19. Bloomberg. Energy Hedge Fund That Shorted Oil Sees Chance for $100 a Barre. 17 April 2020. Available online: https://www.bloombergquint.com/markets/energy-hedge-fund-that-shorted-oil-sees-chance-for-100-again (accessed on 1 May 2020).

 *sustainability*

Article

Let the Game Begin: Enhancing Sustainable Collaboration among Actors in Innovation Ecosystems in a Playful Way

Anastasia Roukouni [1,*], Heide Lukosch [2], Alexander Verbraeck [3] and Rob Zuidwijk

[1] Faculty of Civil Engineering & Geosciences, Delft University of Technology, Stevinweg 1, 2628 CN Delft, The Netherlands

[2] HIT Lab NZ, University of Canterbury, 69 Creyke Road, Christchurch 8041, New Zealand; heide.lukosch@canterbury.ac.nz

[3] Faculty of Technology, Policy and Management, Delft University of Technology, Jaffalaan 5, 2628 BX Delft, The Netherlands; a.verbraeck@tudelft.nl

[4] Rotterdam School of Management, Erasmus University Rotterdam, Burgemeester Oudlaan 50, 3062 PA Rotterdam, The Netherlands; rzuidwijk@rsm.nl

* Correspondence: a.roukouni@tudelft.nl; Tel.: +31-613248875

Received: 18 September 2020; Accepted: 10 October 2020; Published: 15 October 2020

Abstract: Logistics and transport systems are complex systems for which sustainable innovations are urgently needed. Serious games are an acknowledged tool for training, learning, and decision making, as well as for helping to introduce innovative concepts for complex systems. Technological innovations for the transport domain that can improve sustainability are usually heavily dependent on the collaboration among actors. A simulation gaming approach can help these actors in understanding the challenges involved, and in finding solutions in a playful, interactive way. Our research approach includes a thorough literature review on games for innovation and collaboration in transport networks, and the development of two dedicated simulation games addressing sustainability innovations for the Port of Rotterdam, the largest seaport in Europe and one of the largest in the world. The two innovation cases are truck platooning and multi-sided digital platforms for barge transportation, both improving the sustainability of hinterland transportation. The games serve as instruments to reveal interactions and tensions among actors, contribute to the interpretation of their behavior, and eventually help all parties to reach a better understanding on how innovation adoption can be fostered, using an innovation ecosystem perspective. We are convinced that serious gaming, by providing a better understanding of the innovation process, will help the implementation of sustainability innovations in complex systems.

Keywords: transport networks; logistics; simulation games; serious games; innovation ecosystem; Port of Rotterdam; complex systems; truck platooning; barge transport; multi-sided platforms

---

1. Introduction

Bridging the gap between the conception and introduction of an innovation in the context of research and development, and its effective adaptation and diffusion in an industry is a challenge that many sectors are currently facing [1–3]. The development of a technical innovation can be much faster than the capacity of organizations to implement it. For example, sensors used by vehicles to scan their environment are technically very mature, yet the implementation of sensor technology on our streets is still hindered by a lack of investment, trust, and cooperation of actors. The transport and logistics sector is no exception in experiencing a gap between development and implementation of technological innovations [4]. The sector includes a large number of interconnected actors, being highly dynamic and complex [5], which makes the implementation of innovations quite challenging. Nonetheless, the transport and logistics sector is in desperate need for innovations, especially to reduce emissions while offering competitive services, and shift towards more sustainable practices in general.

Transport and logistics communities often consider the adoption of innovative business models as an effective means to enable their participation in a sustainable, agile and collaborative logistics and transportation system. This can involve accelerating the digitization of the sector to enhance supply chain visibility as well as data quality and reliability, often by the use of multi-sided digital platforms [6].

Meanwhile, the concept of an "innovation ecosystem" has recently evolved as a new way to approach and promote innovation [7,8]. An innovation ecosystem is defined as a system in which continuous and dynamic interactions among different actors occur, therefore innovation is not realized as a result of individual efforts but as the outcome of multifaceted communication and synergies among the actors [9,10]. These actors should become aware of the opportunities and risks of technological innovations. Simulation games are evolving, at an accelerated pace, into a recognized means to endorse learning processes in an intriguing setting, that invites diverse actors to work with one another actively towards common goals [11]. They represent a widely accepted instrument in the transportation and logistics domain for many years already [12].

The research presented in this paper attempts to connect the aforementioned topics. The research questions on which our work is based are: (1) Can simulation games be used as an instrument to comprehend the underlying forces around innovation processes in transport and logistics using an "innovation ecosystem" perspective? (2) Can they be used as a means to facilitate knowledge transfer of innovation from theory to practice and contribute in exploring the roles of different actors involved?

The research approach to answer these questions is built around two serious board games that have been developed to study the role of collaboration among stakeholders in the field of transport, logistics and supply chain management towards the effective implementation of innovations. Mobinn (Mobilize Innovation), the first game, is tailored to the case of truck platooning, an innovation to increase sustainability for truck transport that is mature in terms of developed technology, yet unfledged when it comes to implementation. After testing this game through several game sessions, a second game has been developed: Platform4Barge: a Mobinn game. The lessons learnt from the first game have been incorporated into this new version which focuses on another important component of the transport network; the introduction of multi-sided digital platforms and its impact on port environments, especially on barge transport, which is a more sustainable transport means than truck transport when utilized well. Both games are aimed to raise awareness for the challenges that come with implementing such innovative systems, to ensure a viable implementation pathway. Such games can be seen as novel approaches allowing stakeholders to pre-experience future technology effects, which go beyond verbal descriptions or abstract representations [13].

The remainder of this paper is structured as follows: The literature review section provides an overview of the concept of innovation ecosystems, followed by an overview of existing game approaches towards innovation with a focus on transport and logistics. After that, the case context is presented, with insights regarding the challenges faced by actors in port environments, when it comes to introducing truck platooning or multi-sided digital platforms. The two serious games that have been developed are introduced and described in the next section, followed by the results and discussion section that is based on the outcome of the game sessions that have been organized for both games. The paper concludes with some general guidelines and recommendations regarding the use of serious games to enhance the implementation process of innovations in transport and logistics.

2. Literature Review

2.1. Overview of the Concept of Innovation Ecosystems

The idea that innovation is a continuous process, seen more as a journey in time rather than several events that takes place at specific moments is one of the core concepts in innovation management [14]. It has been found that the impact of innovation can also be wider when it is examined as a continuous process rather than a series of discrete events [15]. Innovation can thus be defined as a process through which novel solutions emerge as multiple actors interact and exchange resources to co-create value [16]. Innovation relies upon a complex network of actors and activities, which can be rather heterogeneous [17].

The use of the term ecosystem in a business context was introduced by Moore [18] in his paper of 1993 "Predators and Prey: A New Ecology of Competition", using the similarities he observed between natural ecosystems and the mechanisms related to businesses and innovation as a base for his reasoning. Innovation ecosystems emerge as a result of collaborative interactions, when cooperating actors accomplish a certain level of integration regarding a shared identity, shared strategy and shared objectives [19]. Moreover, an innovation ecosystem stimulates synergies among people and firms in a given geographical space, enhancing the development and commercialization of new ideas [20] and creating and capturing value [21].

Focusing on urban technology innovation ecosystems, Mulas et al. [22], describe these as a number of stakeholders, assets, and the interaction among them in city environments resulting in new technology and ideas. The innovation ecosystem thinking focuses on the creation of social capital and the enhancement of interpersonal relationships which contributes, among others, in the more effective flow of information, financial resources as well as talent worldwide [23]. Under an innovation ecosystem perspective, innovation diffusion is a multi-level and multi-actor phenomenon [2].

According to Su et al. [24], a shift of interest is observed among innovative enterprises lately; they no longer focus on competition in terms of products and services, but in terms of the innovation ecosystems they belong to. Innovation ecosystems vary considerably in terms of their organization and business models, having an impact on the strategic choices made by businesses, both existing and new ones. These choices can shape competition and therefore encourage innovation and eventually transform the innovation ecosystem itself [25].

Nowadays, the nature of government interventions is also changing, moving away from the purely administrative role they used to have. The countries and areas in which government bodies and authorities of all levels introduce new functional roles of facilitators and intermediaries for collaborative interactions within and among innovation ecosystems appear to provide a competitive advantage in creating and maintaining successful innovation ecosystems [19].

Vasconcelos Gomes et al. [26] aimed to provide an answer to the question how to coordinate a complex network of actors in the presence of individual and collective uncertainties in innovation ecosystems. They conducted a comparative analysis of five case studies; all of them start-ups that created and coordinated complex innovation ecosystems for the development and commercialization of radical innovations. They found that many actors in an innovation ecosystem tend to initially consider uncertainties more as individual uncertainties without realizing that other partners may also face the same kind of uncertainty. Moreover, when trying to mitigate or overcome these uncertainties in an isolated way, without taking into account the innovation system partners, they may contribute to the diffusion of those uncertainties through the innovation ecosystem; this way, a specific uncertainty that begins as individual is then integrated into the mindset and decision planning of different actors, hence becoming collective.

Lubik and Garnsey [27] claimed that under an innovation ecosystem perspective is it obvious that value generation is a dispersed process which encompasses co-innovation from other actors in the business environment who have the potential to increase the generated value. In addition to that, according to Pellinen et al. [28], in a business ecosystem, the process of decision making can be influenced by actions, as well as expectations of other actors in the same ecosystem, through a pattern of co-evolution.

Focusing on the logistics sector, Meyer-Larsen et al. [29] emphasize the importance of creating innovation ecosystems; they state that the increase of interaction among the different actors involved is expected to limit the existence of trust issues which could result also in the faster adoption of innovation and help overcome existing barriers. In the relevant literature it is increasingly mentioned that a better understanding of the roles that the different ecosystem actors play in the innovation process should be obtained (e.g., see [14,30]).

In summary, we can see that the innovation ecosystem perspective is one that allows us to look at innovation in a connected way, including actors and their relationships, as well as concerted actions towards innovation implementation. We take over this perspective for the transfer of innovation into a game model used to make actors aware of challenges related to innovation implementation in a sustainable way. In the next section, we describe the instrument of serious games in the domain of transport and logistics to help us define our gaming approach.

2.2. Overview of the Use of Games in Transport, Logistics and Innovation

Serious games are used more and more to enhance awareness and encourage the process of knowledge building in an engaging environment that invites collaboration among different stakeholders [11]. Although the term may sound relatively new, in reality serious games is just a modern expression of approaches and models which date back many centuries ago [31]. The process of playing is inherently connected with human societies, all over the world. In ancient Greece for instance, Plato was the first philosopher who emphasized the educational importance of games in formulating children's attitude; he claimed that through games, children are given the opportunity to get familiarized with habits that are going to be useful for them for the rest of their life [32].

Clark Abt, in his book Serious Games (1970) [33], gave a definition of what makes a game be characterized as a serious game. According to this definition, serious games "have an explicit and carefully thought-out educational purpose and are not intended to be played primarily for amusement. This does not mean that serious games are not, or should not be, entertaining" [33] (p. 9). A few years later, Duke estimated that games were going to be the language of the future and published the homonymous book Gaming: The Future's Language [34]. Approximately half a century later, numerous definitions can be found in relevant literature, from different researchers in different scientific areas and fields. In our research, serious games or simulation games are synonymous and can be described as "experimental, rule-based, interactive environments, where players learn by taking actions and by experiencing their effects through feedback mechanisms that are deliberately built into and around the game" [35] (p. 825).

Serious games use a simplified abstracted model of reality, contributing this way to a better understanding of complex issues [36]. Having a set of clearly defined rules, they encourage interaction, while at the same time providing a model for learning. Moreover, the direct consequences of every action and decision can be observed, therefore the players can realize the impact of their own decisions, as well as of the ones of the other players [37], within a "safe" environment [35]. These characteristics make them valuable tools for exploration and observation of player behavior.

According to Olejniczak et al. [38], the use of serious games by policy makers is a very promising approach in estimating the impact of new complex policies and regulations, due to the fact that they comprise the following strengths:

- They are able to reveal mechanisms such as the initial assumptions of the stakeholders involved, their decisions, as well as the feedback loops that are created by their reactions;
- They demonstrate the effects of these mechanisms over time, which in real life becomes evident over a much longer period, and;
- They create a risk-free environment in which the representatives of the different actors can test hypothetical scenarios of different policies and experiment with new ways of interpreting these and similar actions.

In supply chain management, simulation games have been used to bring attention to the topic of shared resources [29]. When simulation games are employed, the ideas generated to overcome barriers related to innovation implementation tend to be more creative [39], and the impacts of applications which are based on information communication technologies (ICT) or intelligent transportation systems

(ITS) can be examined effectively through them [40]. Simulation games can also be used as an efficient channel for information provision around future situations, associated with a high level of uncertainty, such as climate change [41]. Hidayatno et al. [42] designed a serious simulation game to act as a learning medium of supply chain management for biofuel production. Their game focuses on companies which operate crude palm oil mills. According to the authors the game has proven to be effective in increasing the understanding of players on concepts and complex issues related to sustainable supply chain management. With the learning objective of this game focusing on internal material flow, another simulation game was developed, under the objective of making players familiar with the important role that Industry 4.0 technology can play in production logistics [43].

Focusing on the Dutch railway system, Van den Hoogen and Meijer [44], found that simulation gaming can contribute to the coordinated planning of innovation implementation. Kurapati et al. [45], examined how simulation games can be used to foster sustainability in transportation operations through synchromodal corridor management. The authors developed a digital simulation game which was tested in game sessions with experts in The Netherlands.

Simulation gaming can be thus considered as a robust approach increasing comprehension in large-scale systems with a high degree of integration and with numerous actors dealing with significant uncertainties [36], by examining these systems through different perspectives [46].

3. Case Analysis—Context of the Game-Based Research Studies

Port cities around the world have the increasing tendency to be engaged in various strategies to stimulate innovation through activities such as triple-helix collaborations, innovation ecosystem approaches, urban entrepreneurialism etc. [47].

The city of Rotterdam in The Netherlands is home to the Port of Rotterdam, being the largest container seaport in Europe and 10th largest in the world with state-of-the-art container terminals [48]. In 2019, the total number of deep-sea vessels that docked in the Port exceeded 29,000 and that of inland vessels reached approximately 86,000 [48]. This is translated to almost 8.8 million containers (more than 14.8 million TEU (TEU: "twenty-foot equivalent unit. This is the standard unit for counting containers of various capacities and for describing the capacities of container ships or terminals. One 20 foot ISO (International Organization for Standardization) container equals 1 TEU") [49]).

In 2011, Rotterdam Port Authority, in collaboration with various ministries, businesses and research institutes in The Netherlands, published a strategic development plan called 'Port Vision 2030', in which comprehensive strategies to handle the growing volumes of containers in an efficient and sustainable way were presented. In this vision document, it was stated that by 2030 the aim would be to drastically reduce the road share of hinterland container transport, reaching 35%, so that 65% of containers would be transported by barge or rail [50].

It is undeniable that in terms of transport and logistics, The Netherlands has many unique location advantages among European countries with the Port of Rotterdam being a major multi-modal gateway to the rest of Europe. As a result, as already highlighted in Roukouni et al. [51], a huge strain is put on the road network, while the demand keeps increasing and there is a drastic need for innovations that make use of the current infrastructure more efficiently and reduce congestion, while at the same time enhance traffic safety and limit transport emissions [52]. One promising case to deal with these challenges is truck platooning [51].

Truck platooning can be defined as the concept of "trucks driving automatically in small convoys, a short distance apart, using wireless vehicle-to-vehicle (V2V) communication and advanced driver assistance systems resulting in a smoother traffic flow, higher traffic safety, fuel savings and a reduction in CO_2 emissions" [53] (for more information about the ruck platooning concept see [54,55]). According to Roukouni et al. [51], "for the truck platooning concept to work, many factors have to be aligned, e.g., the technology has to be developed, the legislation and road infrastructure have to be adapted, and the business model has to be accepted by the users". Hence, although the technology has been evolving a lot during the last years, there is still a way to go as certain technological barriers exist, in addition to a conservative attitude of several of the logistics companies involved. Moreover, as multiple actors are involved, there are a lot of stakeholders that have to align in order for the final decisions to be taken. Therefore, it is obvious that truck platooning has all the characteristics of a complex system and represents a sustainable innovation on the forefront of development in The Netherlands [51,56]. All these reasons made it an ideal choice for a demonstration case for the development of our first serious board game, Mobinn.

At the same time, the large and continuously growing volume of containers being transported through the Port of Rotterdam also creates a need for other innovative systems, such as a digital platform that would gather all the required information for booking containers into one single place. Multi-sided platforms (MSP) are "technologies, products or services that create value primarily by enabling direct interactions between two or more distinct customer or participant groups" [57]. The tremendous technological progress of our era, facilitated platform development and lead to an unprecedented volume of data exchange, which magnifies the platform's value for all the parties involved, often resulting in disrupting sectors as a whole, as in the case of Uber or Airbnb platforms for example [51,58].

The potential scalability of digital platforms is a reason why they can cause such disruptions. They cause network effects which can be either positive, resulting to additional value, or negative, having the exact opposite result [59]. A key factor to improving visibility in supply chains and transport networks is information sharing [60]. However, many companies and actors are reluctant to share information as they assume the power of information is diminished when it is shared [61]. Olesen et al. [62] found that the lack of information exchange usually has three causes; limited data availability, lack of trust and system complexity. Being aware of the existence of these challenges, we decided to develop the second serious game, Platform4Barge, to increase actors' awareness of the importance of data sharing and transport network visibility for inland barge transport, which is seen, together with rail, as the sustainable replacement for truck transport.

4. Development of the Serious Games

Serious games represent an interactive and efficient instrument for learning, decision-making, and awareness building and they are an effective tool for simulating dynamic and complex systems, such as transport and logistics systems [35,36]. While the sector has successfully implemented game-based approaches to address several challenges already (see e.g., [44,45,63,64]), the question of innovation implementation has not yet been addressed thoroughly through a game-based approach. Below, we explain how the two games, based on the two illustrated cases, have been designed in order to fill this gap.

4.1. Elements and Game Play

The Mobinn (Mobilize Innovation) game is a board game that represents the relationships among actors involved in setting up a truck platoon in a simplified way. Platform4Barge is a second version of the initial game, which has been tailored to the specific situation and challenges around the use of multi-sided digital platforms in barge transportation. The two games were developed based on an iterative process that used loops of testing and adjustments of early prototypes, and included researchers, game developers, and domain experts.

We followed the philosophy of the triadic game design (TGD) concept, introduced by Harteveld [65]. According to the TGD concept, each serious game comprises three different "worlds"—reality, meaning and play. The first one, reality, as the term implies, refers to the world that is surrounding us and to its reflection to the game we are developing. The world of meaning is associated with purposefulness; as we have already discussed previously in Section 2.2 of the article, serious games have certain defined objectives to fulfil, such as creating awareness towards a specific issue etc. A serious game does not exist just to be played and provide entertainment to the participants. At the same time though, the component of fun should not be overlooked as well, as it is placed in the center of the third world, the one of play. This world includes also the game elements and dynamics, which are the means that can make the game experience captivating and intriguing for the players [65].

Embracing this game design philosophy, we considered the component of meaning and the component of play the same for both our serious games. Meaning is reflected on the objective of our research, which, as stated previously in the introduction, is to raise awareness around the challenges associated with implementing innovative systems in transport and logistics in a sustainable way and explore the role of different actors involved. The game materials and dynamics also do not change. The physical game elements of each game include: one game board, four role cards, five action cards per role, event cards, tokens, and pawns. The roles of the players in the two games correspond to the key actors of the truck platooning ecosystem and of the port container ecosystem for Mobinn and Platform4Barge respectively. What changes is the world of reality for the two cases, and thus the content of all the cards involved in each game. A description of how we obtained the critical insights in the reality of each one of the cases, which helped us then unfold the content of the cards, follows.

For the Mobinn game, in order to develop the role cards, as well as the action and event cards of the game, we first needed to understand who are the main actors that are involved in shaping the truck platooning community and what is "at stake" for each one of them; then we translated this information into content of our three types of cards: role cards, action cards and event cards. we started by a comprehensive literature review of scientific papers about truck-platooning (publications in journals and conference proceedings, working papers, theses', e.g., see [54,55,66]). As truck platooning, being a technology-driven innovation, does not have such a large volume of scientific literature available yet, we also decided to study materials such as white papers, reports (e.g., see [52,67], websites and media sources e.g., see [68,69]. Moreover, The Netherlands has organized a few years ago the so-called European Truck Platooning Challenge, when the country was in the EU presidency [70], and we also reviewed the results of this initiative [71,72]. After the online desk research which helped us obtain a better understanding of the truck platooning world and especially of the actors involved, we discussed our initial ideas with actors within the Port of Rotterdam and we concluded that the main categories of actors involved in truck platooning are the following: the transporters, the freight forwarders, the policy-makers and the developers.

Each actor category then helped us in building a role and creating a role card that "brings this role into life"—from the world of reality into the world of play. Each role card briefly explains who the player is and what is his/her mission in the game. For instance, the developer is a representative of an original equipment manufacturer (OEM) company, which is interested in the truck platooning technology and has already invested a significant amount of resources in the research and development process. The objective of this role is to develop the technology before the company's competitors in the field in the most cost-effective way possible. The absence of EU regulations on the topic is a source of concern for this role. The policy-maker's role in the game is represented by an employee with a high-ranking position in the Dutch Ministry of Infrastructure and Water Management. The main dilemma that this role faces is if they should be patient until the EU publishes regulations and then align the national policy accordingly or work towards making The Netherlands a frontrunner country in this field, by introducing country-level policy that would encourage the deployment of truck-platooning. Figure 1a illustrates an example of a role card in Mobinn.

In the case of Platform4Barge, to re-construct the world of reality around it, we followed exactly the same logic in developing the game cards' content; beginning with a thorough literature review on the field of multi-sided digital platforms, with a focus on digital platforms in transport and logistics, and the container ecosystem in particular. As an extra way to gain meaningful insights in this case, we used also our experience from co-developing a set of four teaching cases and a teaching note around the use of multi-sided digital platforms in transport and logistics [73], together with actors from the Port of Rotterdam as well as from the private sector (we worked on the cases' development in parallel with the game development). We organized a number of try-out sessions with master's students in Rotterdam to test and validate the teaching cases, and the feedback we received was also valuable content wise for the development of the game cards. Following that, we arranged meetings with actors of the Port of Rotterdam who helped us in shaping our ideas and finally deciding the four main categories of actors for this game: the shippers/freight forwarders (in a combined role that represents the demand side), the deep-sea terminal operators, the barge operators and the port authority. For example, the role card of the representative of the port authority implies that the player acts as an orchestrator of the introduction of the digital platform and its operation and as a facilitator of the process. Achieving sustainability through modal shift is of crucial importance for this actor. Figure 1b presents an example of a role card in Platform4Barge.

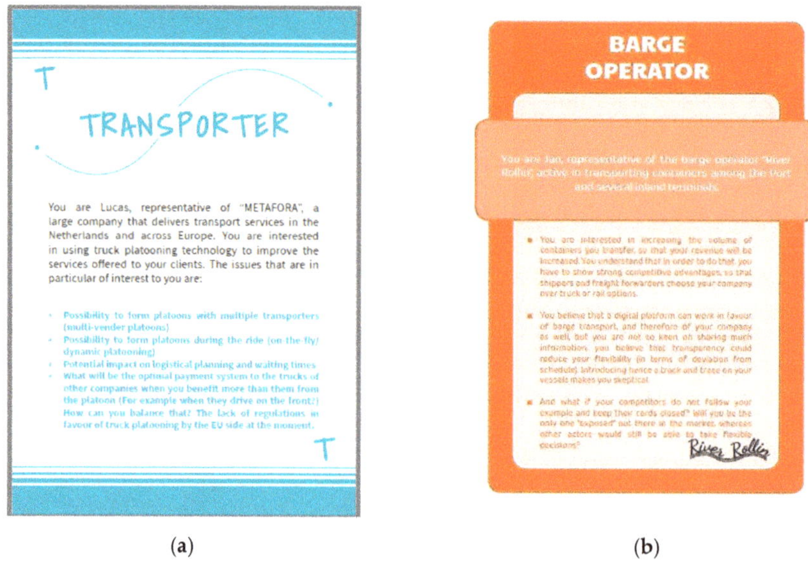

<table>
<tr><td align="center">(a)</td><td align="center">(b)</td></tr>
</table>

Figure 1. Example of a role card in: (**a**) Mobinn and; (**b**) Platform4Barge.

The same process of exploring the world of reality for both our case studies, helped us also in defining the key performance indicators (KPIs) for both the games, as illustrated in Table 1.

Table 1. Key Performance Indicators of Mobinn and Platform4Barge.

Mobinn	Platform4Barge
Technology maturity	Modal shift
Physical/legal infrastructure maturity	Reliability
Innovation adoption	Innovation adoption

The KPIs are illustrated on the game boards, with a 1–9 point scale, using a "traffic light" approach, i.e., dividing the scale in three areas: red, yellow and green, inspired by the work of Van den Ende (2019) (Figure 2a,b). The green area indicates alignment of interests and goals, the yellow area shows only partial alignment, while the red one demonstrates lack of consensus of actors. Each action card, in addition to describing a certain action that can be performed by a certain role, indicates the number of tokens needed to be paid by this role as well as by other(s) in the game in order to be activated. The impact of the card on the three KPIs of the game is also mentioned (Figure 3a,b). The event cards contribute in creating more suspense and fun during the game, as they can indicate either a positive, or a negative event that "disrupts" the players' plan and can make them adapt their strategy (Figure 4a,b).

(**a**)   (**b**)

Figure 2. Game board and key performance indicators (KPIs) in: (**a**) Mobinn and; (**b**) Platform4Barge.

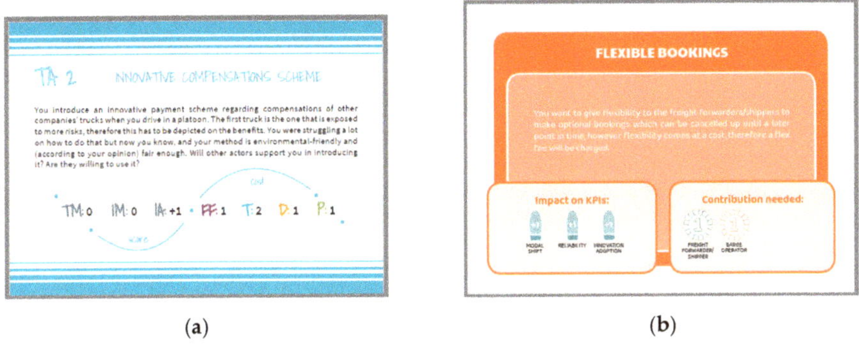

(**a**)   (**b**)

Figure 3. Example of an action card in: (**a**) Mobinn and; (**b**) Platform4Barge.

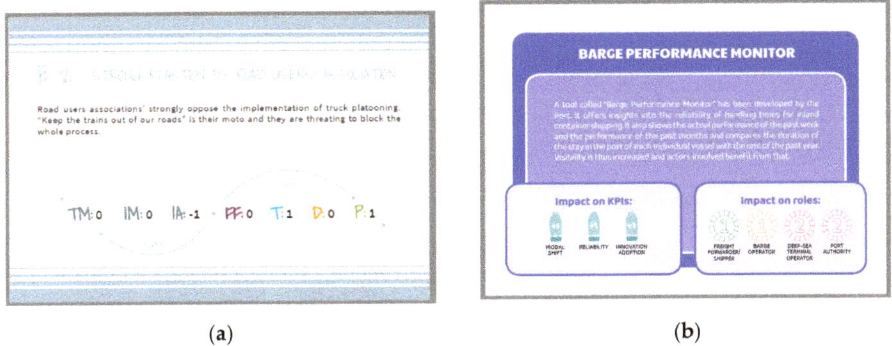

(**a**)   (**b**)

Figure 4. Example of (**a**) a negative event card in: Mobinn and; (**b**) a positive event card in Platform4Barge.

Both games are played in five rounds. At the beginning of a game session, each player gets one role card, all respective actions cards and an initial number of tokens. When a round begins, each player can either bring to the negotiations table one of the action cards or alternatively he/she can choose to not "act" in this round. After every player shares his/her decision, the players who chose to use an action card, simultaneously place them on the board. Each player has to pay the number of tokens that is written on the card for his/her own role before requesting other players to support the action with their tokens.

The card is considered valid and has an impact on the KPIs of the game, only in cases when all the roles mentioned on it agree to contribute with the necessary number of tokens—otherwise, the player who suggested the card loses the invested tokens and he/she can keep the card for future use. At the end of each round, an event card is drawn, and each player gets some extra tokens, representing income from their services. The role of other important actors (in addition to the ones represented by the four main roles), such as the European Commission (in the form of regulations or directives), is introduced in the game through the event cards.

The goal of both games is to "make the innovation happen" (in our case the innovation is aimed at the concepts of truck platooning and the multi-sided digital platform respectively), which is translated in achieving to have all three KPIs in the green area of the game board (points 7–9 in the 9-point scale) at the end of game play.

The main learning goal defined for both games is to understand how important an innovation ecosystem perspective is for a successful adoption of innovation in the transportation domain. Actors, decisions and actions within the innovation ecosystem are highly interdependent. Both games aim to highlight this characteristic and make players more aware of the interrelations of decisions and events. The players have the opportunity to influence the innovation ecosystem by (joint) investments and actions, but they must keep in mind that the innovation is also influenced by external events, limited budget and the existence of only five rounds. The game is played by at least four players (one per role). In cases when there are more participants, teams are formed and two or sometimes three players can together play a single role. This enables participants to first discuss choices of actions and it can be seen as a purposeful game element. Every game session is introduced by a short briefing, during which the facilitator explains the concept and the rules of the game. After the completion of the five game rounds, a debriefing follows. The total time of the game play, including briefing and debriefing is approximately 60 to 90 min for each of the games.

4.2. Play-Test Sessions and Data Collection

In order to test the two games and receive valuable feedback for further improvement, several game sessions were organized in which the game prototypes were played. The Mobinn game was tested between October 2018 and November 2019 at game sessions in The Netherlands and the USA (Figure 5a). The Platform4Barge prototype was tested at sessions in The Netherlands between December 2019 and February 2020. (A fourth game session was planned for March 2020, as part of an alumni-student event organized by Transito, the study association for the Bachelor and Master program Urban, Port and Transport Economics (UPTE) of the Erasmus University Rotterdam. This scheduled event was cancelled due to the measures taken to stop the spread of COVID-19 (Corona Virus Disease 2019) in Europe) (Figure 5b).

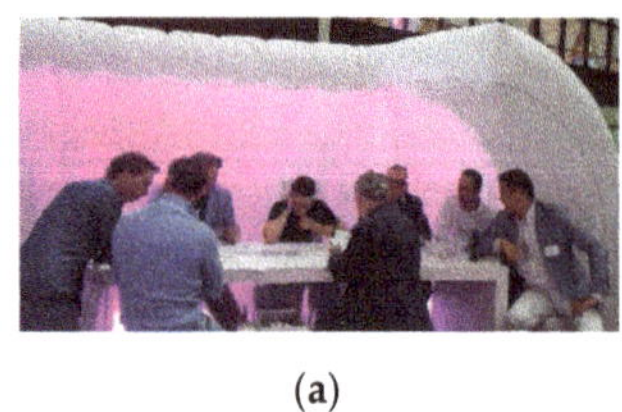

(a)

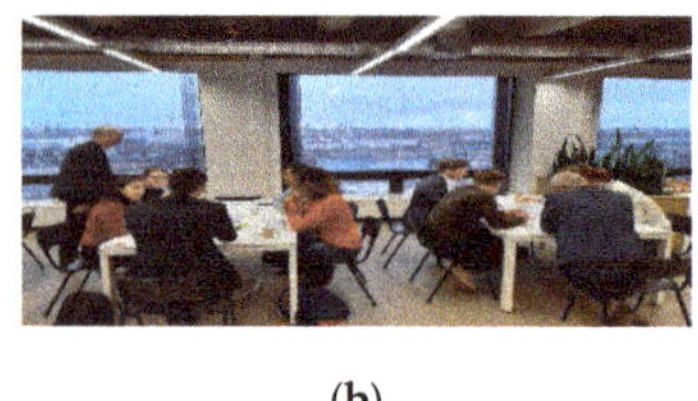

(b)

Figure 5. (**a**) Play-test session of Mobinn, during the event for the 10th year-anniversary of RDM (Rotterdam Dry Docks, in Dutch: Rotterdamsche Droogdok Maatschappij) Campus in Rotterdam, 4 July 2019; (**b**) play-test session of Platform4Barge during the Erasmus Port Campus in Rotterdam, 12 December 2019.

All sessions of both games were facilitated by at least one member of the game development group, who was responsible for the briefing, facilitation, debriefing, as well as for the data collection during and after the game. The data collection took place by audio recording if it was feasible to do so, and by observation through the participating researchers. Usually, one researcher took over the role of process facilitator, and one more observed the session. Beyond dedicated game sessions, the game was also played as part of a larger event, where usually background noise made audio recording difficult. During these sessions, data was collected by observation and notes during the game-play. The notes taken included decisions of the players, the cards they chose to use, as well as their game play behavior. After the end of the game, a debriefing phase was conducted based on semi-structured questions regarding the game experience and the take-aways. (The debriefing questions were adapted from the framework described by Kriz [74], in line with the former adaptation of this framework made for the debriefing and workshops of the Plaitra Game Session, which took place in the context of the iTRACK Project, at TU Delft Campus, The Netherlands, in May 2017). The questions aimed at deriving what types of feelings were experienced by the players during the game play and at revealing the lessons learned (if any) by the game. Moreover, the players were asked what, according to their opinion was working well during the game and which elements might need improvement, and how we can achieve this improvement. These questions were used as a trigger to initiate the discussion with the participants, but they were not in any case exclusive—the players were invited to share any kind of feedback, comment, observation or suggestion for change they considered relevant with the game facilitator. Thus, the debriefing phase focused on the game itself as well as the play experience of the players.

The following tables present the number and type of players, as well as the location where each game session took place (Table 2 for Mobinn and Table 3 for Platform4Barge).

The following section discusses the results of the game sessions.

Table 2. Play-tests sessions that have been organized for Mobinn.

Game Session	Number of Players	Type of Players	Location
1	5	Experts and actors related to the truck platooning community	Futureland, Maasvlakte II, Port of Rotterdam, NL
2	8	Transport and logistics experts from academia and industry	SmartPort (SmartPort is "a neutral knowledge platform, stimulating alliances, financing scientific research and providing public knowledge dissemination. The aim is to speed up innovations in the Port of Rotterdam" [75] premises, Rotterdam, NL.
3	4	Researchers in the field of gaming and/or transportation	TPM Faculty, TU Delft, NL
4	12	MBA program students	Supply Chain Management Center, Robert H. Smith School of Business, USA
5	11	Experts in transport and logistics from the Dutch Railways	Railcenter, Amersfoort, NL
6	5	Actors in the field of transport and logistics	RDM Rotterdam Campus, NL

Table 3. Play-tests sessions that have been organized for Platform4Barge.

Game Session	Number of Players	Type of Players	Location
1	4	Experts from the Rotterdam School of Management community	Erasmus Centre for Urban, Port and Transport Economics, NL
2	8	Participants of Erasmus Port Campus (Erasmus Port Campus is "an initiative aiming to bring the port to the campus. The event was organized by Erasmus UPT and SmartPort@Erasmus to connect the academic world with the latest developments in the port. Theme of the day was: Digitization in ports; Lifting smart logistics to a higher level" [76], students/alumni and professionals/experts)	Erasmus Centre of Entrepreneurship, NL
3	7	Actors/experts working at the Port of Rotterdam on key positions	World Port Center, Port Authority of Rotterdam, NL

5. Results and Discussion

The results discussed in this section represent a summary of the observations of the facilitator and accompanying researcher, collected and analyzed by the researcher team in a qualitative way, based on the in-situ observations during the game play and during the debriefing stage, as well as based on the recordings (when available) and notes for both stages. This approach follows methods as proposed in the field of serious gaming towards solving sustainability-related issues by Den Haan and Van der Voort [77]. Their study found that the majority of publications which fall into the aforementioned category of trying to address sustainability challenges have used qualitative data for the game evaluation, including unstructured observations by the researchers and debriefings.

Both games have been tested by players with a diverse background, knowledge and with different roles; from real actors in the field to professionals and members of the academic and research community; all do have knowledge of the domain of transport and logistics from their respective profession. The facilitator of the game was trying to intervene as little as possible during the game play, to minimize the associated biases and be able to detect and analyze the negotiations and discussions that were taking place among players, and only tried to support the process of play, but not the decision making. High engagement levels were observed during all game sessions, which proved that the games succeeded in creating an enjoyable, comfortable and fun environment for the players.

The negotiations and discussions among players involved more tension when actors with a presence in the field were involved. On the other hand, students and researchers were more curious and keener on learning about the innovation process around truck platooning and multi-sided digital platforms. The professional actors, being already familiar with the concepts, were able to focus more on attributes such as the aspirations of the different roles, and the characteristics of the different actions that could be taken. Students also reported that after the end of the game play, they felt that they obtained a more holistic view and a more comprehensive understanding regarding the challenges for new technologies and innovative concepts to be adopted by the transport and logistics sector. They also realized the value of communication among the stakeholders and organizations in an innovation ecosystem.

As the concept of innovation ecosystem and its importance emerged through the games, the participants appeared afterwards to have realized that nobody can win these games playing alone; it was necessary to discuss, negotiate and collaborate with other parties to increase the KPIs. It was observed that in those game sessions in which the players realized early in time that they needed to collaborate and share information and knowledge with each other to boost the KPIs, almost every game finished with all three KPIs in the green area of the spectrum. On the other hand, in sessions where the players appeared more reluctant to collaborate and unwilling to have increased visibility of their negotiation strategy in the beginning of the game, they struggled towards the end of the game and often failed to reach the objective of having all KPIs in the green area within the five rounds of the game.

A very interesting observation was made during a Mobinn game session with real actors related to the Port of Rotterdam; at one of the rounds, the resources of one player/role were not enough to invest in an action card that that player would like to use. At that moment, another player suggested to pay his part instead, because he said that this action could really contribute to the direction of "making the innovation happen" and thus it would be a win-win for both the roles they represented if they made this joint investment. The noteworthy point here is that this was never mentioned in the game instructions as an option, but the game play motivated the creative, out-of-the box thinking of the player towards the direction of collaboration.

When players had already some background knowledge on truck platooning, they considered it a plus in better understanding the Mobinn game. As for Platform4Barge, even players which did not have any background knowledge on the specific topic of exchange platforms, did not encounter difficulties in understanding, because they were familiar with the concept of digital multi-sided platforms through very widespread platforms of this kind in other sectors, such as Uber and other on-demand transport services for shared mobility, and Airbnb in housing.

A repeated comment during the Mobinn sessions was about the equal distribution of money (tokens) among roles; several players felt that this is often not the case in real-life, as the authorities' budget is usually considerably higher than the one of private actors. For this reason, for Platform4Barge it was decided to start with a different number of tokens per role to better reflect the situation in reality.

Last but not least, feedback from experts included the view that identifying all actors involved in an innovation ecosystem, along with their potential actions, is already a very powerful characteristic of both the games.

6. Conclusions

Actors from diverse backgrounds participated in a number of play sessions with both games; from experts who represented the real actors in the field, to students, researchers and professionals. This led to very interesting observations and proved that both games can act as triggers to make participants aware of the existing challenges that stakeholders involved in transport and logistics face when implementing innovations to increase sustainability. The games provide useful insights in the benefits of collective goal setting and can support and stimulate discussions towards the importance of collaborative action in innovation processes. The games can be used to trigger a valuable discussion and make actors aware of the value of an innovation ecosystem, and the need for collaboration to achieve sustainability goals.

The results of the game sessions highlight the critical role that each actor can play in an innovation ecosystem, by collaborating on data sharing and improving data visibility and transparency. The study found that using serious games to address the challenge of enhancing awareness, communication and collaboration of actors in innovation ecosystems can be highly effective.

Author Contributions: Conceptualization, A.R., H.L., A.V. and R.Z.; methodology, A.R., H.L., A.V. and R.Z. validation, A.R., H.L., A.V. and R.Z.; formal analysis, A.R., H.L., A.V. and R.Z.; investigation, A.R., H.L., A.V. and R.Z.; resources, A.R., H.L., A.V. and R.Z.; data curation, A.R., H.L., A.V. and R.Z.; writing—original draft preparation, A.R., H.L., A.V. and R.Z.; writing—review and editing, A.R., H.L., A.V. and R.Z.; visualization, A.R., H.L., A.V. and R.Z.; supervision, H.L., A.V. and R.Z.; project administration, H.L., A.V. and R.Z.; funding acquisition, H.L., A.V. and R.Z. All authors have read and agreed to the published version of the manuscript.

Funding: Mobinn game was developed as part of the InDeep (Innovation Network Design Enables Excellent Ports) project, funded by The Netherlands Organization for Scientific Research (NWO). The initial version of the Platform4Barge game was developed in the context of the EU Horizon2020 Project SELIS (Towards a Shared European Logistics Intelligent Information Space). The final version was developed in the context of the project "Digital Platforms for Logistics: Flipping the Classroom by Means of Games and Teaching Cases", funded by the Community for Learning and Innovation (CLI) of Erasmus University Rotterdam as an Educational Innovation project.

Acknowledgments: The authors would like to thank the lead developers that were involved in the development of the two games. In the case of Platforrm4Barge, in addition to the authors this includes Donald Baan, Arco Jansen and Maurice Jansen. Game production and visual design was realised by the TU Delft GameLab: for Mobinn, by Linda van Veen and for Platform4Barge, by Linda van Veen and Resy Arts. The authors would also like to thank all the participants of the game sessions for their time and valuable feedback on the game and the anonymous reviewers who helped in improving the manuscript by providing their constructive suggestions and comments.

Conflicts of Interest: The authors declare no conflict of interest. The funders had no role in the design of the study; in the collection, analyses, or interpretation of data; in the writing of the manuscript, or in the decision to publish the results.

References

1. United Nations Department of Economic and Social Affairs. *World Economic and Social Survey 2018: Frontier Technologies for Sustainable Development*; United Nations Department of Economic and Social Affairs: New York, NY, USA, 2018; ISBN 9789211091793.
2. Trischler, J.; Johnson, M.; Kristensson, P. A service ecosystem perspective on the diffusion of sustainability-oriented user innovations. *J. Bus. Res.* **2020**, *116*, 552–560. [CrossRef]
3. Gruenhagen, J.H.; Parker, R. Factors driving or impeding the diffusion and adoption of innovation in mining: A systematic review of the literature. *Resour. Policy* **2020**, *65*, 101540. [CrossRef]
4. Acciaro, M.; Sys, C. Innovation in the maritime sector: Aligning strategy with outcomes. *Marit. Policy Manag.* **2020**, 1–19. [CrossRef]
5. Nilsson, F.R. A complexity perspective on logistics management: Rethinking assumptions for the sustainability era. *Int. J. Logist. Manag.* **2019**, *30*, 681–698. [CrossRef]
6. Roukouni, A.; Zuidwijk, R. *Port of Rotterdam: Booking.com for Container Transport*; Erasmus Research Institute of Management, RSM Case Development Centre: Rotterdam, The Netherlands, 2020; pp. 1–14.
7. Liu, Z.; Stephens, V. Exploring innovation ecosystem from the perspective of sustainability: Towards a conceptual framework. *J. Open Innov. Technol. Mark. Complex.* **2019**, *5*, 48. [CrossRef]
8. Bacon, E.; Williams, M.D.; Davies, G. Coopetition in innovation ecosystems: A comparative analysis of knowledge transfer configurations. *J. Bus. Res.* **2020**, *115*, 307–316. [CrossRef]
9. Toshiyuki, K.; Kazuaki, K. An institutional approach to the creation of innovation ecosystems and the role of law. *Penn St. JL Int'l Aff.* **2015**, *4*, 167.
10. Smorodinskaya, N.; Russell, M.; Katukov, D.; Still, K. Innovation Ecosystems vs. Innovation Systems in Terms of Collaboration and Co-creation of Value. In Proceedings of the 50th Hawaii International Conference on System Sciences, Waikoloa Village, HI, USA, 4–7 January 2017. [CrossRef]
11. Agogué, M.; Levillain, K.; Hooge, S. Gamification of Creativity: Exploring the Usefulness of Serious Games for Ideation. *Creat. Innov. Manag.* **2015**, *24*, 415–429. [CrossRef]
12. Raghothama, J.; Meijer, S.A. A Review of Gaming Simulation in Transportation. In *Proceedings of the Frontiers in Gaming Simulation. ISAGA 2013. Lecture Notes in Computer Science*; Springer: Berlin/Heidelberg, Germany, 2014.
13. Midden, C.; McCalley, T.; Ham, J.; Zaalberg, R. Using persuasive technology to encourage sustainable behavior. In Proceedings of the 6th International Conference on Pervasive Computing, Workshop on Pervasive Persuasive Technology and Environmental Sustainability, Sydney, Australia, 19 May 2008; Volume 113, pp. 83–86.
14. Bessant, J.; Rush, H.; Gray, W.; Hoffman, K.; Ramalingam, B.; Marshall, N. *Development D. for I. Innovation Management, Innovation Ecosystems and Humanitarian Innovation*; University of Brighton: Brighton, UK, 2014.
15. Duncan, L. Innovation as an ongoing process: Developing creative capabilities. *Strateg. Dir.* **2014**, *30*, 13–15. [CrossRef]
16. Vargo, S.L.; Akaka, M.A.; Wieland, H. Rethinking the process of diffusion in innovation: A service-ecosystems and institutional perspective. *J. Bus. Res.* **2020**, *116*, 526–534. [CrossRef]
17. Corsaro, D.; Cantù, C.; Tunisini, A. Actors' Heterogeneity in Innovation Networks. *Ind. Mark. Manag.* **2012**, *41*, 780–789. [CrossRef]
18. Moore, J.F. Predators and Prey: A New Ecology of competition. *Harv. Bus. Rev.* **1993**, *71*, 75–83. [PubMed]
19. Russell, M.G.; Smorodinskaya, N.V. Leveraging complexity for ecosystemic innovation. *Technol. Forecast. Soc. Chang.* **2018**, *136*, 114–131. [CrossRef]

20. Katz, B.; Wagner, J. The Rise of Innovation Districts: A New Geography of Innovation in America. *Harv. Bus. Rev.* **2014**, 1–34. Available online: https://www.brookings.edu/wp-content/uploads/2016/07/InnovationDistricts1.pdf (accessed on 25 September 2020).
21. Dattee, B.; Alexy, O.; Autio, E. Maneuvering in Poor Visibility: How Firms play the ecosystem game when uncertainty is high. *Acad. Manag. J.* **2017**, 1–67. [CrossRef]
22. Mulas, V.; Minges, M.; Applebaum, H. Boosting Tech Innovation: Ecosystems in Cities. *Innovations* **2016**, *11*, 98–125. [CrossRef]
23. Russell, M.G.; Huhtamäki, J.; Still, K.; Rubens, N.; Basole, R.C. Relational capital for shared vision in innovation ecosystems. *Triple Helix* **2015**, *2*. [CrossRef]
24. Su, Y.S.; Zheng, Z.X.; Chen, J. A multi-platform collaboration innovation ecosystem: The case of China. *Manag. Decis.* **2018**, *56*, 125–142. [CrossRef]
25. Zahra, S.A.; Nambisan, S. Entrepreneurship and strategic thinking in business ecosystems. *Bus. Horiz.* **2012**, *55*, 219–229. [CrossRef]
26. de Vasconcelos Gomes, L.A.; Salerno, M.S.; Phaal, R.; Probert, D.R. How entrepreneurs manage collective uncertainties in innovation ecosystems. *Technol. Forecast. Soc. Chang.* **2018**, *128*, 164–185. [CrossRef]
27. Lubik, S.; Garnsey, E. Entrepreneurial innovation in science-based firms: The need for an ecosystem perspective. In *Handbook of Research on Small Business and Entrepreneurship*; Chell, E., Karataş-Özkan, M., Eds.; Edward Elgar: Cheltenham, UK, 2014; pp. 599–633.
28. Pellinen, A.; Ritala, P.; Järvi, K.; Sainio, L.M. Taking initiative in market creation—A business ecosystem actor perspective. *Int. J. Bus. Environ.* **2012**, *5*, 140. [CrossRef]
29. Meyer-Larsen, N.; Hauge, J.; Baalsrud, M.R.; Hamadache, K.; Aifadopoulou, G.; Forcolin, M.; Roso, V.; Tsoukos, G.; Westerheim, H. Accelerating the Innovation Uptake in Logistics. In *Innovative Methods in Logistics and Supply Chain Management*; Blecker, T., Kersten, W., Ringle, C.M., Eds.; Epubli: Berlin, Germany, 2014; Volume 19.
30. Warnke, P.; Koschatzky, K.; Som, O.; Stahlecker, T.; Nabitz, L.; Braungardt, S.; Cuhls, K.; Dönitz, E.; Güth, S.; Plötz, P.; et al. Opening Up the Innovation System Framework Towards New Actors and Institutions. Innov. Syst. Policy Anal. 2016, 49, Fraunhofer ISI, Karlsruhe. Available online: https://www.econstor.eu/handle/10419/129191 (accessed on 9 September 2020).
31. Wilkinson, P. A brief history of serious games. In *Entertainment Computing and Serious Games*; van der Spek, E., Göbel, S., Do, E.Y.-L., Clua, E., Baalsrud, H.J., Eds.; Springer International Publishing AG: Berlin/Heidelberg, Germany, 2016; pp. 17–41.
32. D'Angour, A. Plato and Play: Taking Education Seriously in Ancient Greece. *Am. J. Play* **2013**, *5*, 293–307.
33. Abt, C.C. *Serious Games*; Viking Press: New York, NY, USA, 1970.
34. Duke, R.D. *Gaming: The Future's Language*; SAGE Publications Inc., John Wiley & Sons: New York, NY, USA, 1974.
35. Mayer, I.S. The Gaming of Policy and the Politics of Gaming: A Review. *Simul. Gaming* **2009**, *40*, 825–862. [CrossRef]
36. Lukosch, H.K.; Bekebrede, G.; Kurapati, S.; Lukosch, S.G. A Scientific Foundation of Simulation Games for the Analysis and Design of Complex Systems. *Simul. Gaming* **2018**, *49*, 279–314. [CrossRef]
37. Constantinescu, T.; Devisch, O.; Huybrechts, L. Civic Participation: Serious Games and Spatial Capacity Building. In Proceedings of the MEDIA CITY 5 International Conference, Plymouth, UK, 3 May 2015; pp. 179–263.
38. Olejniczak, K.; Wolański, M.; Widawski, I. Regulation crash-test: Applying serious games to policy design. *Policy Des. Pract.* **2018**, *1*, 194–214. [CrossRef]
39. Patricio, R. A gamified approach for engaging teams in corporate innovation and entrepreneurship. *World J. Sci. Technol. Sustain. Dev.* **2017**, *14*, 254–262. [CrossRef]
40. Oonk, M. Smart Logistics Corridors and the benefits of Intelligent Transportation Systems. I. In *Towards Innovative Freight and Logistics, Science, Society and New Technologies Series, Research for Innovative Transports Set*; Wiley: Hoboken, NJ, USA, 2016.
41. van Pelt, S.C.; Haasnoot, M.; Arts, B.; Ludwig, F.; Swart, R.; Biesbroek, R. Communicating climate (change) uncertainties: Simulation games as boundary objects. *Environ. Sci. Policy* **2015**, *45*, 41–52. [CrossRef]

42. Hidayatno, A.; Zulkarnain; Hasibuan, R.G.; Wardana Nimpuno, G.C.; Destyanto, A.R. Designing a serious simulation game as a learning media of sustainable supply chain management for biofuel production. *Energy Procedia* **2019**, *156*, 43–47. [CrossRef]

43. Blöchl, S.J.; Michalicki, M.; Schneider, M. Simulation Game for Lean Leadership—Shopfloor Management Combined with Accounting for Lean. *Procedia Manuf.* **2017**, *9*, 97–105. [CrossRef]

44. Van den Hoogen, J.; Meijer, S. Gaming and Simulation for Railway Innovation: A Case Study of the Dutch Railway System. *Simul. Gaming* **2014**, *46*, 489–511. [CrossRef]

45. Kurapati, S.; Kourounioti, I.; Lukosch, H.; Tavasszy, L.; Verbraeck, A. Fostering sustainable transportation operations through corridor management: A simulation gaming approach. *Sustainability* **2018**, *10*, 455. [CrossRef]

46. Bekebrede, G.; Lo, J.; Lukosch, H. Understanding Complex Systems Through Mental Models and Shared Experiences: A Case Study. *Simul. Gaming* **2015**, *46*, 536–562. [CrossRef]

47. Witte, P.; Slack, B.; Keesman, M.; Jugie, J.H.; Wiegmans, B. Facilitating start-ups in port-city innovation ecosystems: A case study of Montreal and Rotterdam. *J. Transp. Geogr.* **2018**, *71*, 224–234. [CrossRef]

48. Port of Rotterdam, Facts & Figures: A Wealth of Information. Make It Happen. Available online: https://www.portofrotterdam.com/sites/default/files/facts-and-figures-port-of-rotterdam.pdf (accessed on 25 September 2020).

49. Van der Horst, M.; Kort, M.; Kuipers, B.; Geerlings, H. Coordination problems in container barging in the port of Rotterdam: An institutional analysis. *Transp. Plan. Technol.* **2019**, *42*, 187–199. [CrossRef]

50. Roukouni, A.; Lukosch, H.; Verbraeck, A.; Zuidwijk, R. Mobilise Innovation (Mobinn): A playful Approach applied to the Transport and Logistics Sector. In Proceedings of the 50th ISAGA (International Simulation and Gaming Association) Conference: Simulation & Gaming through Time and across Disciplines. Past and Future, Heritage and Progress, Warsaw, Poland, 26–30 August 2019; pp. 468–479.

51. Van Ark, E.J.; Duijnisveld, M.; van Eijk, E.; Janssen, R.; van Ommeren, C.; Soekroella, A. Value Case Truck Platooning—An Early Exploration of the Value of Large-Scale Deployment of Truck Platooning. TNO: Den Haag, The Netherlands, 2017, 1–100. Available online: https://repository.tudelft.nl/view/tno/uuid%3A770c9b7c-40f0-4036-93e8-caba19c2b311 (accessed on 25 September 2020).

52. Deelen, C. Truck Platooning. Available online: https://smartport.nl/en/project/truck-platooning/ (accessed on 5 October 2020).

53. Tavasszy, L. On the Value Case for Truck Platooning in Europe. Discussion Paper for Truck Platooning Next Level Group Meeting, a Side Event to the ITS European Congress, Glasgow, June 6, 2016. Available online: https://www.researchgate.net/publication/312116529_The_value_case_for_truck_platooning (accessed on 9 September 2020).

54. Bakermans, B.A. *Truck Platooning Enablers, Barriers, Potential and Impacts*; TU Delft: Delft, The Netherlands, 2016.

55. Roukouni, A.; Lukosch, H.; Verbraeck, A. Simulation games to foster innovation: Insights from the transport and logistics sector. In *Neo-Simulation and Gaming Toward Active Learning, Translational Systems Sciences*; Hamada, R., Soranastaporn, S., Kanegae, H., Dumrongrojwatthana, P., Chaisanit, S., Rizzi, P., Dumblekar, V., Eds.; Springer: Singapore, 2019; pp. 157–165.

56. Hagiu, A.; Wright, J. Multi-Sided Platforms. *Int. J. Ind. Organ.* **2015**, *43*, 162–174. [CrossRef]

57. Van Alstyne, M.W.; Parker, G.G.; Choudary, S.P. Pipelines, Platforms, and the New Rules of Strategy. *Harv. Bus. Rev.* **2016**, *94*, 54–62.

58. Evans, D.S.; Schmalensee, R. *The Antitrust Analysis of Multi-Sided Platform Businesses*; National Bureau of Economic Research: Cambridge, MA, USA, 2013.

59. SELIS. *Deliverable D7.21, Living Labs Operation Learning Conclusions and Other SELIS Value Propositions, Final Version*; SELIS, 2019; Available online: https://www.selisproject.eu/uploadfiles/D-721-delLL_Operation_Learning_Conclusions_v3.0.pdf (accessed on 9 September 2020).

60. Christopher, M.; Lee, H. Mitigating supply chain risk through improved confidence. *Int. J. Phys. Distrib. Logist. Manag.* **2004**, *34*, 388–396. [CrossRef]

61. Olesen, P.B.; Hvolby, H.H. D-PI Enabling Information Sharing in a Port. In *Proceedings of the Dvances in Production Management Systems. Competitive Manufacturing for Innovative Products and Services. APMS 2012. IFIP Advances in Information and Communication Technology*; Emmanouilidis, C., Taisch, M., Kiritsis, D., Eds.; Springer: Berlin/Heidelberg, Germany, 2013; pp. 152–159.

62. Macharis, C. Mobility is a serious game: A game to explore the future of mobility. In Proceedings of the Transportation Research Board 97th Annual Meeting, Washington, DC, USA, 7–11 January 2018.

63. König, A.; Kowala, N.; Wegener, J.; Grippenkoven, J. Introducing a mobility on demand system to prospective users with the help of a serious game. *Transp. Res. Interdiscip. Perspect.* **2019**, *3*, 100079. [CrossRef]

64. Harteveld, C. *Triadic Game Design: Balancing Reality, Meaning and Play*, 1st ed.; Springer: London, UK, 2011; ISBN 978-1-84996-157-8.

65. Bhoopalam, A.K.; Agatz, N.; Zuidwijk, R. Planning of truck platoons: A literature review and directions for future research. *Transp. Res. Part B* **2018**, *107*, 212–228. [CrossRef]

66. Janssen, G.; Zwijnenberg, H.; Blankers, I.; de Kruijff, J. *Future of Transportation Truck Platooning*; TNO: Delft, The Netherlands, 2015; pp. 1–36.

67. Commendatore, C. For Truck Platooning to Work, Here's What has to Happen. FleetOwner. 2018. Available online: https://www.fleetowner.com/technology/article/21702545/for-truck-platooning-to-work-heres-what-has-to-happen (accessed on 25 September 2020).

68. Malcolm Wheatley, M.R. Time to Fall in Line. Available online: https://www.automotivelogistics.media/time-to-fall-in-line/21089.article (accessed on 5 October 2020).

69. Dutch Ministry of Infrastructure and Environment. European Truck Platooning Challenge 2016—A Fresh Perspective on Mobility and Logistics. 2016, 1–50. Available online: https://www.government.nl/documents/leaflets/2015/10/06/leaflet-european-truck-platooning-challenge-2016 (accessed on 9 September 2020).

70. Association, E.A.M. First Cross-Border Truck Platooning Trial Successfully Completed. Available online: https://www.acea.be/press-releases/article/first-cross-border-truck-platooning-trial-successfully-completed (accessed on 5 October 2020).

71. Roberts, J. Success of Truck Platooning Challenge Clears Way for Real-Life Convoys—Steve Phillips, CEDR. Available online: https://horizon-magazine.eu/article/success-truck-platooning-challenge-clears-way-real-life-convoys-steve-phillips-cedr.html (accessed on 5 October 2020).

72. Anastasia, R.; Rob, Z. *Multi-Sided Platforms in Europe's Logistics Sector*; RSM Case Development Centre: Rotterdam, The Netherlands, 2020.

73. OECD Glossary. Available online: https://stats.oecd.org/glossary/detail.asp?ID=4313 (accessed on 9 September 2020).

74. Kriz, W.C. A systemic-constructivist approach to the facilitation and debriefing of simulations and games. *Simul. Gaming* **2010**, *41*, 663–680. [CrossRef]

75. SmartPort About SmartPort. Available online: https://smart-port.nl/en/about-smartport-2/ (accessed on 9 September 2020).

76. Erasmus, U. Erasmus Port Campus. Available online: https://www.eur.nl/en/upt/news/first-erasmus-port-campus-launches-focus-digitization-ports (accessed on 9 September 2020).

77. Den Haan, R.J.; van der Voort, M.C. On evaluating social learning outcomes of serious games to collaboratively address sustainability problems: A literature review. *Sustainability* **2018**, *10*, 4529. [CrossRef]

sustainability

MDPI

Article

Efficiency Assessment of Operations Strategy Matrix in Healthcare Systems of US States Amid COVID-19: Implications for Sustainable Development Goals

Aydın Özdemir [1,2,*], Hakan Kitapçı [1], Mehmet Şahin Gök [1] and Erşan Ciğerim [1]

[1] Department of Management, Faculty of Business Administration, Gebze Technical University, Kocaeli 41400, Turkey; kitapci@gtu.edu.tr (H.K.); sahingok@gtu.edu.tr (M.Ş.G.); cigerim@gtu.edu.tr (E.C.)
[2] Besni Ali Erdemoğlu Vocational High School, Adiyaman University, Adiyaman 02300, Turkey
* Correspondence: aydinozdemir17@gmail.com

Abstract: The objective of this study is to assess the efficiency of the operations strategy matrix in the healthcare system of U.S. states amid COVID-19. Output-Oriented Data Envelopment Analysis was used to assess the efficiency of the operations strategy matrix. Strategic Decision Areas (Capacity, Supply Network, Process Technology, and Development and Organization) were considered inputs while competitive priorities (Quality, Cost, Delivery, and Flexibility) were considered outputs. According to results; Alaska, Alabama, Arkansans, Florida, Hawaii, Iowa, Idaho, Louisiana, Minnesota, Missouri, Mississippi, Montana, North Carolina, New Jersey, New York, Oklahoma, South Carolina, South Dakota, Texas, Vermont, Wisconsin, and Wyoming are relatively efficient. Additionally, Connecticut, Louisiana, Minnesota, New Jersey, Rhode Island, Tennessee, Utah, Vermont, Washington, and Wyoming are fully efficient while South Dakota is the state that needs the most improvement in terms of strategic decision areas and competing priorities. On the other hand, inefficient states have larger population and GDP than efficient states. Based on these results, implications for sustainable development goals (SDGs) are drawn.

Keywords: efficiency assessment; operations strategy matrix; data envelopment analysis; healthcare systems; COVID-19

Citation: Özdemir, A.; Kitapçı, H.; Gök, M.Ş.; Ciğerim, E. Efficiency Assessment of Operations Strategy Matrix in Healthcare Systems of US States Amid COVID-19: Implications for Sustainable Development Goals. *Sustainability* **2021**, *13*, 11934. https://doi.org/10.3390/su132111934

Academic Editors: Golam Kabir, Sanjoy Kumar Paul and Syed Mithun Ali

Received: 1 September 2021
Accepted: 20 October 2021
Published: 28 October 2021

Publisher's Note: MDPI stays neutral with regard to jurisdictional claims in published maps and institutional affiliations.

1. Introduction

A novel coronavirus named COVID-19 was identified in Wuhan/China on 31 December 2019, and the World Health Organization (WHO) assessed that COVID-19 could be characterized as a pandemic on 11 March 2020. The first case was confirmed on 22 January 2020 and as of 27 June 2021, due to COVID-19, there were 34,494,677 confirmed cases and 619,424 confirmed deaths in the USA [1].

Centers for Disease Control and Prevention (CDC), a service organization that protects the public's health, has prepared a global response to COVID-19. It includes the goals limiting human-to-human transmission, minimizing the impact of COVID-19 in some states with limited healthcare delivery capacity, and reducing certain threats that pose a risk to the United States' healthcare system [2]. In this context, public health, and social measures have been implemented to prepare and respond to the pandemic [3]. Although there have been many actions implemented to cope with the pandemic, the increasing number of COVID-19 cases poses a major threat to healthcare delivery [4].

WHO has provided comprehensive guidance to community and service providers to minimize the spread of the disease [5]. Since the early experience shows that COVID-19 requires unprecedented mobilization of health systems, it is recommended to assess the health systems to mitigate the outbreak's impact [6].

Although health is only one of the UN Sustainable Development Goals (SDGs), many other health-related goals comprise determinants of health [7]. The SDGs represent a unique opportunity to promote and assess public health [8]. The healthcare assessment process seeks to measure how care affects individuals or populations' health and well-being [9]. The assessment is also vital for achieving the SDGs, especially Goal 3, described as ensuring healthy lives and promoting well-being for all at all ages [10]. because sustainability is a managerial trend that plays an important role [11], and implications drawn based on healthcare assessment help to help policymakers for favorable policymaking [12]. Mentioned assessment depends on the development of standards [13]. There are some standards and tools for assessing healthcare systems such as effectiveness, efficiency, humanity, and equity [14]. As a novel evaluation tool, Operations Strategy Matrix describes the operations strategy as the intersections of an organization's performance objectives (competitive priorities) with its decision areas [15]. According to Slack and Lewis (2017); the matrix can be considered a checklist of the issues required to be assessed and helps operations to be extensive. Additionally, all intersections on the matrix don't have equal importance. Some intersections are more critical than others. This depends on the nature of operations. The matrix has four strategic decision areas (capacity, supply network, process technology, and development and organization), and four competitive priorities (quality, cost, delivery, and flexibility) [15].

The research question of this study is as follows: are US states efficient in the healthcare system amid the COVID-19 pandemic, and how do the efficient or inefficient conditions affect Sustainable Development Goals (SDGs).

The objective of this study is (i) to assess the efficiency of the operations strategy matrix in the healthcare system of US states amid COVID-19. For this purpose, 50 US states were considered as Decision Making Unit (DMU), and output-orientated Data Envelopment Analysis (DEA) was used; and (ii) to draw implications for sustainable development goals (SDGs) based on DEA results. The unique contribution of this study lies in the use of the Operations Strategy Matrix to assess the healthcare system of US states and to draw the implications for Sustainable Development Goals (SDGs). To validate the Operations Strategy Matrix, Data Envelopment Model was used.

The research paper is organized as follows. The following section presents the exiting literature review associated with COVID-19 and the healthcare system. Section 3 describes the DEA methodology. Section 4 presents the results. Section 5 discusses the findings and draws implications for sustainable development goals (SDGs). The final section portrays the conclusion and opportunities for future work.

2. Related Research

The section presents the existing literature about COVID-19 and the healthcare system.

O'Leary et al. (2021) have documented the healthcare policies developed during the initial wave of widespread COVID-19 transmission in Ireland, and they have developed six category headings to describe to focus VOVID-19 policies: (i) infection prevention and control, (ii) residential care settings, (iii) maintaining non-COVID-19 healthcare services, and supports, (iv) testing and contact tracing, (v) guidance for healthcare workers concerning COVID-19, and (vi) treating COVID-19 [16]. Tsai and Yang (2020) have examined the impact of the COVID 19 outbreak on voluntary demand for non COVID-19 healthcare, and their research has indicated a substantial decline in healthcare use is caused by the prevention of measures for COVID-19 [17]. Adwibowo (2020) has recorded the daily COVID-19 cases and has continued with the forecasting of the average daily demand (ADD) of healthcare facilities including beds, ICUs, and ventilators using the ARIMA model, and the model has shown that the healthcare ADD is different in each population [18]. Alshammari et al. (2021) have evaluated the potentiality of emerging technologies for controlling the COVID-19 transmission and ensuring health safety and their study has revealed that most people receive information from social networking sites, health professionals, and television without facing any challenges [19]. Brodie et al. (2021) have explored how adopting a service ecosystem perspective provides insight into the complexity of healthcare systems during times of COVID-19, and they have provided an understanding of the relevance of managerial flexibility, innovation, learning, and knowledge sharing [20]. Arlotti and Ranci (2021) have argued that the negative impact of COVID-19 stems from the poor development of long-term care policy and the marginality of residential institutions within the Italian healthcare system [21]. Rodríguez and Hignett (2021) have presented a model for integrating Human Factors/Ergonomics into healthcare systems to make them more robust and resilient during the COVID-19 pandemic, and they have argued that it is crucial to have a systems vision focused on optimizing the interactions people and other healthcare systems elements [22]. Akiyama et al. (2020) have described how their experience with inpatient care has changed in the wake of COVID-19 and they have determined that COVID-19 hospital discharge rates surpass admission rates [23]. Tasri and Tasri (2020) have analyzed the competency influence of medical records and health information management on the planning and decision making of healthcare services, and they have concluded that the new tools or systems can be an alternative to improve healthcare management [24]. Bel et al. (2021) have built a theoretical model and used it to develop an empirical strategy, analyzing the drivers of policy-response agility during the outbreak, and according to their model, healthcare system capacity and cost-related variables have a significant influence on reaction time; therefore, this situation has negatively affected government strategy [25]. Chamboredon et al. (2020) have presented a brief history of the development of the pandemic in France, including the political decisions that have been taken to combat the outbreak, and they have argued that the history of the development of the pandemic helps explain the establishment of the state of health emergency and containment of the population. [26]. Leite et al. (2020) have evaluated the impact of the COVID-19 pandemic on healthcare systems' demand, resources, and capacity. Then they have discussed that the sustainability of lean post-pandemics is a new modern operational issue [27].

Currently, the world is facing a global health crisis COVID-19; however, ensuring healthy lives and promoting well-being at all ages is essential to sustainable development, and the pandemic provides a milestone for health emergency preparedness and investment in critical 21st-century public services [28]. The SDGs aim to be relevant to all countries to promote prosperity, and almost all of the other 16 goals are related to health, or their achievement will contribute to health implicitly [29]. The SDGs are a tremendous opportunity to improve the health of the people of the world, and to make progress on the SDGs, states/countries must invest more heavily in health services research [30]. To meet the SDG targets and improve health system quality, states/countries need to undertake a measurement agenda [31]. Measuring and assessing the efficiency of health systems has been explored to ensure the sustainability of countries'/states' health and social systems [32]. Furthermore, gaps in the efficiency of the healthcare system will interrupt achieving SDGs [33].

All of these research findings reveal that there is a close relationship between the efficiency of the healthcare system and the implications for sustainable development goals amid the COVID-19 pandemic.

3. Materials and Methods

3.1. Background of Data Envelopment Analysis (DEA)

DEA is a nonparametric method of evaluating the relative efficiency of DMUs that use the same inputs to produce the same outputs, such as firms or public sector agencies [34–36]. DEA is widely used in many areas such as healthcare, banking, education, auditing, market research, and agriculture [37–40].

DEA has been used as a benchmarking and assessment tool in healthcare worldwide [21]. In addition, DEA is considered one of the most effective techniques for relative efficiency assessments, as it can evaluate multiple inputs and outputs simultaneously [22]. According to Ozcan (2009), since it is a multi-dimensional construct, the assessment of efficiency is often troublesome. Nevertheless, DEA addresses the limitations of ratio analysis and regression. Additionally, DEA uses multiple outputs and multiple inputs to identify efficiencies and inefficiencies and also to project how inefficient DMUs can become more efficient, by identifying reference sets (best practices). A reference set function can be built from observed inputs and outputs [23].

DEA models can be divided into two main types; constant returns to scale (CRS), also known as CCR (Charnes-Cooper-Rhodes) model, was initially introduced by Charnes et al. in 1978 [41], and variable returns to scale (VRS), also known as BCC (Banker-Charnes-Cooper) model, was later developed by Banker et al. in 1984 [42]. While the CCR (CRS) model yields an objective assessment of overall efficiency, identifies the sources, and estimates the number of inefficiencies in this way, the BCC (VRS) model distinguishes between technical and scale inefficiencies by estimating pure technical efficiency and identifying possible increasing, decreasing, or constant returns to scale [20]. Unlike CCR (CRS) models, input and output efficiency are equal, BCC (VRS) models yield different input and output efficiencies [21]. In addition, while CCR (CRS) models can assume that economies of scale don't change as the size of service facility increases, BCC (VRS) models cannot assume that economies of scale don't change as the size of service facility increases [26].

DEA models can be either input-orientated or output-orientated. In the input-orientated case, the DEA method defines the frontier by seeking the maximum possible proportional reduction in input usage, with output levels held fixed for each DMU; in the output orientated case, the DEA method seeks the maximum proportional increase in output production, with input levels held constant for each DMU [43,44].

There are four DEA model classifications, also known as Basic Envelopment Models: CCR (CRS)-Input Orientation, CCR (CRS)-Output Orientation, BCC (VRS)-Input Orientation, and BCC (VRS)-Output Orientation [26].

The BCC (VRS) efficiency scores are considered pure technical efficiency scores, while CCR (CRS) efficiency scores are considered technical efficiency scores. On the other hand, scale efficiency can be calculated by dividing CCR (CRS) efficiency scores by BCC (VRS) efficiency scores [28].

The efficiency of each DMU is evaluated relative to optimal production patterns, which are computed from the performance of DMUs with input and output combinations that are the best of any peer DMU [40]. These evaluations result in an efficiency score that represents the degree of efficiency [45]. When the efficiency score for each DMU is a score of 1, it represents efficiency [40]. While DMUs with an efficiency score ranging from zero to one are defined as inefficient in the input-oriented models, DMUs with an efficiency score greater than 1 are defined as inefficient in the output-oriented models [43]. Once the efficiency frontier is set, the performance of inefficient DMUs can be improved by increasing the outputs quantities or decreasing the inputs quantities [46]. For this purpose, a reference set, also known as best practices, is identified for each inefficient DMUs [40,43].

According to Ozcan (2014), the performance of DMUs can be assessed either as fully efficient or weakly efficient. If one DMU's efficiency score is 1 and slack values (both input slack and outputs slacks) are 0, it is considered fully efficient [26]. On the other hand, while the slack value in an input "i" represents an additional inefficient use of input "i", the slack value in an output "r" represents an additional inefficient in the production of output "r" [21].

The basic efficiency measure is the ratio of total outputs to total inputs [17,21]. The mathematical structure and formulations of Basic Envelopment Models can be found in the studies of Cooper et al. (2006), Ozcan (2014), and Emrouznejad and Cabanda (2014) [20,22,24].

3.2. Sample

DMUs used in this study consist of 50 US states. The correct selection of DMUs to be compared is essential and DMUs should be homogeneous. In other words, DMUs should be performing the same tasks with similar objectives using the same inputs and outputs under the same set of market conditions [47]. The sample size is expected to be at least 2 or 3 times larger than the sum of the number of inputs and outputs [34]. This requirement was met, as the study has 50 DMUs, 4 inputs, and 4 outputs. Descriptive statistics of the sample are presented in Table 1.

Table 1. Descriptive statistics of the sample.

DMU	Population	GDP (2021 1st Quarter—$)	Region
Alaska (AK)	731,545	54,629	Far West
Alabama (AL)	4,903,185	238,726	Southeast
Arkansas (AR)	3,017,804	137,312	Southeast
Arizona (AZ)	7,278,717	394,490	Southwest
California (CA)	39,512,223	3,237,389	Far West
Colorado (CO)	5,758,736	413,578	Rocky Mountain
Connecticut (CT)	3,565,287	294,546	New England
Delaware (DE)	973,764	79,124	Mideast
Florida (FL)	21,477,737	1,151,608	Southeast
Georgia (GA)	10,617,423	653,938	Southeast
Hawaii (HI)	1,415,872	92,541	Far West
Iowa (IA)	3,155,070	205,694	Plains
Idaho (ID)	1,787,065	89,826	Rocky Mountain
Illinois (IL)	12,671,821	909,487	Great Lakes
Indiana (IN)	6,732,219	397,134	Great Lakes
Kansas (KS)	2,913,314	184,184	Plains
Kentucky (KY)	4,467,673	222,880	Southeast
Louisiana (LA)	4,648,794	257,593	Southeast
Massachusetts (MA)	6,892,503	611,917	New England
Maryland (MD)	6,045,680	442,858	Mideast
Maine (ME)	1,344,212	69,409	New England
Michigan (MI)	9,986,857	542,566	Great Lakes
Minnesota (MN)	5,639,632	396,994	Plains
Missouri (MO)	6,137,428	340,144	Plains
Mississippi (MS)	2,976,149	122,015	Southeast
Montana (MT)	1,068,778	55,107	Rocky Mountain
North Carolina (NC)	10,488,084	619,595	Southeast
North Dakota (ND)	762,062	58,777	Plains
Nebraska (NE)	1,934,408	137,268	Plains
New Hampshire (NH)	1,359,711	89,605	New England
New Jersey (NJ)	8,882,190	649,829	Mideast
New Mexico (NM)	2,096,829	106,380	Southwest
Nevada (NV)	3,080,156	185,163	Far West
New York (NY)	19,453,561	1,758,071	Mideast
Ohio (OH)	11,689,100	713,507	Great Lakes
Oklahoma (OK)	3,956,971	198,008	Southwest
Oregon (OR)	4,217,737	262,587	Far West
Pennsylvania (PA)	12,801,989	821,117	Mideast
Rhode Island (RI)	1,059,361	63,053	New England
South Carolina (SC)	5,148,714	255,468	Southeast
South Dakota (SD)	884,659	58,878	Plains

Table 1. *Cont.*

DMU	Population	GDP (2021 1st Quarter—$)	Region
Tennessee (TN)	6,829,174	386,444	Southeast
Texas (TX)	28,995,881	1,879,785	Southwest
Utah (UT)	3,205,958	209,203	Rocky Mountain
Virginia (VA)	8,535,519	579,860	Southeast
Vermont (VT)	623,989	34,565	New England
Washington (WA)	7,614,893	651,107	Far West
Wisconsin (WI)	5,822,434	357,365	Great Lakes
West Virginia (WV)	1,792,147	79,690	Southeast
Wyoming (WY)	578,759	39,061	Rocky Mountain

In this table, population data were provided from CovidActNow [48], and gross domestic product (GDP) and region data were provided Bureau of Economic Analysis (BEA) [49].

3.3. Inputs, Outputs, and Data

Since the input-output combination preferred by DMUs should produce an output bundle from the input bundle, it should be technically feasible [35]. In these bundles, both good (desirable) and bad (undesirable) inputs and outputs factors may be available [24]. When evaluating DMUs' efficiency, the desirable and undesirable factors should be addressed differently. Because in the standard DEA model, increases in undesirable inputs and desirable outputs are allowed, similarly, decreases in undesirable outputs and desirable inputs are allowed [50].

The operations strategy matrix mentioned earlier was used as the primary measurement tool.

Data used for inputs and outputs were collected from many databases, public and private agencies, web pages, etc. in the US for 17 months (from 22 January 2020 to 27 June 2021). The data source of each input and output is mentioned in its description.

There are four inputs used in the analysis; these are Decision Areas/Strategic Decision Areas in the Operations Strategy Matrix adapted from Slack and Lewis (2017) and Karuppan et al. (2016) as the main input framework [15,51].

Capacity (i1): An operation's capacity is the maximum level of value-added activity that it can perform under normal conditions [15]. Capacity measures can be based on outputs or the availability of inputs [52]. Most healthcare facilities' capacity indicators are composed of volume measures per reporting period [53]. The capacity of healthcare systems is determined by the capacity of each server and the number of servers being used [40]. According to the CDC, as an essential disease control measure used by local and state health department personnel for decades, contact tracing is a key strategy for preventing the further spread of COVID-19. Thus, it requires people with the training [54]. On the other hand, Donabedian's Model has argued that personnel is an input measure of healthcare quality [13]. The Capacity, therefore, is represented by the "Contact Tracer Capacity Ratio" in this study. Data were provided from CovidActNow [48].

Supply Network (i2): Since a supply network perspective is defined as setting an operation with all the other operations, it interacts with materials, parts, ideas, information, data, knowledge, and people all flow through the supply network formed by all these operations [55]. The Supply network involves the design and management of seamless inter and intra-organizational processes [39]. Moreover, material, information, and capital flows must be sustainable [40]. However, COVID-19 has created trouble to manage a sufficient supply network [41]. Since COVID-19 vaccines reduce the risk of people spreading the virus, vaccinations are vital to get population immunity [56–59]. COVID-19 vaccines and ancillary supplies are procured and distributed by the US federal government at no cost to jurisdictions (states, territories, tribes, and local entities) [60,61]. The transportation ecosystem provides the technologies, services, and processes necessary to facilitate market penetration [44]. According to CDC, distribution is the process of shipping vaccines to provider locations, as directed by jurisdictions, federal agencies, and pharmacy partners who are enrolled in the COVID-19 Vaccination Program. Vaccine delivery is the last part of the distribution process. Deliveries represent the vaccine doses that have arrived at their destination [62]. The Supply Network, therefore, is represented by the "Distributed Vaccine Amount/Per Capita" in this study. Data were provided from CovidActNow [48].

Process Technology (i3): Process technology refers to the machines, devices, and equipment that create and deliver products and services [55]. Process technologies can be classified into three types: material processing technologies such as flexible manufacturing systems, information processing technologies such as optical character-recognition machines, and customer processing technologies such as medical equipment/devices [15]. The successful delivery of health care depends on medical devices; therefore, the Management of Medical Equipment (MME) is considered the most critical component within Health Systems [63]. The SEIR model attempts to predict how a disease will evolve in a population. Accordingly, all of the COVID-19 deaths have progressed from ICU cases [64]. On the other hand, Donabedian's Model has argued that facilities and equipment are input healthcare quality measures [13]. The Process Technology, therefore, is represented by the "Intensive Care Unit (ICU) Bed Capacity/Per 100,000 Population" in this study. Data were provided from CovidActNow [48].

Development and Organization (i4): This input is concerned with a broad and long-term set of decisions governing how the operation is run continuingly due to different employees having different tolerance for risk and ambiguity [15]. Additionally, it includes the workforce organization, planning and control, and improvement activities as an infrastructural decision [55]. Since COVID-19 has swept across the USA, it became clear that certain factors—such as healthcare systems—make some states more vulnerable to its impact [65]. Vulnerability is a state of sensitivity to disaster [66]. States with higher vulnerability have pre-existing economic, social, and physical conditions that may make it hard to respond to the COVID-19 outbreak [67]. According to Surgo Ventures, the COVID-19 Community Vulnerability Index (CCVI), developed by Surgo Ventures, evaluates how well any state could respond to the health, economic and social consequences of COVID-19 without appropriate response and additional support. It ranges from 0 to 100 and constitutes seven themes: Socioeconomic Status, Minority Status & Language, Household & Transportation, Epidemiological Factors, Healthcare Systems Factors, High-Risk Environments, and Population Density. Theme 5, Healthcare System Factors, measure the capacity, strength, accessibility, and preparedness of the healthcare system to respond to the COVID-19 outbreak [68]. Development and Organization, therefore, is represented by the "The COVID-19 Community Vulnerability Index (Theme 5: Healthcare System Factors)" in this study. Data were provided from Surgo Ventures (Precision for COVID) [68].

There are four outputs, two of which undesirable output, used in the analysis: these are Competitive Priorities or Performance Objectives in the Operations Strategy Matrix adapted from Slack and Lewis (2017) and Karuppan et al. (2016) as the main output framework [15,51].

Quality (o1): Since quality usually means the high specification of a product or service [15], it can be defined as consistent conformance to customers' expectations [15,55]. Because quality has two dimensions, high-performance design and, goods and services consistency [69], in the healthcare area, quality means that patients receive the most appropriate treatment and that their treatment is carried out correctly [55]. Adopting the most effective prevention and treatment practices for the leading causes of mortality, and enabling healthy living are healthcare systems' priorities [51]. On the other hand, Donabedian's Model has argued that patient experience, restoration of function, recovery, and survival are the output measure of healthcare quality [13]. Therefore, as an undesirable output, Quality, is represented by the "Deaths/1M Population" in this study. Data were provided from Worldometers [1].

Cost (o2): As a competitive priority, cost means offering a product or a service at a low price relative to the substitute's product or service prices [69]. Even organizations that compete on things other than price are interested in keeping their costs low [15]. Economic efficiency considers whether a given output is produced at a low cost [13]. In the healthcare area, the total cost includes the costs of resources such as material, equipment, people, facilities, and the costs inherent in the transformation process [51]. According to USASpending.Gov; in early 2020, the U.S. Congress appropriated funds in response to the COVID-19 pandemic. These funds were made possible through the Coronavirus Aid, Relief, and Economic Security (CARES) Act and other supplemental legislation. In March 2021, additional funds were appropriated through the American Rescue Plan Act. Once the federal government has determined that an individual, organization, business, or state, local, or tribal government will receive an award, the money is obligated (promised) and then outlay (paid) according to the terms of the contract or financial assistance [70]. Therefore, as an undesirable output, the cost is represented by the "Award Outlays/Per Capita Spending" in this study. Data were provided from USASpending.Gov [70].

Delivery (o3): Delivery, as a competitive priority, refers to the ability to provide services on time (on-time delivery) or to deliver the services faster (rapid delivery) than the competitors [51]. Rapid delivery means how quickly an order is received, while on-time delivery means how often deliveries are made on time [69]. Healthcare delivery includes many repetitious workflows, such as filling prescriptions, reporting laboratory test results, and completing radiology images [71]. Donabedian's Model has argued that preventive management, coordination, and continuity of care, acceptability of care to the recipient are the process measure of healthcare quality [13]. Immunization with a safe and effective COVID-19 vaccine is a critical component of the United States strategy to reduce COVID-19-related conditions [60]. Vaccines are widely accessible in the United States and are available for everyone at no cost [59]. Percent vaccinated is the percentage of the total population of a state that has started the vaccination process by receiving at least their first dose [67]. Therefore, delivery is represented by the "Vaccinated Rate (+1 Dose)" in this study. Data were provided from CovidActNow [48].

Flexibility (o4): Flexibility means being able to change the operation anywise [55]. Competing on flexibility refers to responding to changes with minimal penalties [51]. One operation that can exhibit a wide range of abilities is more flexible than another [15]. As customer demand is increasing, organizations are increasingly promoting different models of their products and service [72]. A flexible system can rapidly increase or decrease the quantity of production to meet demand fluctuations [69]. Some operational uncertainties can be balanced by systematic flexibility [73]. In the healthcare area, excess capacity or a flexible workforce is often required to meet demand fluctuations [52]. An essential metric for coping with demand fluctuations caused by COVID-19 is whether hospitals can handle the increased load of new COVID-19 cases without resorting to crisis standards of care [64]. Flexibility, therefore, is represented by the "Available ICU Capacity (by percentage)" in this study. Data were provided from CovidActNow [48].

According to the explanations above, it is assumed that the input bundle used by DMUs can produce the output bundle in the Operations Strategy Matrix.

Summary statistics of inputs and outputs are presented in Table 2.

Table 2. Summary statistics of inputs and outputs.

Inputs	Min.	Max.	Mean
Capacity (i1)	0.200	8.5700	1.8162
Supply Network (i2)	0.8705	1.4739	1.1311
Process Technology (i3)	14.25	41.10	25.38
Development and Organization (i4)	0.0000	1.0000	0.5212
Quality (o1)	364	2976	1692
Cost (o2)	2316	25,947	4063
Delivery (o3)	0.3590	0.7350	0.5244
Flexibility (o4)	0.1800	0.6000	0.3368

3.4. Analysis Design

The analysis consists of two stages. In the first stage; since there are two undesirable (bad) outputs in the dataset, BCC (VRS)-Output Orientation DEA model was employed in this study. The model proposed by Seiford and Zhu (2002) was applied for undesirable outputs [50]. Then, DMUs were evaluated using pure technical efficiency scores, a reference set was identified for each inefficient DMUs, and improvement options of inefficient DMUs were calculated. In addition to this, fully efficient DMUs are determined based on efficiency scores and slack values. All analyses were performed using the deaR package in the R project [74].

In the second stage; an independent-samples *t*-test was conducted to compare DMUs' populations in the efficiency and inefficient conditions and an independent-samples *t*-test was conducted to compare DMUs' gross domestic product (GDP) in the efficiency and inefficient conditions.

4. Results

Pure technical efficiency scores and inputs/outputs values of US states are shown in Table 3.

Table 3. Pure technical efficiency scores and input/output values.

DMU	Efficiency Score	i1	i2	i3	i4	o1	o2	o3	o4
Alaska	1	1.64	1.1338	17.50	0.24	502.00	6287.97	0.484	0.27
Alabama	1.01025	0.44	0.9793	33.37	0.96	2312.00	2655.36	0.394	0.18
Arkansas	1	0.54	0.9400	32.18	0.70	1953.00	2316.43	0.416	0.29
Arizona	1	0.16	1.1121	30.36	0.82	2461.00	2695.86	0.492	0.50
California	1.01127	2.09	1.2408	17.58	0.90	1609.00	3544.20	0.608	0.31
Colorado	1.00959	0.62	1.2188	22.21	0.44	1211.00	3741.60	0.576	0.30
Connecticut	1	2.65	1.3218	29.51	0.12	2321.00	3701.11	0.666	0.51
Delaware	1.03175	2.29	1.3115	19.51	0.10	1740.00	4226.13	0.577	0.25
Florida	1.00605	0.33	1.1561	29.71	0.92	1759.00	3190.01	0.531	0.25
Georgia	1.06975	0.97	1.0511	26.01	0.98	2015.00	3963.08	0.425	0.25
Hawaii	1	1.48	1.3699	15.33	0.28	364.00	4628.79	0.694	0.42
Iowa	1.02645	1.18	1.0937	21.39	0.52	1943.00	3253.92	0.512	0.34
Idaho	1	0.51	0.9325	17.18	0.38	1200.00	3009.66	0.393	0.35
Illinois	1.04749	3.09	1.1558	26.52	0.88	2023.00	4596.03	0.588	0.40
Indiana	1.01591	1.12	0.9919	33.51	0.62	2053.00	2782.81	0.443	0.34

Table 3. *Cont.*

DMU	Efficiency Score	i1	i2	i3	i4	o1	o2	o3	o4
Kansas	1.02255	0.69	1.0558	29.11	0.40	1767.00	3419.46	0.489	0.31
Kentucky	1	1.88	0.9963	41.10	0.48	1612.00	2522.23	0.492	0.40
Louisiana	1	0.54	0.8705	39.71	0.50	2307.00	3374.14	0.378	0.38
Massachusetts	1	8.57	1.4103	19.91	0.36	2610.00	4225.28	0.700	0.27
Maryland	1.00622	5.14	1.3954	21.68	0.20	1610.00	3448.70	0.610	0.31
Maine	1	0.71	1.3775	24.10	0.16	638.00	3779.09	0.661	0.23
Michigan	1.02937	1.31	1.1600	24.62	0.68	2099.00	3277.50	0.512	0.27
Minnesota	1	2.27	1.1547	16.05	0.72	1358.00	25,947.46	0.567	0.28
Missouri	1	0.02	0.9961	29.25	0.78	1611.00	2979.07	0.445	0.18
Mississippi	1	0.31	0.8937	28.26	0.60	2485.00	2554.42	0.359	0.30
Montana	1	0.78	1.0382	21.05	0.14	1555.00	4520.29	0.475	0.40
North Carolina	1	0.71	1.1117	22.88	1.00	1279.00	2540.43	0.450	0.23
North Dakota	1	7.14	0.9361	25.59	0.26	2005.00	6125.47	0.437	0.42
Nebraska	1	5.83	1.0791	29.05	0.18	1168.00	3829.86	0.514	0.37
New Hampshire	1.01757	1.32	1.3409	20.74	0.66	1008.00	3984.12	0.618	0.36
New Jersey	1	2.56	1.2890	34.05	0.56	2976.00	3848.45	0.644	0.60
New Mexico	1	0.89	1.1268	20.36	0.64	2067.00	2876.87	0.616	0.38
Nevada	1	0.31	0.9952	28.38	0.66	1840.00	3124.91	0.489	0.33
New York	1.05499	6.26	1.2284	27.51	0.94	2774.00	4278.89	0.596	0.39
Ohio	1.01952	1.28	1.0585	35.56	0.46	1735.00	3061.39	0.480	0.37
Oklahoma	1.02923	0.71	1.0137	24.59	0.34	1866.00	3644.81	0.446	0.35
Oregon	1	0.62	1.3349	19.28	0.58	655.00	3305.07	0.582	0.33
Pennsylvania	1.00539	1.70	1.2188	28.75	0.54	2168.00	3185.67	0.624	0.28
Rhode Island	1	2.14	1.4247	14.25	0.04	2575.00	4531.06	0.642	0.26
South Carolina	1	1.56	1.0255	22.53	0.42	1907.00	2346.02	0.438	0.33
South Dakota	1.08609	4.95	1.0824	26.34	0.32	2295.00	5348.51	0.503	0.44
Tennessee	1	2.52	0.9224	35.08	0.76	1838.00	2735.26	0.413	0.30
Texas	1.03728	0.59	1.0966	23.33	0.86	1811.00	3566.55	0.478	0.18
Vermont	1	0.73	1.0031	19.65	0.80	735.00	3117.00	0.481	0.40
Washington	1	2.00	1.2100	22.37	0.00	1335.00	3184.50	0.586	0.33
Wisconsin	1	2.32	1.4739	16.51	0.84	410.00	5277.58	0.735	0.27
Washington	1	0.92	1.2303	16.44	0.74	781.00	3369.97	0.607	0.28
Wisconsin	1	1.45	1.0489	27.62	0.30	1391.00	3251.07	0.534	0.46
West Virginia	1	0.82	1.0545	37.33	0.06	1605.00	2627.55	0.431	0.29
Wyoming	1	0.15	0.8911	24.02	0.22	1279.00	5343.58	0.390	0.60

According to Table 3, Alaska, Arkansans, Arizona, Connecticut, Hawaii, Idaho, Kentucky, Louisiana, Massachusetts, Maine, Minnesota, Missouri, Mississippi, Montana, North Carolina, North Dakota, Nebraska, New Jersey, New Mexico, Nevada, Oregon, Rhode Island, South Carolina, Tennessee, Utah, Virginia, Vermont, Washington, Wisconsin, West Virginia, and Wyoming have a score of 1, and they are relatively efficient.

Based on Ozcan (2009) and Ozcan (2014), a reference set, also known as best practices, was identified for each inefficient DMUs to become efficient [40,43].

Improvement options for the inefficient states are shown in Table 4.

Table 4. Improvement options for the inefficient states.

DMU	i1	i2	i3	i4	o1	o2	o3	o4
Alabama	0.00	0.00	−1.78	−0.24	−216.39	−238.75	0.04	0.16
California	−1.15	−0.03	0.00	−0.22	−456.29	−252.56	0.01	0.01
Colorado	0.00	0.00	0.00	0.00	−67.33	−212.93	0.01	0.03
Delaware	−0.44	−0.06	0.00	0.00	−39.28	−689.72	0.02	0.06
Florida	0.00	0.00	−2.16	−0.29	−7.37	−137.73	0.00	0.10
Georgia	0.00	−0.03	0.00	−0.35	−147.96	−1533.55	0.03	0.06
Iowa	0.00	−0.01	0.00	0.00	−27.35	−600.25	0.01	0.01
Illinois	−1.89	0.00	−4.02	−0.31	−45.31	−1429.02	0.03	0.02
Indiana	0.00	0.00	−1.81	−0.04	−159.90	−368.61	0.01	0.01
Kansas	0.00	0.00	−1.12	0.00	−27.28	−508.00	0.01	0.03
Maryland	−2.62	−0.16	0.00	0.00	−8.50	−139.93	0.00	0.02
Michigan	−0.09	−0.08	−3.18	−0.15	−111.96	−665.94	0.02	0.09
New Hampshire	−0.36	−0.05	−2.09	−0.17	−34.59	−385.83	0.01	0.01
New York	−4.89	−0.05	−4.90	−0.43	−641.48	−1191.63	0.03	0.02
Ohio	0.00	−0.01	0.00	0.00	−24.24	−446.69	0.01	0.01
Oklahoma	0.00	0.00	−0.08	0.00	−343.73	−651.85	0.01	0.01
Pennsylvania	−0.27	−0.05	−6.80	0.00	−39.73	−122.62	0.00	0.12
South Dakota	−3.60	0.00	0.00	0.00	−648.11	−1773.55	0.04	0.04

According to Table 4:

- The DMU that needs the most improvement in terms of "i1" is New York.
- The DMU that needs the most improvement in terms of "i2" is Maryland.
- The DMU that needs the most improvement in terms of "i3" is Pennsylvania.
- The DMU that needs the most improvement in terms of "i4" is New York.
- The DMU that needs the most improvement in terms of "o1" is South Dakota.
- The DMU that needs the most improvement in terms of "o2" is South Dakota.
- The DMUs that need the most improvement in terms of "o3" are South Dakota and Alabama.
- The DMU that needs the most improvement in terms of "o4" is Texas.

As shown in Table 5, independent samples *t*-tests were conducted to compare DMUs' populations and gross domestic product (GDP) in the efficiency and inefficient conditions.

Table 5. Independent sample *t*-test results.

Dependent Variable	Efficiency Condition	N	Mean	Std. Deviation	P
Population	Efficiency	31	3,988,512.03	2,803,864.465	0.011
	Inefficiency	19	10,731,047.42	10,285,057.50	
GDP	Efficiency	31	252,090.90	199,971.078	0.018
	Inefficiency	19	735,539.84	802,756.118	

According to Table 5:

- There was a significant difference in the scores for efficiency (M = 3,988,512.03, SD = 2,803,864.465) and inefficiency (M = 10,731,047.42, SD = 10,285,057.50) conditions; t (48) = −2.795, p = 0.011 (in terms of population)
- There was a significant difference in the scores for efficiency (M = 252,090.90, SD = 19,9971.078) and inefficiency (M = 735,539.84, SD = 802,756.118) conditions; t (48) = –2.577, p = 0.018 (in terms of GDP)

5. Discussion

COVID-19 pandemic is a complex and unprecedented public health crisis worldwide. It is crucial to evaluate health systems for mitigating the impact of the COVID-19 pandemic [6].

Although all countries are affected by the pandemic, the United States is the country with the highest number of COVID-19 cases and deaths reported as of 27 June 2021. Therefore, both the CDC and each state's health department are committed to stopping its spread.

CDC's global COVID-19 response works toward the goals mentioned above by meeting the following objectives: strengthening healthcare system capacity to prevent, detect, and respond to local COVID-19 cases; mitigating COVID-19 transmission in the community, across borders, and in healthcare facilities; contributing to the scientific understanding of COVID-19; ensuring readiness to implement and evaluate vaccination programs [2].

Since healthcare systems rely heavily on scientific analysis to find results that can be generalized to a larger context [9], the evaluation should be both critical and as objective as possible [14]. Operation Strategy Matrix is a novel and objective evaluation tool. The matrix deal with how Strategic Decision Areas (Capacity, Supply Network, Process Technology, Development and Organization) affect Competitive Priorities (Quality, Cost, Delivery, Flexibility) [15].

DEA has been used widely in assessing healthcare efficiency in the United States and around the world at different levels of decision-making units [43]. Therefore, each U.S State's Operation Strategy Matrix was evaluated using DEA in the context of the COVID-19 pandemic. During the analysis, Strategic Decision Areas (Capacity, Supply Network, Process Technology, and Development and Organization) were considered as inputs while competitive priorities (Quality, Cost, Delivery, and Flexibility) were considered as outputs. Capacity was represented by "Contact Tracer Capacity Ratio". Supply Network was represented by "Distributed Vaccine Amount/Per Capita". Process Technology was represented by "Intensive Care Unit (ICU) Bed Capacity/Per 100,000 Population". Development and Organization was represented by "The COVID-19 Community Vulnerability Index (Theme 5: Healthcare System Factors)". Quality was represented by "Deaths/ 1M Population". The cost was represented by "Award Outlays/Per Capita Spending". Delivery was represented by "Vaccinated Rate (+1 Dose)". Flexibility was represented by "Available ICU Capacity (by percentage)". Additionally, since there are two undesirable (bad) outputs in the dataset, BCC (VRS)-Output Orientation Data Envelopment Analysis model was employed. Inputs (Capacity, Supply Network, Process Technology, and Development and Organization) are essential components of any operations including healthcare operations while outputs (Quality, Cost, Delivery, and Flexibility) are results of the inputs.

According to the results, 31 states are efficient, while 19 states are inefficient. Additionally, Connecticut, Louisiana, Minnesota, New Jersey, Rhode Island, Tennessee, Utah, Vermont, Washington, and Wyoming are fully efficient DMU based on efficiency scores and slack values.

A reference set was identified for each inefficient DMUs to become efficient. In the reference sets:

- New Mexico appeared eighteen times.
- Arizona appeared seven times.
- Oregon, Maine, and Idaho appeared six times.
- South Carolina appeared five times.
- Wyoming, West Virginia, Hawaii, and Arkansas appeared four times.

- Wisconsin, Massachusetts, and Connecticut appeared three times.
- Washington, Virginia, North Carolina, New Jersey, Missouri, and Kentucky appeared two times.
- Rhode Island, Nevada, and Mississippi appeared one time.

Additionally, the DMUs that need the most improvement are New York, Maryland, Pennsylvania, South Dakota, Alabama, and Texas.

It is relatively hard limiting the human-to-human transmission of COVID-19 in large communities because independent samples t-test results demonstrate that inefficient states have more population than efficient states. Moreover, 62.25% of the total US population lives in inefficient states, and 9 out of 12 states in the 3rd Population Quartile (7,845,049.50) are inefficient, while 9 out of 12 states in the 1st Population Quartile (1,790,876.50) are efficient. Furthermore, only 10.70% of the total US population lives in fully efficient states.

Economic and human mobility is maximum in the states with high GDP. For this reason, the spreading of the virus is fast. Hence, independent samples t-test results demonstrate that inefficient states have more GDP than efficient states. Moreover, 64.14% of the total US GDP is formed by inefficient states and 8 out of 12 states in the 3rd GDP Quartile (587,874.25) are inefficient, while 9 out of 12 states in the 1st GDP Quartile (91,862.25) are efficient. Additionally, only 13.69% of the total U.S GDP is formed by fully efficient states.

From the perspective of BEA region grouping [49]:

- 5 out of 6 states (83.3%) are efficient in the New England Region.
- 1 out of 5 states (20.0%) is efficient in the Mideast Region.
- 1 out of 5 states (20.0%) is efficient in the Great Lakes Region.
- 4 out of 7 states (57.1%) are efficient in the Plains Region.
- 9 out of 12 states (75.0%) are efficient in the Southeast Region.
- 2 out of 4 states (50.0%) are efficient in the Southwest Region.
- 4 out of 5 states (80.0%) are efficient in the Rock Mountain Region.
- 5 out of 6 states (83.3%) are efficient in the Far West Region.

New England and Far West are the most efficient regions while Mideast and Great Lakes are the least efficient regions by percentages. These results are in line with the population and GDP findings mentioned above. Additionally, three out of six states are fully efficient in the New England Region and 1 of 6 states is fully efficient in the Far West Region.

The SDGs are a universal agenda taking various aspects in development into account and applying them to all countries [75–77]. According to the United Nations Department of Economic and Social Affairs, SDG 3 is described as ensuring healthy lives and promoting well-being for all at all ages, and COVID-19 has shortened life expectancy [10]. United Nations has called the COVID-19 pandemic, beyond being a health emergency, as a systemic crisis that is already affecting economies and societies in unprecedented ways [72]. United Nations has created so that the countries cope with the pandemic, and the dashboards include some indicators such as the capacity of the healthcare system, vulnerability, and poverty [73]. These indicators are also included in our research model with the same or close names.

This study had some limitations. Since COVID-19 is a global pandemic, it is difficult to assess its effect on health systems. Additionally, data used for inputs and outputs were collected from many databases, public and private agencies, web pages, etc., in the US mentioned above. As such these databases were considered to be correct. Furthermore, data include 17 months from 22 January 2020 to 27 June 2021.

6. Managerial Implication and Conclusions

This study provides the first overview of the healthcare systems of U.S. states in the context of COVID-19, based on the first 17 months of the pandemic. The use of a novel evaluation tool named Operation Strategy Matrix constitutes the unique aspect of the study. Thus, the study is important for healthcare operations research literature.

The study results might contribute to both the federal and state health authorities; while the inefficient states can become efficient by applying our improvement suggestions, the federal government can develop new strategies to mitigate the COVID-19 pandemic considering our findings especially inductive statistical findings regarding population, GDP, and region.

Since SGDs address the global challenges we face, it could be yielded some critical projections for the future from the study's findings. Moreover, our research findings might help to ensure healthy lives in the context of COVID-19. The reference sets identified for each inefficient DMUs to become efficient can be used for both the US and other countries to achieve SDG Goal 3 amid COVID-19. Especially, fully efficient states (Connecticut, Louisiana, Minnesota, New Jersey, Rhode Island, Tennessee, Utah, Vermont, Washington, and Wyoming) can be served as a model to achieve Targets 3.8, 3. c, and 3.d.

Further studies can be developed new models applying different DEA approaches (Fuzzy DEA, Two-Stage DEA, Network DEA, etc.) to prioritize issues, challenges, drivers, and barriers related to healthcare systems and SDGs in the context of the COVID-19 pandemic.

Author Contributions: Conceptualization, A.Ö., H.K., M.Ş.G., and E.C.; methodology, A.Ö., H.K., M.Ş.G., and E.C.; software, A.Ö., H.K., M.Ş.G., and E.C.; validation, A.Ö., H.K., M.Ş.G., and E.C.; formal analysis, A.Ö., H.K., M.Ş.G., and E.C.; investigation, A.Ö., H.K., M.Ş.G., and E.C.; resources, A.Ö., H.K., M.Ş.G., and E.C.; data curation, A.Ö., H.K., M.Ş.G., and E.C.; writing—original draft preparation, A.Ö., H.K., M.Ş.G., and E.C.; writing—review and editing, A.Ö., H.K., M.Ş.G., and E.C.; visualization, A.Ö., H.K., M.Ş.G., and E.C. All authors have read and agreed to the published version of the manuscript.

Funding: This research received no external funding.

Institutional Review Board Statement: Not applicable.

Informed Consent Statement: Not applicable.

Data Availability Statement: Not applicable.

Conflicts of Interest: The authors declare no conflict of interest.

References

1. Worldometer. Coronavirus Update. 2021. Available online: www.worldometers.info/coronavirus (accessed on 27 June 2021).
2. Centers for Disease Control and Prevention (CDC). CDC COVID-19 Global Response. 2020. Available online: www.cdc.govcoronavirus2019-ncovglobal-covid-19global-response.html (accessed on 21 November 2020).
3. Al-Taweel, D.; Al-Haqan, A.; Bajis, D.; Al-Bader, J.; Al-Taweel, A.R.M.; Al-Awadhi, A.; Al-Awadhi, F. Multidisciplinary academic perspectives during the COVID-19 pandemic. *Int. J. Health Plan. Manag.* **2020**, *35*, 1295–1301. [CrossRef]
4. Ogunkola, I.O.; Adebisi, Y.A.; Imo, U.F.; Odey, G.O.; Esu, E.; Lucero-Prisno, D.E. Rural communities in Africa should not be forgotten in responses to COVID-19. *Int. J. Health Plan. Manag.* **2020**, *35*, 1302–1305. [CrossRef]
5. Gökmen Kavak, D.; Öksüz, A.S.; Cengiz, C.; Kayral, İ.H.; Şenel, F.Ç. The importance of quality and accreditation in health care services in the process of struggle against Covid-19. *Turk. J. Med. Sci.* **2020**, *50*, 1760–1770. [CrossRef] [PubMed]
6. World Health Organization (WHO). Strengthening the Health System Response to COVID-19 Recommendations for the WHO European Region Policy Brief. 2020. Available online: https://apps.who.int/iris/bitstream/handle/10665/333072/WHO-EURO-2020-806-40541-54465-eng.pdf?sequence=1&isAllowed=y (accessed on 8 April 2020).
7. Aftab, W.; Siddiqui, F.J.; Tasic, H.; Perveen, S.; Siddiqi, S.; Bhutta, Z.A. Implementation of health and health-related sustainable development goals: Progress, challenges and opportunities-a systematic literature review. *BMJ Glob. Health* **2020**, *5*, e002273. [CrossRef]
8. Barcelona Instute for Global Health. The Sustainable Development Goals (SDGs) and Global Health. 2021. Available online: https://www.isglobal.org/en/-/global-health-inequities (accessed on 10 October 2021).
9. Block, D.J. *Healthcare Outcomes Management: Strategies for Planning*; Jones and Bartlett Publishers: Burlington, MA, USA, 2006.
10. United Nations Department of Economic and Social Affairs Sustainable Development. The 17 Goals. 2021. Available online: https://sdgs.un.org/goals (accessed on 29 August 2021).
11. Bappy, M.M.; Ali, S.M.; Kabir, G.; Paul, S.K. Supply chain sustainability assessment with Dempster-Shafer evidence theory: Implications in cleaner production. *J. Clean. Prod.* **2019**, *237*, 117771. [CrossRef]
12. Alam, S.T.; Ahmed, S.; Ali, S.M.; Sarker, S.; Kabir, G.; Ul-Islam, A. Challenges to COVID-19 vaccine supply chain: Implications for sustainable development goals. *Int. J. Prod. Econ.* **2021**, *239*, 108193. [CrossRef]
13. Donabedian, A. Evaluating the Quality of Medical Care. *Milbank Q.* **2005**, *83*, 691–729. [CrossRef] [PubMed]

14. Jenkinson, C. *Assessment and Evaluation of Health and Medical Care: A Methods Text*; Open University Press: Buckingam, PA, USA, 1977.
15. Slack, N.; Lewis, M. *Operations Strategy*, 5th ed.; Pearson Education Limited: Harlow, UK, 2017.
16. O'Leary, N.; Kingston, L.; Griffin, A.; Morrissey, A.-M.; Noonan, M.; Kelly, D.; Doody, O.; Niranjan, V.; Gallagher, A.; O'Riordan, C.; et al. COVID-19 healthcare policies in Ireland: A rapid review of the initial pandemic response. *Scand. J. Public Health* **2021**, *49*, 713–720. [CrossRef]
17. Tsai, Y.; Yang, T. COVID-19 Outbreak and Voluntary Demand for Non-COVID-19 Healthcare: Evidence from Taiwan. *SSRN Electron. J.* **2020**. [CrossRef]
18. Adwibowo, A. COVID 19 healthcare facility demand forecasts for rural residents. *medRxiv* **2020**. [CrossRef]
19. Alshammari, N.; Sarker, M.N.I.; Kamruzzaman, M.M.; Alruwaili, M.; Alanazi, S.A.; Raihan, M.L.; AlQahtani, S.A. Technology-driven 5G enabled e-healthcare system during COVID-19 pandemic. *IET Commun.* **2021**, 1–15.
20. Brodie, R.J.; Ranjan, K.R.; Verreynne, M.-L.; Jiang, Y.; Previte, J. Coronavirus crisis and health care: Learning from a service ecosystem perspective. *J. Serv. Theory Pract.* **2021**, *31*, 225–246. [CrossRef]
21. Arlotti, M.; Ranci, C. The Impact of COVID-19 on Nursing Homes in Italy: The Case of Lombardy. *J. Aging Soc. Policy* **2021**, *33*, 431–443. [CrossRef]
22. Rodríguez, Y.; Hignett, S. Integration of human factors/ergonomics in healthcare systems: A giant leap in safety as a key strategy during Covid-19. *Hum. Factors Ergon. Manuf.* **2021**, *31*, 570–576. [CrossRef] [PubMed]
23. Akiyama, M.J.; Arnsten, J.H.; Roth, S. New normal: Caring for hospitalised patients in the Bronx, New York, during COVID-19. *Intern. Med. J.* **2021**, *51*, 288–290. [CrossRef] [PubMed]
24. Tasri, Y.D.; Tasri, E.S. Improving clinical records: Their role in decision-making and healthcare management–COVID-19 perspectives. *Int. J. Health Manag.* **2020**, *13*, 325–336.
25. Bel, G.; Gasulla, Ó.; Mazaira-Font, F.A. The Effect of Health and Economic Costs on Governments' Policy Responses to COVID-19 Crisis under Incomplete Information. *Public Adm. Rev.* **2021**, 1–16. [CrossRef]
26. Chamboredon, F.; Roman, C.; Colson, S. COVID-19 pandemic in France: Health emergency experiences from the field. *Int. Nurs. Rev.* **2020**, *67*, 326–333. [CrossRef] [PubMed]
27. Leite, H.; Lindsay, C.; Kumar, M. COVID-19 Outbreak: Implications on Healthcare Operations. *TQM J.* **2021**, *33*, 247–256. [CrossRef]
28. United Nations. Goal 3: Ensure Healthy Lives and Promote Well-Being for All at All Ages. 2021. Available online: https://www.un.org/sustainabledevelopment/health/ (accessed on 18 October 2021).
29. World Health Organization (WHO). Sustainable Development Goals (SDGs). 2021. Available online: https://www.who.int/health-topics/sustainable-development-goals#tab=tab_1 (accessed on 18 October 2021).
30. Rajan, S.; Ricciardi, W.; McKee, M.; McKee, M. The SDGs and health systems: The last step on the long and unfinished journey to universal health care? *Eur. J. Public Health* **2020**, *30*, i28–i30. [CrossRef]
31. Kruk, M.E.; Gage, A.D.; Arsenault, C.; Jordan, K.; Leslie, H.H.; Roder-DeWan, S.; Adeyi, O.; Barker, P.; Daelmans, B.; Doubova, S.V.; et al. High-quality health systems in the Sustainable Development Goals era: Time for a revolution. *Lancet Glob. Health.* **2018**, *6*, e1196–e1252. [CrossRef]
32. Gavurova B, Kocisova K, Sopko, J. Health system efficiency in OECD countries: Dynamic network DEA approach. *Health Econ. Rev.* **2021**, *11*, 1–25.
33. Jordi, E.; Pley, C.; Jowett, M.; Abou Jaoude, G.J.; Haghparast-Bidgoli, H. Assessing the efficiency of countries in making progress towards universal health coverage: A data envelopment analysis of 172 countries. *BMJ Glob. Health* **2020**, *5*, e002992. [CrossRef] [PubMed]
34. Ramanathan, R. *An Introduction to Data Envelopment Analysis a Tool for Performance Measurement*; Sage Publications India Pvt Ltd.: New Delhi, India, 2003.
35. Ray, S.C. *Data Envelopment Analysis Theory and Techniques for Economic and Operations Research*; Cambridge University Press: New York, NY, USA, 2004.
36. Tone, K. Radial DEA Models. In *Advances in DEA Theory and Applications with Extensions to Forecasting Models*; Tone, K., Ed.; John Wiley & Sons Ltd.: Hoboken, NJ, USA, 2017.
37. Charnes, A.; Cooper, W.W.; Lwein, A.Y.; Seiford, L.M. *Data Envelopment Analysis: Theory, Methodology, and Application*; Springer: New York, NY, USA, 1994.
38. Emrouznejad, A.; Cabanda, E. Managing Service Productivity Using Data Envelopment Analysis. In *Managing Service Productivity Using Frontier Efficiency Methodologies and Multicriteria Decision Making for Improving Service Performance*; Emrouznejad, A., Cabanda, E., Eds.; Springer: Berlin/Heidelberg, Germany, 2014.
39. Gok, M.S.; Sezen, B. Analyzing the Efficiencies of Hospitals: An Application of Data Envelopment Analysis. *J. Glob. Strat. Manag.* **2011**, *2*, 137. [CrossRef]
40. Ozcan, Y.A. *Quantitative Methods in Health Care Management: Techniques and Applications*, 2nd ed.; John Wiley & Sons, Inc.: San Francisco, CA, USA, 2009.
41. Charnes, A.; Cooper, W.W.; Rhodes, E. Measuring the efficiency of decision making units. *Eur. J. Oper. Res.* **1978**, *2*, 429–444. [CrossRef]
42. Banker, R.D.; Charnes, A.; Cooper, W.W. Some Models for Estimating Technical and Scale Inefficiencies in Data Envelopment Analysis. *Manag. Sci.* **1984**, *30*, 1078–1092. [CrossRef]

43. Ozcan, Y.A. *Health Care Benchmarking and Performance Evaluation an Assessment Using Data Envelopment Analysis (DEA)*; Springer: New York, NY, USA, 2014.

44. Emrouznejad, A.; Tavana, M.; Hatami-Marbini, A. The State of the Art in Fuzzy Data Envelopment Analysis. In *Performance Measurement with Fuzzy Data Envelopment Analysis*; Emrouznejad, A., Tavana, M., Eds.; Springer: Berlin/Heidelberg, Germany, 2014.

45. Cooper, W.W.; Seiford, L.M.; Tone, K. *Introduction to Data Envelopment Analysis and Its Uses with DEA-Solver and References*; Springer: New York, NY, USA, 2006.

46. Hua, Z.; Bian, Y. DEA with Undesirable Factors. In *Modeling Data Irregularities and Structural Complexities in Data Envelopment Analysis*; Zhu, J., Cook, W.D., Eds.; Springer: New York, NY, USA, 2007.

47. Golany, B.; Roll, Y. An application procedure of DEA. *Omega* **1989**, *17*, 237–250. [CrossRef]

48. CovidActNow. Covid Act Now API (v2.0.0-Beta.1). 2021. Available online: https://apidocs.covidactnow.org/ (accessed on 27 June 2021).

49. Bureau of Economic Analysis (BEA). Gross Domestic Product 1st Quarter 2021. 2021. Available online: https://www.bea.gov/sites/default/files/2021-06/qgdpstate0621.pdf (accessed on 18 October 2021).

50. Seiford, L.M.; Zhu, J. Modeling undesirable factors in efficiency evaluation. *Eur. J. Oper. Res.* **2002**, *142*, 16–20. [CrossRef]

51. Karuppan, C.M.; Dunlap, N.E.; Waldrum, M.R. *Operations Management in Healthcare Strategy and Practice*; Springer: New York, NY, USA, 2016.

52. McLaughlin, D.B.; Hays, J.M. *Healthcare Operations Management*; Health Administration Press: Chicago, IL, USA; AUPHA Press: Washington, DC, USA, 2008.

53. Bergeron, B.P. *Performance Management in Healthcare: From Key Performance Indicators to Balanced Scorecard*, 2nd ed.; CRC Press: Boca Raton, FL, USA; Taylor & Francis Group: Milton Park, UK, 2018.

54. CDC. Contact Tracing. 2020. Available online: https://www.cdc.gov/coronavirus/2019-ncov/daily-life-coping/contact-tracing.html (accessed on 18 November 2020).

55. Slack, N.; Chambers, S.; Johnston, R. *Operations Management*, 5th ed.; Pearson Education Limited: Essex, UK, 2007.

56. Sakib, N.; Ibne Hossain, N.U.; Nur, F.; Talluri, S.; Jaradat, R.; Lawrence, J.M. An assessment of probabilistic disaster in the oil and gas supply chain leveraging Bayesian belief network. *Int. J. Prod. Econ.* **2021**, *235*, 108107. [CrossRef]

57. El Amrani, S.; Ibne Hossain, N.U.; Karam, S.; Jaradat, R.; Nur, F.; Hamilton, M.A.; Ma, J. Modelling and assessing sustainability of a supply chain network leveraging multi Echelon Bayesian Network. *J. Clean. Prod.* **2021**, *15*, 302.

58. Shahed, K.S.; Azeem, A.; Ali, S.M.; Moktadir, M.A. A supply chain disruption risk mitigation model to manage COVID-19 pandemic risk. *Environ. Sci. Pollut. Res.* **2021**, 1–16.

59. CDC. Key Things to Know About COVID-19 Vaccines. 2021. Available online: https://www.cdc.gov/coronavirus/2019-ncov/vaccines/keythingstoknow.html (accessed on 30 June 2021).

60. CDC. COVID-19 Vaccination Program Interim Operational Guidance Jurisdiction Operations. 2020. Available online: https://www.cdc.gov/vaccines/imz-managers/downloads/COVID-19-Vaccination-Program-Interim_Playbook.pdf (accessed on 30 June 2021).

61. Boucetta, M.; Hossain, N.U.I.; Jaradat, R.; Keating, C.; Tazzit, S.; Nagahi, M. The architecture design of electrical vehicle infrastructure using viable system model approach. *Systems* **2021**, *9*, 19. [CrossRef]

62. Centers for Disease Control and Prevention (CDC). About COVID-19 Vaccine Delivered and Administration Data. 2021. Available online: https://www.cdc.gov/coronavirus/2019-ncov/vaccines/distributing/about-vaccine-data.html (accessed on 30 June 2021).

63. Pan American Health Organization (PAHO). Health Technology Management. 2020. Available online: https://www.paho.org/hq/index.php?option=com_content&view=article&id=11582:health-technology-management&Itemid=41686&lang=en (accessed on 12 November 2020).

64. CovidActNow. Introduction and Contents (Update 26 May 2020). USA. 2020. Available online: https://covidactnow.org/ (accessed on 30 June 2021).

65. Surgo Ventures. Vulnerable Communities and COVID-19: The Damage Done, and the Way Forward. 2021. Available online: https://surgoventures.org/resource-library/report-vulnerable-communities-and-covid-19 (accessed on 30 June 2021).

66. Maru, H.; Haileslassie, A.; Zeleke, T.; Esayas, B.; Rescia, A.; Gmada, S.S. Analysis of Smallholders' Livelihood Vulnerability to Drought across Agroecology and Farm Typology in the Upper Awash Sub-Basin, Ethiopia. *Sustainability* **2021**, *13*, 9764. [CrossRef]

67. CovidActNow. Risk Levels & Key Metrics. 2021. Available online: https://covidactnow.org/covid-risk-levels-metrics#vulnerability (accessed on 30 June 2021).

68. Surgo Ventures. Precision for COVID. 2021. Available online: https://precisionforcovid.org/ccvi (accessed on 27 June 2021).

69. Reid, R.D.; Sanders, N.R. *Operations Management an Integrated Approach*, 5th ed.; John Wiley & Sons: Hoboken, NJ, USA, 2013.

70. USASpending.Gov. COVID-19 Spending. 2021. Available online: https://www.usaspending.gov/disaster/covid-19?publicLaw=all (accessed on 27 June 2021).

71. Buchbinder, S.B.; Shanks, N.H. *Introduction to Health Care Management*, 3th ed.; Jones & Bartlett Learning: Burlington, MA, USA, 2017.

72. Mamun, A.A.; Khaled, A.A.; Ali, S.M.; Chowdhury, M.M. A heuristic approach for balancing mixed-model assembly line of type i using genetic algorithm. *Int. J. Prod. Res.* **2012**, *50*, 5106–5116. [CrossRef]

73. Duan, J.; Li, H.; Zhang, Q. Multiobjective optimization of buffer capacity allocation in multiproduct unreliable production lines using improved adaptive NSGA-II algorithm. *Kuwait J. Sci.* **2020**, *48*, 37–49. [CrossRef]

74. Coll-Serrano, V.; Benítez, R.; Bolós, V. *Data Envelopment Analysis with deaR, R Package Version 1.2.0*; R Core Team: Vienna, Austria, 2018.

75. Tasaki, T.; Tajima, R.; Kameyama, Y.; Randall, A. Measurement of the Importance of 11 Sustainable Development Criteria: How Do the Important Criteria Differ among Four Asian Countries and Shift as the Economy Develops? *Sustainability* **2021**, *13*, 9719. [CrossRef]
76. United Nations Development Programme (UNDP). *2020 Human Development Perspectives COVID-19 and Human Development: Assessing the Crisis, Envisioning the Recovery*; UNDP: New York, NY, USA, 2020. Available online: http://hdr.undp.org/sites/default/files/covid-19_and_human_development_0.pdf (accessed on 29 August 2021).
77. United Nations Development Programme. COVID-19: New UNDP Data Dashboards Reveal Huge Disparities among Countries in Ability to Cope and Recover. 2020. Available online: https://www.undp.org/press-releases/covid-19-new-undp-data-dashboards-reveal-huge-disparities-among-countries-ability (accessed on 29 August 2021).

MDPI

St. Alban-Anlage 66

4052 Basel

Switzerland

Tel. +41 61 683 77 34

Fax +41 61 302 89 18

www.mdpi.com

Sustainability Editorial Office

E-mail: sustainability@mdpi.com

www.mdpi.com/journal/sustainability